T0251999

GANGA-BRAHMAPUTRA-MEGHNA WATERS

ADVANCES IN DEVELOPMENT AND MANAGEMENT

GANGA-BRAHMAPUTRA-MEGHNA WATERS

ADVANCES IN DEVELOPMENT AND MANAGEMENT

MAHESH CHANDRA CHATURVEDI

CRC Press
Taylor & Francis Group
Boca Raton London New York

CRC Press is an imprint of the
Taylor & Francis Group, an **informa** business

CRC Press
Taylor & Francis Group
6000 Broken Sound Parkway NW, Suite 300
Boca Raton, FL 33487-2742

First issued in paperback 2019

© 2013 by Taylor & Francis Group, LLC
CRC Press is an imprint of Taylor & Francis Group, an Informa business

No claim to original U.S. Government works

ISBN-13: 978-1-4398-7376-2 (hbk)
ISBN-13: 978-1-138-38215-2 (pbk)

This book contains information obtained from authentic and highly regarded sources. Reasonable efforts have been made to publish reliable data and information, but the author and publisher cannot assume responsibility for the validity of all materials or the consequences of their use. The authors and publishers have attempted to trace the copyright holders of all material reproduced in this publication and apologize to copyright holders if permission to publish in this form has not been obtained. If any copyright material has not been acknowledged please write and let us know so we may rectify in any future reprint.

Except as permitted under U.S. Copyright Law, no part of this book may be reprinted, reproduced, transmitted, or utilized in any form by any electronic, mechanical, or other means, now known or hereafter invented, including photocopying, microfilming, and recording, or in any information storage or retrieval system, without written permission from the publishers.

For permission to photocopy or use material electronically from this work, please access www.copyright. com (http://www.copyright.com/) or contact the Copyright Clearance Center, Inc. (CCC), 222 Rosewood Drive, Danvers, MA 01923, 978-750-8400. CCC is a not-for-profit organization that provides licenses and registration for a variety of users. For organizations that have been granted a photocopy license by the CCC, a separate system of payment has been arranged.

Trademark Notice: Product or corporate names may be trademarks or registered trademarks, and are used only for identification and explanation without intent to infringe.

Visit the Taylor & Francis Web site at
http://www.taylorandfrancis.com

and the CRC Press Web site at
http://www.crcpress.com

The book is dedicated to my gurus

Dr. A.C. Mitra, Dr. A.N. Khosla, and Dr. K.L. Rao

Who taught me real life engineering,

Prof. Hunter Rouse

Who taught me the science of water, and

Prof. Roger Revelle

Who was an inspiration about revolutionizing the development

of the Ganga–Brahmaputra–Meghna basin

Contents

Part III   Proposed Revolutionary Policy

List of Figures

List of Tables

Preface

The Ganga–Brahmaputra–Meghna (GBM) basin, inhabited by about 10% of humanity, once the leading region of human well-being, is currently the most backward one. Paradoxically, the highest level of socioeconomic development is urgently required to bring this outstandingly large component of humanity into the mainstream of development and to ensure the sustainability of the region. A very large scale of environmental development, in which the management of water is crucial on account of the hydrologic–climatic conditions of South and East Asia, will be required. The challenge is further made more difficult on account of climate change. This poses severe challenges. Development is being undertaken by the governments of the countries of the basin but, in this author's judgment, several advances are urgently required.

Having been associated with the development of the GBM system for a long time, the author believes that its development can be revolutionized. There are three central themes that we are trying to emphasize. One, it is considered by the author that in view of the unique geophysical and hydrologic characteristics of the GBM basin, several novel technologies that the author has developed are possible, and the development of India's waters, in which GBM has the central role, can indeed be revolutionized. Two, technology has to be considered as part of the societal–economic system, particularly when such a large-scale development, as one involving the GBM system, is being considered. The socioeconomic system has also to be continually advanced. Three, India will be developing the world's largest water resources and over the course of time, India, along with China, will be emerging to achieve her due role in human affairs. Paradoxically, in terms of the ends and means, it will be of leadership, as it had been in history and will logically be in the future.

There is another feature of the proposed revolutionary development of the GBM basin. The environmental system is not conscious of the political subdivisions that humanity has made. Its scientific development has to transcend these divisions. This is particularly applicable in terms of management of water. Water does not recognize political boundaries. Nepal and Bangladesh have, thus, to be considered as integral parts of the GBM basin for its scientific development.

Real life collaboration is a different story. Attempts at collaborative development of India's and Nepal's waters have been made, but it is a story of distrust and inaction. Similarly, development of GBM waters involves Bangladesh integrally. That is another story of a long dispute and inaction. We believe that the physiographic–hydrologic characteristics of the GBM basin present an opportunity of developing certain unique technological developments, which we have developed, as brought out in the book. This will lead to revolutionizing the development of India's waters, generating collaborative development between India and Nepal, and between India and Bangladesh, each addressing the development with enthusiasm as it contributes to the best development in each country's interest.

The study is part of the recent three-volume study of India's waters. They are related but are also independent. One relates to presentation of the development of India's water as it has been undertaken over time. Another one brings out the proposed revolution in concepts, policies, technologies, and management of water. The third, this one, brings out the development of the Ganga–Brahmaputra–Meghna system, which is one of the world's largest hydrologic systems and the most populated region of the world, accounting for about 10% of the total human population!

There is a correlation of the present studies with earlier work and studies, which may therefore be mentioned. Advances have been made in the art and science of development and management of water since about the mid-twentieth century. The Ford Foundation actively and generously supported the author in promoting the advancement of the scientific development of water in India, through a number of activities. First was communicating to senior policy makers of India that a vast advancement in water resources of India will be required as a second India will be born by the turn of the century. Second was organizing studies of the subject by some Indian scientists in collaboration with the scientists of Harvard University in this area at Harvard. Third was development of the capability of some Indian professionals working for advanced research, with the author at the Indian Institute of Technology Delhi. Three books by the author, in the context of each of these three activities, have come out.

Advances continue to be made, with new concepts and analytical technologies being formulated. Water resources development has to be undertaken in a new context. The author has tried to bring out these advances in book form, entitled *Societal Environmental Systems Management*. It will be published soon.

This book has been written in the context of the author's longtime involvement in the development of the GBM waters, with the start of his professional career in His Majesty's Service in the British regime, which continues today. A limitation in writing about GBM may be mentioned. Much of the information, particularly regarding water quantity estimates, is classified. Therefore, there is a possibility of errors, which is acknowledged and regretted.

The author had the privilege of learning from some of the leaders of engineering and science, in India and internationally—Dr. A.C. Mitra, Dr. A.N. Khosla, Dr. K.L. Rao, Prof. Hunter Rouse, and Prof. Roger Revelle—and later working with them. This is gratefully acknowledged by dedicating the book to them.

Acknowledgment is due to a large number of professionals and doctoral scholars with whom the author worked. The book, as the other two books, owes much to the support of a number of younger colleagues. To mention a few, S.K. Pathak, B.S. Mathur, K.N. Duggal, V.K. Srivastava, D.K. Srivastava, Rema Devi, S.D. Khepar, Y.C. Arya, Ranveer Singh, R.K. Prasad,

U.C. Chaube, M.K. Munshi, B.N. Asthana, D.K. Gupta, (Late) P.K. Bhatia, Lata Singh, A.V. Chaturvedi, E.A.S. Sarma, Pramod Deb, and C. Thangraj, former doctoral scholars and later senior academicians or real life engineers, contributed much to the author's education. The author owes much to his colleagues from Harvard University, (Late) Prof. Roger Revelle, (Late) Prof. J.J. Huntington, and Prof. Peter Rogers. Acknowledgment is duly made of their support. Acknowledgment is also due to all the scholars whose work is referred to in the book, as the author's understanding of the subject owes much to them.

The book has been made possible thanks to the support of Dr. B.N. Asthana, who has been kind enough to undertake the editing of the book. His generous support is acknowledged with greatest appreciation. Finally, Shri D.C. Thakur took over the responsibility of editing the drawings and undertaking the publication process with the publishers, which is acknowledged and deeply appreciated.

Last but not the least, my biggest debt is to my wife, (Late) Prof. Vipula Chaturvedi, who was a close associate in life, education, and research. We shared some of the most exciting time as students together in the USA and later as faculty members at several U.S. universities. She encouraged and supported me, even while I neglected her during the course of my engagement on the task of writing several books.

Mahesh C. Chaturvedi
New Delhi, India
Cambridge, MA, USA

Part I

System Characteristics

1

Introduction

1.1 Introduction

The river Ganga, its tributaries, and the flat and fertile plains through which they flow are one of the world's richest natural resources. For thousands of years, the fertile land and abundant water have provided the foundation for an increasingly civilized society based on agriculture, and have led to the development of one of the world's largest human populations. Large empires were established. The Ganga basin was one of the most developed regions of the world until the end of the eighteenth century, attracting the attention of the growing Western world powers (Spear, 1978).

The prosperity, however, was only part of a creamy layer, with the bulk of the people suffering from mass rural poverty (Gailbraith, 1989). For a variety of reasons, the society was trapped into a quasi-steady socioeconomic–environmental state. Agriculture, which was the essential economic activity, remained traditional and at a subsistence level, with little surplus. Irrigation was essential for agriculture in view of the hydrologic conditions of the arid–monsoon climate, and it has been practiced for millennia, chiefly to protect against the uncertainties of the monsoon rains. Deeply embedded cultural, social, and economic practices inhibited modernization of irrigation and agriculture. As a result, the population remained overwhelmingly rural and desperately poor.

A different world history was being written by the Western world from the fifteenth century, for several basic reasons (Landes, 1999). Conditions of agriculture were also improving (Boserup, 1965, 1981). They were not constrained by the hydrologic characteristics as in India. They started on a new human path of development. In this process, they obtained most of the new world and conquered or subjugated the old world. The British conquered India, starting from the Battle of Plassey on June 23, 1757, in which "the British won, and winning changed Indian history" (Landes, 1999, p. 160). The entire country was won by 1818 (Spear, 1978). It led to the establishment of the British rule in India, contributing much to the degradation of India on one hand and the development of Great Britain and the Western world on the other. The Ganges basin, which was the centerpiece of the British Indian Empire, suffered worst. As the industrial revolution progressed in Western Europe from the eighteenth century, it started changing the face of earth and people. With its unfolding, the Western world increasingly became more powerful and affluent, conquering or dominating the rest of the world.

The schism in power and affluence continues to widen. Although the Earth is only one, the world has been fractured into a First and Third World, one glorious and dominant, the other prostrate and bleeding. They are poles apart in terms of socioeconomic–technological and environmental development. The rich, a miniscule proportion, are developed socially and economically, and they rule and enjoy life and the world. The rest, dominant in numbers but abjectly poor, just strive to survive. In terms of income and disparities, the world of today is in the shape of a champagne glass (Figure 1.1). China and India, once leaders of the human society, are currently among its poorest components, each accounting for about 20% of the total. The process continues, increasing at an exponential scale and poses serious challenges for humanity and the environment.

The Ganga–Brahmaputra–Meghna (GBM) basin, once the richest region of the world, holds a major share of these deprived and poor people. Some salient statistics about the conditions in the GBM basin are brought out in

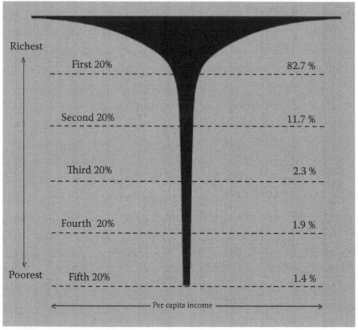

Source Extracted from Human Development Report, UNDP (2005)
Website: http://hdrundp.org

FIGURE 1.1
Distribution of world income and disparities.

Table 1.1. The statistics, however, do not bring out the tremendous challenges—social, economic, and environmental. The developing countries have first to meet their basic needs of drinking water, food, shelter, sanitation, and so on, even which are not available to most of the people. The economy of these societies is predominantly agricultural, in which management of water is crucial in view of the hydrologic characteristics of monsoons. It is currently at the lowest level of survival. The energy needs of the poor are primarily met from biomass, putting severe pressure on the forests. For traction, in agriculture and transportation in rural areas, animal power is the main source, requiring maintenance of pasturelands, which are dwindling. Although poverty has been the prevailing condition of most of the people throughout the centuries, in earlier times, it did not inevitably cause unacceptable and lasting environmental damage because populations, by and large, were small and stable and lived within the confines of their existing resource base, maintaining a balance.

With the rapid growth of population in the last several decades, the age-long delicate balance is increasingly being destroyed. The growth forces more and

TABLE 1.1

GBM Basin: Some Development Indicators (in Order of Population)

Item	India	Bangladesh	Nepal	Bhutan
Estimated population (1998) (in millions)	987	128	24	1.7
Population annual growth rate (1995–2000) (%)	1.8	2.2	2.5	2.4
Life expectancy at birth (1997) (years)	62	58	57	53
Infant mortality rate (1997)	71	81	75	87
Under 5 mortality rate (1997)	108	109	104	121
Without access to safe water (1995) (%)	19	21	52	42
Without access to sanitation (1995) (%)	71	65	80	30
Malnourished children under 5 years (%)	53	56	47	38
Illiterate adults (%)	48	62	72	58
Population below poverty line (1989–1994) (%)[a]	53	29	53	?
Per capita GNP (1997) ($)[b]	370	360	220	430
Real per capita GNP (1995) (PPP$)[c]	1422	1382	1145	1382
Human development index (1995)	0.451	0.371	0.351	0.347
Gender development index (1995)	0.424	0.342	0.327	0.330
Arable land (1994–1996) (hectare/capita)	0.17	0.07	0.13	?
Per capita commercial energy use (KgOE)[d]	476	197	320	—
Per capita electricity consumption (KWh)	347	97	43	—

Source: Ahmad, Q.K., A.K. Biswas, R. Rangachari, and M.M. Sainju (eds.), *Ganges–Brahmaputra–Meghna Region: A Framework for Sustainable Development,* University Press, Dhaka, Bangladesh, 2001. With permission.
[a] $1 per day (1989–1994).
[b] GNP, gross national product.
[c] PPP, purchasing power parity.
[c] Annual, 1996. KgOE, kilogram of oil equivalent.

more extensive use of land, water, and forest resources. It thrusts people into more marginal and environmentally sensitive lands and leads to increasing elimination of pasturelands and deforestation. There is an increasing shortage of cultivable land and water. It even uproots men, women, and children from their natural rural habitats, filling the cities with relentlessly growing multitudes of human beings who have no skills, no jobs, and no resources, causing even more environmental degradation, misery, and social tension. Poverty is the prevailing condition in both cities and the countryside, and poverty compels the people to extract from the even shrinking resource base, destroying it in the process in the viscous cycle of poverty, population explosion, and environmental degradation, even in the process of mere survival.

A further frightening scene unfolds with the climate change continuing at a rapid rate. Greenhouse gases will be making summers hotter, melting the glaciers and raising sea levels. Temperatures are projected to rise by between 1.1°C and 6.4°C and sea levels by between 18 and 59 cm this century. While this macabre drama is unfolding all over the world, the countries most affected also tend to be the poorest of the poor and, in some cases, paradoxically, once advanced and prosperous regions. Apparently, they had already strained their natural resources to the limit of their technosocial capability and reached highest sustainable population pressures. The GBM basin, embracing Bangladesh, Nepal, and Northeastern India, is a classic example.

The GBM basin has been one of the world's most populated regions over history and continues as such. The population has been estimated to have remained almost stable at about 100 million from the invasion of Alexander in 324 BC to the end of Akbar's reign in AD 1605 (Thapar, 1977). It was still about 150 million in the beginning of the twentieth century when the first census was carried out. It was about 600 million by 1999 and is still growing, estimated to stabilize at about a billion by the middle of the twenty-first century. The population density is already about 343 persons per square kilometer. With large areas as desert and uninhabitable mountains, the effective population is already about 600 persons per square kilometer. Agriculture, which is still primitive, remains the dominant activity, with about 75% of the population engaged in it.

The GBM is characterized by endemic poverty—being home to about 40% of the total number of the abysmally poor people of the world, with about 40% earning less than a dollar per day and about 80% less than about two dollars per day (WB, 2010). The performance of the region with respect to the social indicators such as economic growth, education, and health is one of the worst in the world. Per capita gross national product (GNP) for Bangladesh, India, and Nepal, the three riparian countries, is (as of 1998) US$350, US$430 and US$210 respectively, even much lower than the global average of US$4890 for the low-income countries. Adult illiteracy, about 50%, is much higher even compared to the average of 29% for all developing countries. The situation is even worse in the case of women compared to men. The countries of the region have put very low emphasis on social aspects of development. The public expenditure on education is only about 3% compared to the global average of 4.8%, and on health only about 1% compared to the global average of 2.5%, even in the developing countries. Access to safe drinking water is very poor and to sanitation abysmal.

Although the region is remarkably rich in natural resources, with nearly 45% of the land of GBM being intensely arable, the per capita availability is very small—about one-tenth of a hectare, which is half of the global average for developing countries. In the energy sector, the GBM countries have a very low dependence on utilization of commercial energy. Against the per capita energy use of about 1680 KgOE globally and 5340 KgOE for high-income countries, the corresponding figures for Bangladesh and India are 197 and 320 KgOE, respectively.

The region is already synonymous with abject poverty, filth, and frightening man-made and natural disasters. It is difficult to imagine life in the future when even now most of the people live, if it can be called living, without adequate food, potable drinking water, clothing, and shelter toiling under a broiling sun from morning till dusk for miserable rewards without revolting, which may not be ruled out if current trends continue. A foreboding future's societal and environmental scenario can be visualized with the increasing natural resources requirement.

These societies, some of which like India and China, have been leaders of human civilization and are not going to stay at the threshold of modernity, peeping at it. They are justifiably trying to join the mainstream of humanity's development. However, this poses a tremendous challenge for the management of the environment. With the average per capita GNP of the developed and developing countries at $21,000 and $1000, respectively, and with the doubling of the latter's population, a 42-fold increase in natural resources requirement works out (Daly, 1993). Even allowing for improvement of technology over the spectrum of the production–consumption cycle and reuse and structural changes in the economy, a very disturbing picture of the future's natural resource scarcity and environmental degradation emerges. Signs of this disturbing scenario are already visible. Rivers have been increasingly polluted, almost being transformed into polluted drains. Scarcity of water is increasingly leading to conflicts about meeting the diverse demands among the states, much less the riparian countries. This has serious sociopolitical implication for the developing countries and even for the developed countries in view of the emerging global village.

In addition to lack of resources—poverty being both a cause and a consequence of environmental damage—reasons for this include lack of awareness, the myopic view of decision makers, lack of technological capability, institutional overlaps and bureaucratic inefficiencies, political and financial vested interests, insensitivity about the interlinkage of environmental management and socioeconomic development, and the sheer difficulty of dealing with a myriad of relatively small-scale natural resource–using activities. Taken together, all these are responsible for the bulk of environmental degradation.

This need not be so. The GBM basin, to reiterate, is one of the world's richest regions in terms of environmental resources, specifically land, water, and forests. It has the most fertile land and has been called the world's richest agricultural estate (Ruttan, 1987). It has one of the world's highest water availabilities per unit of land, second only to the Amazon, but on a comparatively much smaller land area. With a hydroelectric potential that is estimated currently at 150,000 MW at 60% load factor (which is a totally wrong and poor estimate, as we will demonstrate later), most of which is undeveloped, it has one of the world's richest potential of renewable, nonpolluting energy.

However, the foreboding challenge of environmental management cannot be met if "business as usual" continues. New concepts, technologies, and planning and management will have to be developed (Agenda 21, 1992; Chaturvedi, 1976, 2001, 2011a, 2011b, 2011c). Development of the GBM or the Greater Ganga, as it has been called, is a great challenge to the art and science of development and management of the environment. We will demonstrate that it can be revolutionized, meeting all the challenges comfortably, in contrast to the current dismal predictions. The environmental management revolution will be at two levels—macro and micro. Our focus will be on the macro level because we are particularly emphasizing a technological revolution in the management of the GBM basin. It may be emphasized that the micro level revolution is equally, if not more, important. It will not be discussed in this book, but it has been covered elsewhere (Chaturvedi, 2011a, 2011c).

The problem of environmental management is interlinked with the overall problem of socioeconomic development, and advances on several fronts will be required to meet the serious challenge. Population control, education, and rapid economic growth are some of the leading issues in a host of activities. Physical environment provides the foundation for the social activities. Our center of attention, however, will be the issue of environmental management, further focusing on water, as it is the key to the sustainable development of the environment, particularly under the hydrologic climatic conditions of the GBM, but it has to be set in the context of societal development challenge. Even water resource development has to be undertaken in a wide context of conjunctive, technological, managerial, and social advancement (Vaidyanathan, 1999, 2010; Pasucal et al., 2009; Kumar, 2010; Damodaran, 2010).[1]

The study thus has four objectives. First, it is hoped that the study would contribute to the sustainable development of the GBM environment to meet the urgent challenge of its rapid socioeconomic development. The study is only an introduction to detailed studies, which hopefully will be urgently taken. Some action has ensued, but much work remains to be done.[2] Second, it is demonstrated that the challenge has to be met creatively. New concepts, developmental policies, and creative approaches are required. It is demonstrated that the physiographic–hydrologic characteristics of the region offer the possibility of some novel technologies that will almost double the available water of the region and of India and increase the hydroelectric potential several fold and that too on an interstate–intertemporal basis. Most importantly, the water will be put in the hands of the farmer, making him independent of the officials and the Gods! Third, it has been argued that a revolutionary change in the science of the management of the environment, called *societal–environmental systems management*, according to which the physical and social–environmental systems have to be considered integrally, backed by participatory transparent modeling, has to be adopted to meet the challenge (Chaturvedi, 2011a). The study, focusing on the GBM basin, becomes a case study for the advancement of this new science. Fourth, the study sets the starting point for implementation and more detailed study with contribution to the art and science of the subject, which hopefully is undertaken on a continuing basis to contribute to indigenous capability at the highest level in India.

Interest has centered on the study of the GBM basin of late by some generalists, but as the organizers themselves clarified, they were merely descriptive.[3] We develop a new revolutionary perspective and novel technologies to revolutionize the development of GBM waters. It is important that a total societal–environmental perspective be kept in view. A revolution has to be brought out in the concepts, technology, and management. The presentation of the GBM basin is made in this perspective. Another clarification may be made. Our focus is on the development of the Greater Ganga and we are not discussing any controversy or some recent efforts on collaborative development between India and China.

1.2 Overview

To enable action, the current development will be reviewed in sufficient detail, and then the needed revolutionary changes will be demonstrated. The study is therefore organized into three parts. Part 1 deals with an introduction of the socioeconomic and environmental scene. Part 2 gives an overview of the development and current state. Part 3 deals with futures challenges and the proposed revolution in the management of the environment.

Our basic thesis is that a comprehensive development approach is required, and we focus on environmental systems management in that context. Although an integrated view of the environmental system is the central theme, the emphasis is on water because the major focus has to be on water in view of the climatic–hydrologic conditions and the agrarian economy. Even in the future, management of water will be of major concern, but it will have to be undertaken with a changed perspective of societal–environmental systems management and through the proposed revolutionary technologies.

In Part 1, after a brief introduction to the subject, an overview of the environmental system is undertaken in Chapter 2. The river basin has a unique geological history leading to a unique physiographic–hydrologic environment that poses challenging opportunities and dilemmas. The physical environment is considered in reasonable detail to bring out appreciation of developmental activities of the past and some novel technologies and proposals for the future.

Management of the environment is embedded in the cultural scene. These are made daunting with the socioeconomic backwardness and the population explosion. This is considered in Chapter 3, with an overview of the current societal political and economic scene at the outset in the chapter. The cultural environment is also presented in reasonable detail, as it is important to appreciate it to formulate appropriate developmental policy and technological options for environmental management.

Environmental management, like all technological and policy matters, has historicity, and therefore, an overview of the historical and current state of development is undertaken in Part 2. First, in Chapter 4, development until the end of the British colonial regime is studied, focusing on India, as there was little development in the independent Nepal or Bhutan, which were the other constituent countries. With the independence, a division of the basin into India and East Pakistan took place. East Pakistan achieved a separate entity shortly and was renamed Bangladesh. The development and current state in each country, Nepal and Bhutan, India, and Bangladesh, are studied in Chapters 5, 6, and 7, respectively, starting with the uppermost countries. Next the intercountry conflicts and their resolution efforts are studied. The Indo–Nepal interaction is analyzed in Chapter 8 and that for Indo–Bangladesh in Chapter 9. Technology is a facet of the social system. It is managed through institutional setups that reflect the culture of the socioeconomic milieu. Institutional setup and management are studied in Chapter 10.

With this background, we proceed to the creative part of the study in Part 3. Future perspectives of resource demand and environmental degradation, if the present policies, technologies, and management efforts continue to be followed, are studied in Chapter 11. Advances that have been made recently in the art and science of water management have been kept in view. A concept of integrated water resources management has been proposed. This has been further advanced by a concept of societal–environmental systems management. These are briefly discussed in Chapter 12, before presenting the proposed revolutionary technologies and policies. (They are separately based on a detailed presentation; Chaturvedi, 2011a.) New policies and technologies will be required to meet the formidable challenge of managing the ecosystem in the context of a large-scale increase in the economy and the population. These aspects are covered in Chapter 13. These advances will be required to be backed by systems studies. Considerable work in this context has been done, focusing on the GBM basin (Chaturvedi, 1987). It is briefly presented. Further advances have been made, and these are also presented in Chapter 13. It is also emphasized that while planning is an important component of the technological management cycle, the real issue is action. The subjects of institutional modernization and suggestions for comprehensive management of water have been presented in Chaturvedi (2011a, 2011c) and are therefore not repeated.

Although environmental processes do not understand political divisions, human perception is based on them. The considerations so far have therefore been in terms of the political boundaries, which have caused both inefficient development and management of the natural resources and conflicts, further compounding the inefficient management of the environmental resources and processes. From both humanistic and scientific considerations, it is logical that integrated development of the GBM basin is undertaken. The physical and hydrological characteristics of the GBM basin particularly dictate that an integrated consideration, rising above the political bickering is taken. This is undertaken in Chapter 14. It will be demonstrated that paradoxically, this is synonymous with country-oriented perception.

A brief conclusion follows in Chapter 15.

Notes

1. It may be clarified that these studies, generally by economists, are unfortunately completely innocent of the technological aspects and their implications.

2. On the recommendation of the author, the Ministry of Water Resources and the Ministry of Environment and Forests, Government of India, have each sanctioned projects under which scientists from all the Indian Institutes of Technology (IIT) and sponsored engineers from the GBM basin official departments shall be working together on a detailed study of the development of the GBM basin. Secondly, a recommendation was made by the author to his alma mater, The University of Iowa, Iowa City, IA, USA, to study the GBM basin in terms of their annual course entitled "International Perspectives in Water Resources Science and Management." They have kindly accepted the suggestion, and they will be sending a group of about 20 students to India in late December 2011 to early January 2012, for about a fortnight to study the GBM basin, with the course organized in collaboration with the IITs, with which the author is also associated.

3. Varghese (1990), a reputed journalist, has carried out a very detailed study of the GBM basin, which gives valuable information. Attempts have been made by him for the region to be studied by a collaboration of interested people from the three countries so that the governments of these countries undertake collaborative development of the region and that the current mutual bickering is resolved (Ahmad et al., 2001). However, as the organizer of one set of these studies clearly stipulated, "these studies are addressed to the lay reader and are not intended to be technical treatises" (Panandikar, 1992).

2

System Environmental Characteristics

2.1 Introduction

The Ganges–Brahmaputra–Meghna (GBM) basin is a unique environmental system and one of the world's richest natural resource estates (Table 2.1). It is the world's largest river system in terms of population and the second largest river system in terms of discharge intensity ($m^3/s/km^2$), which is almost the same as that of the Amazon and way above any other major river system.

The environmental characteristics—the physiography, geology, climate, hydrology, land and soil, fauna and flora, and the people—dictate the developmental policy and therefore have been given in reasonable detail. An engineering perception of the GBM instead of the usual journalistic presentations often made is being emphasized. Secondly, a macro as well as micro development of the GBM basin is being emphasized, as detailed later. Therefore, the environmental system is described in considerable detail.

The biotic characteristics have undergone considerable change with the growing population and economic development. The abiotic factors, land and water, have also been considerably perturbed. We will first consider them in their natural state. Some salient characteristics of the GBM basin, in the perspective of the challenge of development, have already been given at the outset in Table 1.1.

TABLE 2.1

Some Major Characteristics of Some Major World Rivers (Arranged according to Drainage Area)

River	Drainage Area (1000 km²)	Length (km)	Discharge Characteristics		
			Mean (1000 m³/s)	Rank Order	m³/s/km²
Amazon	7050	6400	180	1	0.0255
Mississippi–Missouri	3221	6020	18	8	0.0052
Yangtze	1959	6300	34	4	0.0174
Ganga–Brahmaputra	1621	2897	38	2	0.0237
Indus	1166	2900	5	—	0.0047
Rhine	160	1392	2	—	0.0137

Source: *Encyclopaedia Britannica*, 1982. With permission.

2.2 Location and Physical Geography

The GBM basin occupies a dominant position, located in the northern and northeastern part of the Indian subcontinent between north latitudes 21°58' and 31°30' and east longitudes 73°30' and 97°50'. It is a long, narrow basin surrounded by mountains on all sides except for a low, unnoticeable watershed on the west, which separates the Indus basin from the Ganges basin, and the southeastern tip where the rivers, after confluence, meet the ocean at the Bay of Bengal.

The basin is one drainage system, with the Ganges and Brahmaputra arising from locations close by in the northwest end each flowing in a well-defined region and both come to confluence shortly before they merge in the ocean through several branches in one of the world's biggest deltas. River Meghna, or Barak as it is called in the upstream reaches, meets the two, after their confluence, at the extreme tail end. The GBM basin is shown in Figure 2.1. It is also often called the Ganga system, as River Ganga is dominant and has historically been referred to as such.

The basin covers five countries or parts thereof—India, Nepal, Bangladesh, Bhutan, and China (Tibet), covering an area of 1.75 million km². The latter is almost entirely in the Brahmaputra basin, is isolated and uninhibited presently, and has almost no bearing on the management. Some salient statistics are given in Table 2.2. In area as well as population and development

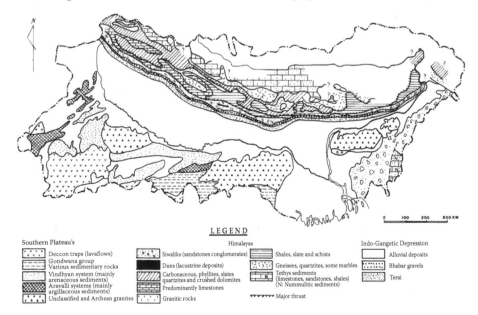

LEGEND

Southern Plateau's

- Deccon traps (lavaflows)
- Gondwana group
- Various sedimentary rocks
- Vindhyan system (mainly arenaceous sediments)
- Aravalli systems (mainly argillaceous sediments)
- Unclassified and Archean granites

Himalayas

- Siwaliks (sandstones conglomerates)
- Duns (lacustrine deposits)
- Carbonaceous, phyllites, slates quartzites and crushed dolomites
- Predominantly limestones
- Granitic rocks

- Shales, slate and schists
- Gneisees, quartzites, some marbles
- Tethys sediments (limestones, sandstones, shales) (N: Nummulitic sediments)
- Major thrust

Indo-Gangetic Depression

- Alluvial deposits
- Bhabar gravels
- Terai

0   100   200   300 KM

FIGURE 2.1
Ganges–Brahmaputra–Meghna basin. (From Bruijnzeel, L.A., C.N. Bremmer, "Highland–Lowland Interactions in the Ganges–Brahmaputra River Basin: A Review of Published Literature," ICIMOD Occasional Paper No. II, Kathmandu, 1989. With permission.)

TABLE 2.2

GBM Salient Statistics

Item	Unit	Ganga			Total or Average Figure for Basin	Brahmaputra		Total or Average Figure for Basin	Barak/Meghna		Total or Average Figure for Basin	Total for Greater Ganga Basin	Total in Each Country		
		India	Nepal	Bangladesh	Basin	India	Bangladesh	Basin	India	Bangladesh	Basin		India	Nepal	Bangladesh
A. Land															
Geographical area	Mha	86.14	14.08	6.74	106.96[a]	18.71	4.7	23.41[b]	4.4	3.62	8.02	138.39	109.25	109.25	15.06
Cultivable area at present cultivation	Mha	60.30	3.98	3.64	67.92	12.15	3.0	15.15	1.11	2.87	3.98	87.05	73.56	3.98	9.51
B. Population															
1971 Census	Million	221.19	11.29	27.03	259.51	17.65	24.83	42.48	5.35	19.10	24.43	326.42	244.17	11.29	70.96
Population density	Per km²	257.00	80.00	401.00	243	94	528	181	121	523	305	236	223	80	471
Agr. population	Million	149.74	10.34	19.06	179.14	11.93	17.5	29.43	3.60	13.46	17.06	222.03	165.27	10.34	50.02
Agr. population	%	67.6	91.6	70.5	69.03	67.6	70.5	69.28	67.6	70.5	69.83	68.02	67.69	91.58	70.5
C. Water															
Mean annual rainfall	cm/yr	60–200	100–250	150–212	120	212	200	212	450	240	350	—	—	—	—
Total annual runoff	mhm	55.01[c]	—	Neg	55.01	5.25	10.25	61.5	6.0	5.1	11.1	127.61	112.26	—	—
Irrigation potential (present)	Mha	17.81	1.202	2.95	21.96	2.3	1.46	3.76	0.115	0.47	0.585	26.305	20.225	1.202	4.88

(continued)

TABLE 2.2 (Continued)

GBM Salient Statistics

| Item | Unit | Ganga | | | Total or Average Figure for Basin | Brahmaputra | | Total or Average Figure for Basin | Barak/Meghna | | Total or Average Figure for Basin | Total for Greater Ganga Basin | Total in Each Country | | |
		India	Nepal	Bangladesh		India	Bangladesh		India	Bangladesh			India	Nepal	Bangladesh
D. Energy															
Hydropower potential	Mkw	13.274	85.00	Neg	98.27	13.43	Neg	13.43	2.5	Neg	2.5	114.2	29.2	85	Neg
Hydropower installed	Mkw	1.83	0.41	Neg	2.24	00.18	Neg	0.18	0.08	Neg	0.08	2.5	2.09	0.41	Neg
Hydropower installed	%	13.79	0.49	Neg	2.33	1.34	0.0	1.34	3.2	Neg	3.2	2.23	7.16	0.49	Neg
E. Unit figures															
Land per capita	Ha	0.39	1.25	0.25	0.4	1.06	0.189	0.55	0.82	0.189	0.33	0.42	0.45	1.25	0.18
Utilizable water/unit cultivable land	m³	116.00	—	—	—	171.0	—	—	379.0	—	—	—	—	—	—
Cropping intensity	%	125	—	—	—	118.0	—	—	124.0	—	—	—	—	—	—
Irrigation intensity	%	23	—	—	—	21.3	—	—	13.0	—	—	—	—	—	—
Utilizable and utilized water resource															
Surface	mhm	18.5/70.4	—	—	—	0.61/70	—	—	0.20/27	—	—	—	19.21	—	—
Ground	mhm	10.64/39.4	—	—	—	0.02/1.1	—	—	0.75/11.3	—	—	—	10.735	—	—
Total	mhm	29.14/59.4	—	—	—	2.68/23.5	—	—	1.03/9.7	—	—	—	32.85	—	—

Note: Neg = Negligible; Agr. = Agricultural. These figures are derived from the Irrigation Commission Report. The figures have been taken from various published sources and are only indicative.

a 2.9 Mha in Tibet not included in this figure.

b 29.3 Mha in Tibet and 5.29 Mha in Bhutan not included in this figure.

c Inclusive of runoff of Nepal.

opportunities, India and Ganges have the central focus (Table 2.3). India's population and area share are 76% and 63%, respectively, while the corresponding share of Bangladesh is 21% and 7%; Nepal, whose almost entire territory is within the Ganga basin, has an 8% share of the area and a 3.5% share of the population. Ganga basin has a prominent place in India (Table 2.4).

The basin has an interesting geological history, which has shaped its unique environmental characteristics. The Indian subcontinent is an individual tectonic plate, which, in the geologic period of 60 million years ago, was moving roughly north at a rate of about 15 to 20 cm per year (Molnar, 1986). About 40 million years ago, the Indian plate began to meet and pass underneath what is now Tibet and its movement slowed to about 5 cm a year. The Himalayas and the Indo-GBM basin developed as a consequence of this

TABLE 2.3

GBM–Indian Characteristics as Part of India

Item	Total for Greater Ganga Basin	Share for India
Land		
Geographical area (10^6 ha)	138.39	109.25
Cultivable area (10^6 ha)	87.05	73.56
Population		
1971 Census (million)	326.42	244.17
Population density (per km^2)	80—645 with an average of 240	223.0
Agricultural population (million)	222.03	165.27
Agricultural population (% of total)	68.02	67.69
Water		
Mean annual rainfall (cm/yr)	60–450 with an average of 350	—
Total annual runoff (m ha m)	127.61	112.26
Irrigation potential (m ha)	26.3	20.2
Energy		
Hydropower potential (MW)	114.2	29.2
Hydropower installed	2.5	2.09
Unit figures		
Land per capita (ha)	0.42	0.45
Utilizable water/unit culturable land (m^3)	116–379	—
Cropping intensity (%)	—	118–125
Irrigation intensity (%)	—	13–23
Utilizable water (m ha m)		
Surface	—	19.21
Ground water	—	10.735

Source: Chaturvedi, M.C., *Water Resources Systems Planning*, Tata McGraw-Hill, New Delhi, 1987. With permission.

TABLE 2.4

Ganga as Part of India

Item	Ganga Basin	Total India	Ganga (% of Total)
Land			
Geographical area (10^6 ha)	86.14	328.06	26.3
Cultivable area (10^6 ha)	60.30	199.92	30.1
Population			
1971 Census (million)	221.19	546.9	40.4
Population density (per mi.^{2})	286–1180	470	—
Rural population (%)	80–93	—	—
Cultivators and agricultural labor (%)	55 + 22	of total labor force	
Water			
Mean annual precipitation (cm/yr)	60–200	118	—
Potential evapotranspiration (cm/yr)	155–175	228	93
Total annual runoff (m ha m)	55.01	188.12	29.3
Groundwater (% of total)	26.6	—	—
Utilizable water resources			
Surface	18.50	66.60	27.8
Surface as percent of total	34	—	—
Groundwater	10.64	26.10	41.0
Total	29.14	92.70	31.4
Unit figures			
Land/Capita (ha)	0.27	0.37	
Utilizable water/unit culturable land (m^3)	116.0	115.00	100%
Cropping intensity	125	—	—
Irrigation intensity	23.0	20.6	—
Hydropower potential (MW)	13,285	—	—
Hydropower installed (MW)	1,830	—	—

Source: Chaturvedi, M.C., *Water Resources Systems Planning,* Tata McGraw-Hill, New Delhi, 1987. With permission.

intercontinental collision between the Eurasian plate in the north and the Indian plate in the south, and its aftermath over a geological time (Wadia, 1975, 1986; Jhingran, 1982; Molnar, 1986). The intercontinental collision is believed to have taken place along the ocean floor and the line of subduction is considered to be along the Indus–Tsango (Brahmaputra) Suture where the two plates joined. As the Southern Gondwanaland kept pressing forward, the Tibetan Plateau was heaved out of the sea to the north of the Tibetan Himalaya, damming rivers and streams into remnant tectonic lakes.

In response to successive thrusts, Himalayas developed as a series of several, more or less, parallel or converging ranges, trending northwest to southeast. The width of the Himalayas is between 160 and 400 km and the length of the central range is about 2500 km from west to east between Nanga Parbat (8126 m) in Jammu and Kashmir and Namcha Barwa (7736 m) in Tibet, where Brahmaputra takes an abrupt southern turn. The Himalayan mountain ranges consist essentially of the scrapings from the forward edge of the Indian plate forced backward (i.e., to the south) over the advancing mass. The individual ranges generally present a steep face toward the south and a gentler inclined slope toward the north. The extremities of the range are generally considered to be marked by the two "syntaxes" or great bends of the Indus in Kashmir and the Brahmaputra in upper Assam.

On the southern end of the Himalayas is the Indian peninsular shield consisting of Aravali, the Vindhyan ranges and Southeastern plateau. They traverse almost the whole width of peninsular India. Together with the Satpura range, which runs almost parallel to it, the Vindhyan range forms the major watershed of the subcontinent, dividing north India from Deccan. The Vindhyan ranges are geologically very old, mostly granite and basalt rocks and have wide valleys. While the average elevation of the Vindhyas is only 300 m, several peaks in Satpura rise above 1000 m. On the western flank of the Vindhyas, the Aravali Mountains bridge the gap between the Vindhyas and the Rajasthan desert. The Aravali Mountains are an ancient, heavily eroded chain of folded mountains, which are up to 1300 m high.

Over the geological period, Indus and Brahmaputra flowed west and east, respectively, along the trough, while the Ganga and its tributaries, and the tributaries of Indus, cut through the Himalayan ranges to flow south. It is interesting to note that the world's three biggest rivers, in terms of discharge, have almost the same starting point, each however flowing in different directions. Over the geologic time, severe climatic and hydrologic changes took place. The Himalayan mountain system was cut, heavily eroded, and the sediment load deposited in the trough in between Himalayas and the Vindhyas leading to the Great Indo-Gangetic plain. These alluvial plains are outcome of these hydrologic processes and geological activities, resulting in rich, fertile land consisting of unconsolidated riverborne sediments thousands of meters deep. Though the nature of the alluvial sediment is generally complex and there are quick alternations of fine and coarse stratum, it is usually possible to group a large number of layers into a limited number of distinct groups, each having a family likeness, to lead to meaningful fence charts as discussed later. The region is very rich in groundwater of purest quality. A large network of large rivers and smaller channels serves the plains.

The Ganges basin with a drainage area of 1,087,400 km^2 gives the characteristic shape and character to the basin. The basin covers practically whole of the northern India between the Himalayas and Vindhyas except the states of Jammu, Kashmir, and Punjab. It covers more than one fourth (26.3%) of India and is inhabited by almost half of the Indian population. It is the center of India's history and civilization. Several major tributaries of river Ganges from north and south drain into it.

The Brahmaputra basin extends over an area of nearly 580,000 km². It lies between east longitude 82°0′ and 97°50′ and north latitudes 25°10′ and 31°30′. The basin is irregular in shape and can be considered in two parts. The first is the upper portion lying on the north of the Great Himalaya, principally in China (Tibet), in which the river flows from west to east. This region comprises mostly mountain ranges and narrow valleys, almost desolate, with little biotic components or human population. The second portion is after river Brahmaputra takes a sharp, almost 180° bend, at the eastern most end and enters India, with its tributaries draining India and Bhutan. In this portion, the region is mostly hilly with a narrow river flowing in between. The width is only about 80 km on the average with hills, forests, and tea gardens. It widens gradually, takes an abrupt south turn, and enters Bangladesh meeting the Ganges shortly later forming the great Ganga–Brahmaputra delta. This part is bounded by the Himalayas and the ridge separating it from the Ganga basin on the north and west, by Patki, Naga, and Lushai hills running along the Assam–Burma border on the east and on the south by the Assam range of hills.

The Meghna or Barak basin is located at the lowermost part of the GBM basin. It extends over an area of about 7800 km². It is bounded on the north by the Barail range, which separates it from Brahmaputra, on the east by the Naga and Lushi hills, on the south and west, the delta. Originating from the southern hills of Nagaland, Barak flows through Manipur and Cachar in Assam in India to enter Bangladesh, where it is known as Meghna. There are two physiographic regions in the basin, namely, the hilly regions and the plains. The plains are very thickly populated and cultivated.

Mention may be made of river Saraswati, as its discovery is a remarkable scientific and technological achievement, which provides considerable geophysical and geological information. Saraswati River is referred to and described in the Hindu Vedas as a mighty river taking off from the Himalayas from about the same source as the trinity of Indus, Ganges, and Brahmaputra and flowing down to the Arabian Sea, in between the Indus and the Ganges system. There has been no description of the river or trace in history. Recent geological and satellite imagery investigations have confirmed the geological existence of the river. It followed the course of the present Ghaggar in Punjab, Haryana, and Rajasthan. A rich civilization developed in the Saraswati basin as archeological findings have confirmed. The river disappeared for three reasons: (1) diversion of its tributary Yamuna toward the east for rendezvous with river Ganges at Allahabad, (2) diversion of its tributary Sutlej toward the west at Ropar for a rendezvous with river Beas, and (3) seepage of the remaining river water in underground paleochannels (Parpola, 1994; Valdiya, 1996; Kalaynaraman, 2000). Two practical implications of the findings need to be noted: (1) consciousness of the geological–climatic forces and (2) consciousness of the hydrogeological implications of this history.

2.3 Physiography and Geology

The basin can be considered in terms of three characteristic features: the young and active Himalayan ranges in the north, the old and geologically stable shield area in the south, and the Indo-Gangetic depression in between. The main geostructural features of the GBM basin are shown in Figures 2.2 and 2.3 and are discussed briefly.

2.3.1 Physiography of the Himalayas

From south to north, the Himalayan ranges can be grouped into four parallel longitudinal belts of varying width each having distinct physiographic features and its own geological history. They are designated as (i) the Outer or Sub-Himalayas, (ii) the Lesser or the Lower Himalayas, (iii) the Great or Higher Himalayas, and (iv) the Tethys or Tibetan Himalayas. Further north lie the Trans-Himalayas in Tibet proper as an eastward continuation of some of the most northerly Himalayan ranges. The boundaries between the four major units are mainly of a geological nature. Two important features in this respect are the so-called "main boundary thrust" (MBT) and the "main central thrust" (MCT). From west to east, the Himalayas are divided broadly into three mountainous regions: Western, Central, and Eastern, with the first one in the Indus basin and the latter two in the GBM basin. The physiographic zones of the Nepal Himalaya, characteristic of the Himalayas, are shown in Figure 2.4.

The Outer Himalayas comprise flat-floored structural valleys and the Shivalik Hills (also known as Chunia hills in Nepal), which border the Himalayan mountain system to the south. Except for small gaps in the east, the Shivaliks run for the entire length of the Himalayas with a maximum width of 100 km in the Indian State of Himachal Pradesh. In general, the 280-m contour line marks their southern boundary; they rise to another 760 m to the north. The Shivalik zone is an extremely rough landscape. The main Shivalik range has steeper southern slopes facing the Indian plains and descends gently northward to flat-floored basins called *Dun*. The best known of these is the Dehra Dun, in Uttaranchal, which is the mountainous part of former Uttar Pradesh (UP). There is a very abrupt transition, generally referred to as the Himalayan mountain front, between the Shivaliks hills and the great Gangetic Plain to the south. As the slope changes abruptly, the bed load of boulders and coarse sediment is deposited in the so-called Bhabar zone. As the slope becomes abruptly flat, the ground water table almost meets the ground levels leading to wet marshy lands known as Terai.

Northward, the Shivalik Range abuts against an 80 km wide massive mountainous tract, the Lesser Himalayas where mountains rising to 4500 m and valleys with altitudes of 900 m run in different directions. It is an intricate network of valley and ridges generally following the main west–northwest

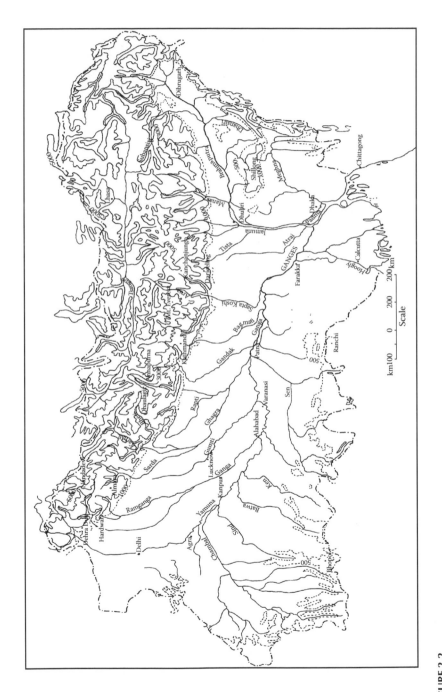

FIGURE 2.2

Generalized geological map of GBM basin. (From Bruijnzeel, L.A., C.N. Bremmer, "Highland–Lowland Interactions in the Ganges–Brahmaputra River Basin: A Review of Published Literature," ICIMOD Occasional Paper No. II, Kathmandu, 1989. With permission.)

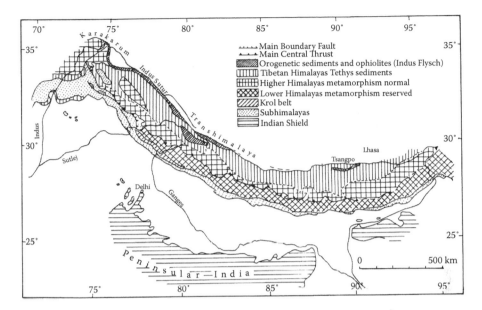

FIGURE 2.3
Main geostructural features of the GBM basin. (From Bruijnzeel, L.A., C.N. Bremmer, "Highland–Lowland Interactions in the Ganges–Brahmaputra River Basin: A Review of Published Literature," ICIMOD Occasional Paper No. II, Kathmandu, 1989. With permission.)

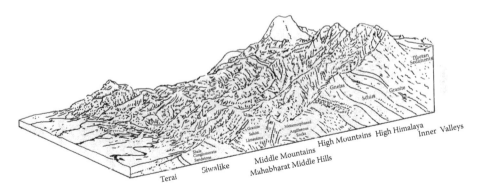

FIGURE 2.4
Physiographic zones of the Nepal Himalaya. (From Bruijnzeel, L.A., C.N. Bremmer, "Highland–Lowland Interactions in the Ganges–Brahmaputra River Basin: A Review of Published Literature," ICIMOD Occasional Paper No. II, Kathmandu, 1989. With permission.)

to east–southeast terrain of the country. There is a general conformity of altitude among neighboring summits, which creates the appearance of a highly dissected plateau. It is an area of steep relief, high rainfall, and very high erosion. In Uttarakhand, the Lesser Himalayas is a homogeneous physiographic unit. In Nepal it can be subdivided into the Lesser Himalayas proper, grading to the north into the ranges of High Himalayas; the midland valleys and basin, the most populous part of Nepal with elevations in the range of 600 to 2000 m; and the Mahabharat ranges which rise to altitudes of up to 3000 m. Rivers of the Nepalese Himalayas, which generally follow the east–west trend of the Midland valleys, cut through the Mahabharat in very narrow, deep gorges.

The backbone of the Himalayan system is formed by the Great Himalayas, a single high range rising above the line of perpetual snow. Their average height extends to 6000 m. The Great Himalayan Range rises to its maximum height in Nepal, having in that section nine of the 14 highest peaks of the world. From west to east they are Dhaulagiri (8172 m), Annapurna (8078 m), Manashi (8156 m), Kao-sent-tsan Feng (Gosainath, 8014 m), Cho Oyu (8153 m), Mt. Everest (8848 m), Makalu (8481 m), Lhotse (8511 m), and Kangchenjunga (8598 m). Further east, the range changes from a southeasterly to an easterly direction as it enters Sikkim. After this, it runs eastward for another 320 km through Bhutan and the eastern part of Arunachal Pradesh as far as the peak of Kangto (7096 m) and finally turns northeast, terminating in the Namcha Barwa.

There is no sharp boundary between the Great Himalayas and the ranges, plateaus, and basins lying to the north of the Great Himalayan Range, generally grouped together under the name of Tethys Himalayas and extending northward into Tibet. In Kashmir, the Tethys are its widest, forming the Spiti Basin and the Zaskar mountains, the highest peaks of which to the southeast are Leo Pargil (6719 m), rising north of the Sutlej River opposite Shipki La (pass) and Shilla (7026 m).

2.3.2 Geology of the Himalayas

A study of the geological history of the Himalayas reveals that marine sediments of the Paleozoic and Mesozoic eras (between about 65,000,000 and 570,000,000 years ago) deposited on the floor of the ancient Thethys Sea, the frontal part of the crystalline massif (mountain mass) of peninsular India, estuarine deposits along the flanks of the embryonic mountains, and, finally, products of surface erosion of the rising mountains all contributed to the formation of the present-day range (*Encyclopaedia Britannica*, 1984). The uplift of the Himalayas took place in at least three distinct and widely separated phases. The first phase of the major mountain-building movement took place at the close of the Eocene Epoch (about 38,000,000 years ago), although the beginning of the main Himalayan uplift started in Middle and Upper Cretaceous times (from about 100,000,000 to 65,000,000 years ago), with the advancement of the crystalline massif of peninsular India toward the Plateau of Tibet. This movement caused the rise of the Tethys Himalayas, along with the greater part of the Great Himalayas. In the second phase of upheaval, which occurred in the Miocene Epoch (7,000,000 to 26,000,000 years ago), the estuarine deposits and the Indian Massif formed the ranges of the Lesser Himalayas. The final mountain-building phase started at the end of the Tertiary Period (about 7,000,000 years ago), lifting the detrital deposits accumulating at the base of the Himalayas to form the Shivalik Range, the foothills of the Outer Himalayas. Since the middle Pleistocene Epoch (about 1,500,000 years ago), the Himalayas have risen at least 1372 m, an occurrence witnessed by the early man. That the Himalayas still continue to rise is evidenced by the upheaval of the younger river terraces. It is estimated that India moves about 5 cm per year under Tibet, the central Great Himalayas rise about 1 cm per year in altitude, and slippage in this process generates "a roughly annual moderate earthquake and the rare but all too frequent great earthquakes" (Molnar, 1986, p. 154).

Precambrian metamorphic rocks (rocks formed by heat and pressure from 570,000,000 to 4,600,000,000 years ago) form the bulk of the Himalayas. These rocks represent the frontal part of the Indian Shield, which, according to the theory of continental drift, pushed northward, uplifting the Himalayas as it presses against the Asian landmass. Only in the Spiti Basin and in a few other localities can large outcrops of marine sediments from Paleozoic and Mesozoic times be seen.

Plutonic rocks (formed deep down from a molten state), such as granites and granodiorites of pre-Miocene age (i.e., more than 26,000,000 years old), outcrop in extensive areas in north Kashmir. The intrusion during post-Miocene times (i.e., within the last 7,000,000 years) of tourmaline granite into an older series of gneisses (rocks formed by heat and pressure and made of bands that differ in color and composition) and schists (crystalline rocks, the constituent minerals of which are usually arranged in a foliated or parallel pattern), aided by up-thrusts in many areas, has given rise to many of the high peaks of the Himalayas such as Makalu, Manaslu, and Nanga Parbat, which are typical examples. Mt. Everest and its two associated peaks, Lhotse and Cho Oyu, are, however, formed of limestone and pelitic (clay and mudstone) rocks, the latter dipping toward the north. It is possible that the entire formation of Mt. Everest is thrust up over a foundation of gneiss, as nappes or overturned folds.

The nappe structure of the Himalayas can also be seen elsewhere in the ranges. The Krol Nappes of the Simla region, in Himachal Pradesh, and the Garhwal Nappes of Uttarakhand, are typical. These nappes are also evident in Nepal in the Nawakot and Kathmandu areas, which were formed along the line of the main central thrusts. This thrust zone borders the Great Himalayas to the south, rising abruptly in height and showing changes in the sequence of geological beds. The Pir Panjal Range, for example, owes its origin to thrust faulting; in the Simla region, Krol limestones of the Carboniferous Period (i.e., from 280,000,000 to 345,000,000 years old) are overthrust onto much younger deposits of Pliocene Epoch (from 7,000,000 to 2,500,000 years old).

The geological structure is much simpler in the Outer Himalayas, where the foothills are mainly composed of tertiary formations (from 2,500,000 to 65,000,000 years old), grouped under the Lower, Middle and Upper Shivaliks. These consist mostly of freshwater deposits, such as sandstone, shale, and conglomerates. The Lower Shivaliks are from 550 to 1800 m thick, the middle Shivaliks from 900 to 1400 m thick, and the Upper Shivaliks from 1400 to 1800 m thick. At the same time that the Shivalik deposits were occurring, lacustrine (lake) deposits known as Karewas (flat-topped terraces) were being formed in the Vale of Kashmir. Both the Karewas and Shivaliks show evidence of glaciation during Pleistocene times from 10,000 to 2,500,000 years ago.

2.3.3 Physiography and Geology of the Vindhyas

The southern plateau consists largely of very old (Panaezoic and older) crystalline and sedimentary rocks and especially in the southwestern part of the basin by the hard lava flows known as the Deccan traps. Although the shield occasionally attains maximum heights of about 1000 m, most of the area lies between 300 and 600 m. This not only represents the mature character of the region, but also moderate subrecent geological uplift associated with Himalayan Orogeny. As such, rivers emanating from the areas have somewhat higher gradients than would be expected from such an old landscape. In addition, these rivers have flashy flow regimes and have shifted their beds upon entering the Ganges plain for thousands of years. However, despite very serious surface erosion in the area, sediment loads carried by most southern tributaries of the Ganges are only a fraction of those carried by the northern tributaries.

2.3.4 Indo-Gangetic Depression

The Indian plate has been bent downward by the weight of the Himalaya, thus creating a major structural basin, filled with debris from the mountains in the north, and to a lesser extent from the southern plateau. The northern deposits are often grouped into a piedmont (footslope) zone and more low-lying alluvial deposits. The entire complex is commonly referred to as the Terai, although the upper part of the piedmont is known as the Bhabar zone. The Bhabar formation essentially consists of very coarse-textured alluvial fan deposits topped by a thin layer of finer textured material.

The thickness of the sediment is at a maximum close to the Shivaliks and may reach 5000 m (Carson et al., 1986). Soils are well- to excessively-drained and pose drought problems. Forest clearance has been limited up to now, mainly for this reason, but where it has occurred, serious and gully erosion has been the result. Because of the dramatic reduction in river gradients upon leaving the Shivaliks, streams, which are heavily laden with sediment during the monsoon, assume a braided pattern. Riverbeds can be up to several kilometers wide and erosion is lateral rather than vertical. Indeed, the piedmont zone south of the Himalaya is one of the most impressive (and active) of any such system on earth.

Much of the water infiltrating into the coarse Bhabar deposits emerges a few kilometers downstream, producing a sudden change in river morphology from braided to meandering. This coincides with a knick point in river gradient, with the deposits becoming increasingly finer textured to the south. Also, drainage density increases considerably below this 'line of saturation.' From here onward (in the Terai proper), a system of sandy levees and more clayey depressions ("basins") is found, which bears no relationship to the present-day drainage pattern, although the depressions get flooded occasionally. Rice is widely cultivated in these "basins," whereas the (higher) levees are used for settlements and roads. In between the old basin/levee system and the active river channels, another subrecent system exists, called the Meander Flood plain. Its channels are only flooded during extreme events.

In contrast to all other physiographic zones described above, the Terai is a zone of deposition rather than erosion. In addition, it suffers greatly from the rapid and unpredictable shifting of riverbeds (Carson, 1985). The subtropical forests of the Terai are being converted to agricultural fields at a rapid rate.

The alluvial deposits of the Ganges plain are mainly composed of unconsolidated beds of sand and gravel (former channel beds), silt and clay (former depressions), and their mixture in varying proportions. Although the alluvial filling on an average is 1300–1499 m deep, decreasing gradually southward, a zone of over 8000 m depth runs along the foot of the Himalayas (Singh and Verma, 1987).

A distinction is made between the Bhagar, or high (greater than 15 m above the plains) interfluvial zones above the general limit of flooding, and the khadar, the more low-lying riverine tracts whose sandy to clayey deposits are annually renewed.

2.3.5 Brahmaputra River Basin

Excluding the deltaic parts for the moment, the Brahmaputra River Basin can be subdivided into four major physiographic units; viz. the old Meghalaya tableland and the younger Patkai–Naga (or Purvanchal) ranges in the south, the central Assam Valley depression, and the eastern extension of the Himalaya in the north.

The Meghalaya plateau rises to elevations of 600 m in the west to about 2000 m in the center and consists mainly of very old (Pre-Cambrian) hard crystalline rocks (granites and the like). Along the western and southern margins, flat-lying sedimentary rocks (dominated by sand and limestone of Mesozoic to tertiary age) are exposed. Whereas, the central and eastern parts form a true plateau, the western and northern fringes are highly dissected, forming a series of irregular low hills down to the Assam valley (Das et al., 1987).

To the east of the plateau and separated from it and the Assam valley by a major fault, the Purvanchal is found. It consists of a strongly fractured series of N–S to NE–SW trending ridges. Rocks comprise a variety of tertiary sediments; topography is steep and river valleys narrow. In addition, the area is extremely unstable in terms of seismicity due to its proximity to the main transform fault along which the Indian and Southeast Asian tectonic plates rub shoulders (Haroun-er Rashid, 1977). Elevations range from a low 150 m in the southwest to over 3000 m in the extreme northeast. Most of the area, however, is found between 900 and 2100 m.

The Assam Valley is an almost flat plain underlain by some 1500 m of alluvium. Its width ranges from about 90 km at the upstream end to about 50 km lower down. Within the plain a number of isolated granitic hillocks that have become detached from the Meghalaya plateau are found. As can be expected from the contrasts in geology between the mountains in the north and in the south, the physiography of the two riverbanks differs markedly, especially in the western part of the valley.

In the north, a situation similar to that described earlier for the Terai exists, with braided streams that start to meander upon passing the line of saturation. However, before joining the Brahmaputra, these rivers run almost parallel to the main stream as they encounter its levees. In the south, on the other hand, the valley is much less wide and the small tributaries flowing from the Meghalaya plateau run in much less meandering courses.

The Eastern Himalaya runs through Sikkim, Bhutan, and Arunachal Pradesh. The general division of Shivaliks (locally called Duars), Middle and Great Himalaya and Tibetan Marginal Range still applies here, although the topography of the Middle Himalayas now rises steadily and merges with the Great Himalaya. As such, the latter do not stand out as much as they do in Nepal and further west. Also, the depression associated with the Central Midlands is hardly developed here.

Instead, there is a growing tendency for spurs from the Great Himalaya to radiate southward (e.g., the Black Mountains in Bhutan) as one moves toward the east. This is in line with the general change of direction in the axis of the mountains as they approach the eastern end of the Indian tectonic plate.

The Bengal Basin has been filled with sediments washed down from the surrounding highlands, mostly since Pleistocene times. Morphologically speaking, the area consists of the flood plains of the rivers traversing it, with all the features common to such systems ending up in a major delta. Although the basin as a whole is a zone of deposition, it is nevertheless (like the Indo-Gangetic depression) subject to tectonic movements, with some parts actively sinking and others rising. Indeed, the change in the course of the Brahmaputra in 1787 during a single flood event has been described as having become possible because of such movements (Morgan and McIntire, 1959). Estimates of the rates of sinking vary from 2 to 6 cm/yr. (Haroun-er Rashid, 1977). Needless to say, such areas have become even more liable to flooding than before.

2.4 Climate and Hydrology

Two features which strongly influence the climatic and hydrologic characteristics of the basin are the Himalayas, as a great climatic divide affecting air and water circulation system and the monsoon climate. By its situation and the stupendous height, the Great Himalayan Range obstructs the passage of cold continental air from the north into India in the winter and also forces the southwest monsoon rain bearing winds to give up most of the moisture before crossing the range northward thus causing heavy precipitation of rain and snow on the Indian side but arid conditions in Tibet.

2.4.1 Climate

The basin is in the tropics and the climate is temperate, except for the areas in the Himalayan Mountains, which can be extremely cold. In the plains, particularly in the northern regions adjoining the Himalayas, the winters are quite cold, with the temperatures falling as low as 1°C. As one moves south even the winters become milder being moderated by the presence of large water bodies on the land and the proximity of the sea on the eastern side. In the plains in the Vindhyan region, it becomes very warm, with the maximum temperature rising up to 48°C.

2.4.2 Precipitation

One of the most distinguishing characteristics of the basin, besides the geological formation, is the monsoon climate of the country. The rainfall, though varying from region to region, is dominated by the southwest monsoon from June to October, which accounts for almost 80%–90% of the precipitation. Secondly, even during this period, it is not a continuous period of rainfall, but the precipitation is often concentrated in short spells of over a few days and the timing and amount varies widely. The variability increases in regions having lesser precipitation. This seasonality of rainfall, with 50% of the precipitation falling in just 15 days and over 90% of the river flows on just four months, has shaped the society and history.

The GBM basin is abundant in water, with an overall average of 1500 mm of rainfall a year. However, annual rainfall decreases as one moves west from 3000 mm at the Bangladesh coast to 600 mm in Rajasthan to the extreme west and 450 mm in Lhasa, Tibet in the rain shadow of the Himalayas. The mean annual and seasonal rainfall (June–September) is shown in Figure 2.5. The coefficient of variation of annual rainfall is shown in Figure 2.6.

Rainfall over much of the area is concentrated during the monsoon and the isohyetal patterns for these months and the annual totals are quite similar. The eastern part of the river basin is subjected to the influence of the monsoon for a longer time and it receives considerably higher rainfall. The coefficient of variation of annual rainfall, therefore, is much less.

Apart from this overall trend of lesser rainfall toward the west, there are marked topographical effects producing strong local variations throughout

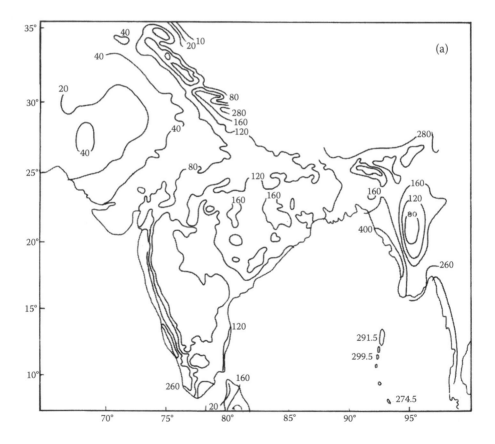

FIGURE 2.5
Mean annual (a) and mean seasonal (June–September) (b) rainfall over the Indian subcontinent. (From Bruijnzeel, L.A., C.N. Bremmer, "Highland–Lowland Interactions in the Ganges–Brahmaputra River Basin: A Review of Published Literature," ICIMOD Occasional Paper No. II, Kathmandu, 1989. With permission.)

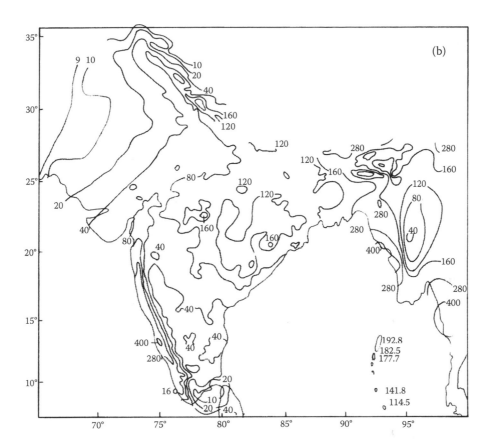

FIGURE 2.5 (Continued)

the entire river basin. For example, the world's highest rainfall of about 11,000 mm per annum has been recorded at Cherrapunji on the southern slopes of the Meghalaya plateau, where the moist air from the Bay of Bengal suddenly rises by 1200 m. However, in Shillong, situated only 50 km to the north on the same plateau, rainfall decreases to about 2400 mm/year and further declines steadily to 1600–1800 mm/year as one descends into the Assam valley. It rises again to 4000 mm/year on the lower Himalayas in the slopes of Arunachal Pradesh.

While the regional variation in rainfall is considerably uniform over the plains, the various longitudinal physiographic zones of the Himalayas, with their strong contrasts in elevation, experience widely varying amounts of rainfall.

The rainfall is quite variable from year to year as brought out by the coefficient of variation shown in Figure 2.6. It does not, however, give the total

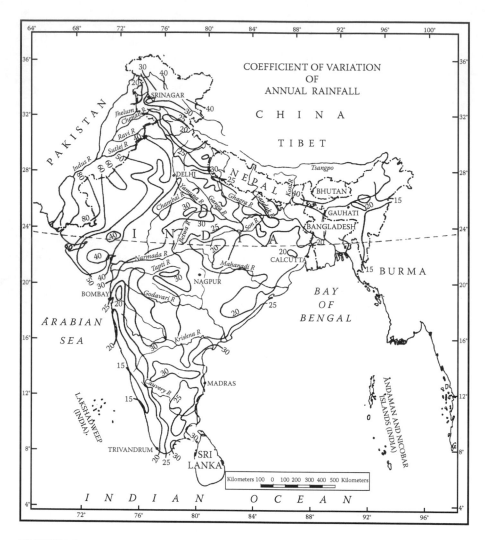

FIGURE 2.6
Coefficient of variation of annual rainfall.

picture. In four years (1877, 1899, 1918, and 1972), more than 40% of India suffered a severe drought (over 70% in 1991) (Mooley and Parathasarthy, 1983). Famines have been a regular feature in India before the introduction of irrigation in the British regime (Bhatia, 1967).

Similarly, excessive rainfall was widespread over India in 1878, 1892, 1938, and 1961 with about 40% of the country being affected during the extreme event of 1892 (Mooley and Parthasarthy, 1983). The drought and excessive rainfalls are not regionally correlated. For instance, droughts and excessive rains in western UP would often show up in eastern UP as well, but not necessarily in sub-Himalayan Bengal and generally not at all in Assam.

The occurrence of a very wet or dry year seems to be related to the degree to which depressions are able to penetrate toward the west. In addition, the regional distribution of rainfall appears to be strongly related to the location of the monsoon trough, a zone of relatively low pressure normally running between southern Bengal and northwestern Rajasthan. The trough may shift toward the foothills of the Himalaya, producing a marked decrease in rainfall over India to the south of the trough, but a distinct increase over the Himalayas (Dhar et al., 1982; Ramaswamy, 1962) has described the synoptic of this situation, which is commonly referred to as a "break" in the monsoon.

Extreme amounts of rain falling over periods of one to several days are obviously of great practical significance because of their role in generating major floods. This is especially so during the height of the summer monsoon, when soils all over the river basin are quite wet. The highest daily rainfall ever observed in the Ganga basin amounted to 823 mm, recorded at Nagina (UP) in September 1880. The associated 2-day total exceeded 1 m of water, viz. 1042 mm.

The corresponding maximum daily rainfall in the Brahmaputra basin amounted to 1036 mm at Cherrapunji (Holeman, 1968). The world's highest rainfall is also recorded at Cherrapunji, where 3.81 m of water fell during one 5-day period in August 1841 (Rogers et al., 1989). It should be realized, however, that such extreme values often represent the core of a much larger "field" of rain with much lower amounts falling as one moves away from the center.

Within the delta and adjoining coastal areas to the east, extreme rainfall may also be associated with typhoons. These occur on average about six times a year, arriving either in early summer (April, May) or in (September–October), during which time much of Bangladesh is already inundated anyway. Cyclones often generate waves that may be 3–8 m high, causing enormous damage, especially when they coincide with high tide (Haroun-er Rashid, 1977).

2.4.3 Evaporation

In addition to rainfall, several other climatic parameters, such as temperature and humidity, solar radiation, and wind spread, will affect the hydrologic behavior of an area. The four parameters are combined into a single variable, evaporation.

The seasonal variation in monthly Penman reference evaporation over the Indian subcontinent is shown in Figure 2.7. By far the highest values are recorded for the hot and dry premonsoon period (April, May).

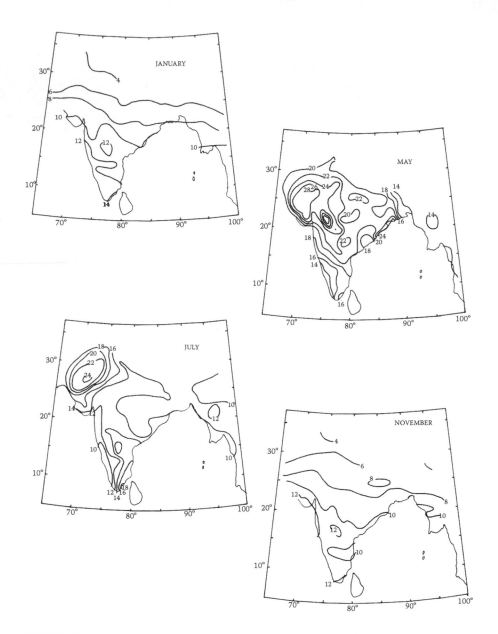

FIGURE 2.7
Mean annual potential evapotranspiration according to Penman (cm) over the Indian sub-
continent. (From Bruijnzeel, L.A., C.N. Bremmer, "Highland–Lowland Interactions in the
Ganges–Brahmaputra River Basin: A Review of Published Literature," ICIMOD Occasional
Paper No. II, Kathmandu, 1989. With permission.)

2.5 Hydro System

The hydro system—the surface and groundwater system—evolves in terms of the physiographic and meteorological characteristics, interacting with the land characteristics. This is particularly brought out in the GBM. Rivers Ganga and Brahmaputra, are one of the world's largest rivers in terms of the annual discharge, and together with River Meghna, the GBM system is second largest with a total annual discharge of, second only to River Amazon as shown in Table 2.1. Comparatively, the drainage basin is also small, with the result that the discharge intensity (discharge/unit area) is one of the highest in the world, second only to River Irrawady. Since the precipitation is concentrated to a few monsoon months, in which there are also periods of heavy precipitation, the peak discharge intensity is one of the highest.

Physiographically, Ganga basin is a long narrow region with very high mountains, which are geologically very young and unconsolidated on the north and also a geologically old and stable mountain system on the south. Very flat alluvial plains constituted by the scouring of these mountain formation lie at the foot of these high mountains. These physiographic and hydrologic features result in several unique characteristics. One, the rivers rush in the mountains and hills and then slow down abruptly on coming into the plain so that each has two very distinct sections, a steep, rapidly flowing one and a slow-moving one. Water's capacity to carry sediment depends largely on the speed of its flow; thus, the pattern of the subcontinental rivers is that they make large deposits, inland deltas, at the point where they suddenly decelerate in the plains. Second, the rivers cause very high floods and have very high sediment loads. These are among the world's highest, respectively for the rivers Ganges and Brahmaputra. River Ganga and its tributaries are also therefore likely to meander over large areas. River Brahmaputra has even more heavy concentrated precipitation and flooding characteristics. After the confluence of the two, shortly followed by the delta, one of the most severe hydraulic conditions of heavy floods, erosion and siltation, and meandering ensues. River Meghna debouches another tremendous amount of discharge at the tail end to turn this veritable fury into the world's worst floods almost every year.

Besides the extreme hydrologic characteristics, there is also phenomenal spatial and temporal variability. In contrast to the high flows during the three or four monsoon months, the flows during the rest of the year are severely reduced. During the summers, the minimum flows even at the tail end of Ganges and Brahmaputra can almost be negligible, with many of even the major southern tributaries of Ganges almost drying up. The annual hydrograph of the two rivers is given in Figure 2.8. The monsoons come almost overnight after a long period of scorching summer when the entire region is absolutely parched, and almost overnight, the land gets inundated with heavy flows or even floods. At the same time, scorching heat

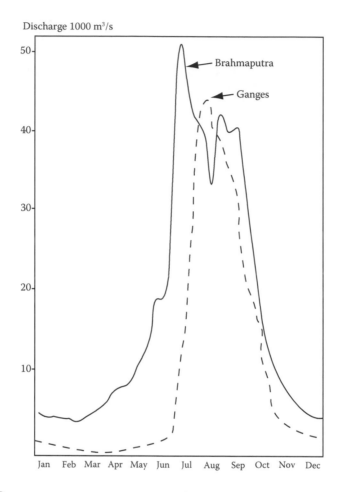

FIGURE 2.8
Annual hydrograph of Ganges and Brahmaputra. (From Bruijnzeel, L.A., C.N. Bremmer, "Highland–Lowland Interactions in the Ganges–Brahmaputra River Basin: A Review of Published Literature," ICIMOD Occasional Paper No. II, Kathmandu, 1989. With permission.)

still parches the land on the western regions. The monsoons may fail, and drought and famine may be experienced instead of floods.

The geological history has important bearing on the drainage and hydro system of the basin. The Great Himalayan Range, which normally would form the main water divide throughout its entire length, functions as such only in limited areas. This situation exists because the major Himalayan rivers, such as the Indus, Brahmaputra, Sutlej, and at least two headwaters of the Ganges—the Alaknanda and Bhagirathi—are older than the mountains they traverse. It is believed that the Himalayas were uplifted so slowly that the old rivers had no difficulty in continuing to flow through their channels and with the rise of the Himalayas, even acquired a greater momentum, which enabled them to deepen their valleys more rapidly. The elevation of the Himalayas and the deepening of the valleys thus proceeded simultaneously, with the result that the mountain ranges emerged with a completely developed river system cut into deep transverse gorges, ranging in depth from about 1500 m to 5000 m and in width from 10 to 40 km. The earlier origin of the drainage system explains the peculiarity that the major rivers drain not only the southern slopes of the Great Himalayan Range but also, to a large extent, the northern slopes, the water divide being north of the crest line.

In the cold heights, there are large deposits of water in the form of ice. Permanent ice covers are estimated 30,000 km^2 in the Himalayas, almost 17% of the total area of the range. Glaciers play an important role in draining the higher altitudes and in feeding the Himalayan Rivers. Several glaciers occur in Uttarakhand, of which the largest, Gangotri, is 40 km long. The Mahalangor himal ("snowfield"), with its Khumbu Glacier, drains the Everest region in Nepal. The rate of movement of the Himalayan region glaciers varies considerably; in the neighboring Karakoram ranges, for example, the Baltoro Glacier moves about 2 m a day, while others, such as the Khumbu, moves only about half m daily. Most of the Himalayan glaciers are in retreat.

The region was densely forested, which mitigated the fury of floods, erosion, meandering, and even the scorching heat. Over a period of time, population has been increasing and, of late, exploding, with the result of forests being almost completely cleared, leading to an increasing incidence of the flood–erosion syndrome. While the region has one of the world's highest per unit discharge intensities, paradoxically, it has also one of the lowest per capita water availabilities, which is worsening every year as population continues to increase mercilessly. As has been observed, the region is a supersaturated solution of land, men, and water during monsoons and at the same time a thirsty piece of parched land over most of the year. With the deforestation, even the world's wettest spot, Cheerapunji, is turned into a desert. While these characteristics apply generally, conditions vary in the GBM from tributary to tributary. The three rivers may therefore be studied in slightly greater detail.

2.5.1 Ganges

The Ganges, which is the anglicized name, is locally known as Ganga and is recognized and venerated by people. The Ganga basin is usually divided into eight or nine subbasins: (i) the Main Ganga (including Ramganga), (ii) the Yamuna, (iii) the Chambal, (iv) the Tons, Karmnasa and others between the Chambal and the Son, (v) the Son, (vi) the right-bank tributaries east of the Son, (vii) the Gomti, the Ghagra and others between them, (viii) the Gandak and other left-bank tributaries. Sometimes, the Ramganga is treated separately. Salient statistics of the main rivers and tributaries of the GBM basin are given in Table 2.5.

The Ganga rises in the southern Himalayas on the Indian side of the Tibet border. Its five head streams—the Bhagirathi, the Alakhnanda, the

TABLE 2.5

Salient Statistics of Main Rivers and Tributaries of GBM

Sl. No.	Subbasin	Land				Water				Utilizable		Utilized % of Utilizable			People			Unit Figures			
		Catchment Area (inside India)	Culturable Area (% of 1)	Net Sown Area (% of 2)	Cropping Intensity	Precipitation	Evaporation	Annual Runoff	Surface Runoff	Annual Groundwater Recharge	Surface Runoff (% of 7)	Groundwater (% of 7)	Surface (% of 9)	Groundwater (% of 10)	Total Utilized % of Utilizable	Population	% of Total Population	Density (km²)	Culturable Area/Capita	Water/Capita	Water/Culturable Land	% Gross Sown Area Irrigated
		m ha	%	%	%	cm	cm	m ha m	m ha m	m ha m	%	%	%	%	%	m	%	No/km²	Ha	m³	%	%
		1	2	3	4	5	6	7	7a.	8	9	10	11	12	13	14	15	16	17	18	19	20
1	Main Ganga (including Ram Ganga)	13.98	64.1	86.9	132	128	175	6.12	1.33	3.0	21.7	35.1	80.0	37.2	74.0	49.69	9.25	353	0.18	1830	101.80	18.3
2	Chambal	13.95	72.6	64.4	110	79	275	2.43	1.80	1.24	74.2	35.8	46.1	64.0	52.0	15.35	2.91	110	0.66	2400	36.15	13.3
3	Yamuna	22.68	73.8	77.8	125	87	260	6.56	1.43	3.05	21.8	33.7	109.2	43.0	70.0	45.66	8.60	201	0.37	2100	57.25	23.6
4	Tons, Karamnasa, and others between Yamuna and Son	2.86	66.3	77.9	131	106	225	1.41	1.04	0.47	73.8	26.5	36.5	29.5	35.10	6.83	1.37	239	0.28	2750	99.0	25.0
7	Gandak and other left bank attributes	5.73	76.4	87.7	137	154	180	17.47	3.77	1.36	21.5	5.9	69.04	11.0	56.6	26.22	4.85	457	0.17	7200	431.0	9.0
	Right bank tributaries east of Son	9.71	69.3	69.2	117	130	175	4.51	3.34	2.52	74.3	38.7	85.0	9.2	60.0	33.27	6.30	342	0.20	2100	104.60	37.1
	Total	86.15	68.6	76.0	125.5	116.39	211.8	55.01	18.50	14.62	33.6	19.3	71.0	39.4	59.1	221.29	41.20	259	0.32	3630	145.26	21.0

Source: Chaturvedi, V., and M.C. Chaturvedi, "Development, Technology and Education," Proceedings of the 12th Comparative Education Society of Europe, Antwerp, July, 1985. With permission.

Mandakini, the Dhauli Ganga, and the Pindar—all rise in the Uttarakhand division of the state of UP. Of these, the two main head streams are the Alakhnanda (the longer of the two) and the Bhagirathi. The Bhagirathi, which is traditionally known as the source of the Ganga, rises in India from the Gangotri glacier in the Himalayas at an elevation of about 7010 m above mean sea level. After its confluence with the Alakhnanda at Dev Prayag, the river assumes the name Ganga. After draining the middle ranges of the Himalayas, the river debouches into the plains at Hardwar. From Hardwar down to Allahabad where the Yamuna joins it on the right bank, there is a distance of about 720 km. It generally flows in a south–southeasterly direction. Lower down, the river flows eastward and past Varanasi, the Ganga is joined by a number of tributaries on both banks. Of the left-bank tributaries in the upstream reaches, prior to Varanasi, the Ramganga and the Gomti are the most important. The Yamuna has a number of important tributaries like the Chambal, the Sind, the Betwa, and the Ken joining it from the south. The Tons and the Karamnasa are other right-bank tributaries in UP.

After leaving UP the Ganga forms the boundary between UP and Bihar for a length of about 110 km and in this reach the Ghagra, which flows down from Nepal, joins it near Chapra. The river then enters Bihar below Ballia and flows more or less through the middle of the state. During its course of nearly 445 km in Bihar, the river flowing eastward is joined by a number of major tributaries on both banks. The Great Gandak, the Bagmati and the Kosi, and the Burhi Gandak join it on the left bank. The first three flow down from Nepal into North Bihar. The Son, the Pun Pun, the Kiul, the Chandan, the Gerua, and others join the Ganga on the right bank.

In West Bengal, the last Indian state that the Ganga enters, the Mahananda joins it from the north. The river then skirts the Rajmahal Hills to the south and flows southeast to Farakka. The delta of the Ganga can be said to start from Farakka. The river divides into two arms about 40 km below Farakka. The left arm, known as the Padma, flows eastward into Bangladesh while the right arm, known as the Bhagirathi–Hooghly, continues to flow in a southerly direction in West Bengal. Two tributaries flowing in from the west, the Damodar and the Rupnarayan, join the Hooghly itself. Until some 300 years ago, the Bhagirathi–Hooghly constituted the main arm of the Ganga, carrying the bulk of the flow. Thereafter, the Padma arm opened up more and more, leaving Bhagirathi a mere spill channel of the Ganga flowing mostly during high stages of flow. The river ultimately flows into the Bay of Bengal, about 145 km downstream of Calcutta. The length of the river (measured along the Bhagirathi and the Hooghly) during its course in West Bengal is about 520 km. The Ganga (Padma) in Bangladesh flows past Kushtia and Pabna until the Brahmaputra (locally called Jamuna) joins it at Goalkunda.

The united stream of the Brahmaputra and the Ganga beyond Goalkunda continues to flow southeast under the name Padma. At Chandpur, 105 km below Goalkunda, the Padma is again joined on the left bank by the Meghna, whose source is in the high mountains, which are subjected to intense rainfall. From this confluence downward, the river known as lower Meghna becomes a very broad estuary, making its exit into the Bay of Bengal.

Dacca, the principal city of Bangladesh, stands on the Burhi Ganga, a tributary of Dhaluwari. Apart from the Hooghly and the Meghna, the other distributary streams which form the Ganges delta are as follows: in West Bengal, the Jalangi, and in Bangladesh, the Metabanga, the Bhairab, the Kobadak, the Gorai (Madhumati), and the Arial Khan.

The precipitation over the Ganga basin is brought about by the southwest monsoon as well as by cyclones originating over the Bay of Bengal. The average annual rainfall in India varies from 35 cm on the western end of the basin to about 200 cm near the delta. The river drains an area of 106.96 m ha (excluding Tibet) and the average annual flow of the Ganga at Farakka is about 55.01 m ha m. The river network is shown in Figure 2.2. The tributaries are themselves big rivers from international standards. Although they are all perennial, for about eight winter and summer months, discharge is meager compared to the four monsoon months and for most of the course until the Ghagra and the Son meet it. Ganga is a sluggish wide stream. The annual discharge in the three periods is Rabi (winter), (ii) summer, and (iii) Kharif (monsoon) is shown in Figure 2.9.

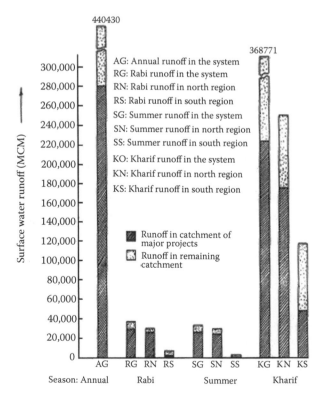

FIGURE 2.9

Annual discharge in Ganges in the three periods: Rabi (winter), summer, and Kharif (monsoon).

The slope of the river is very steep in the Himalayas but changes very abruptly as it enters the plains near Hardwar. Even in the piedmont zone at Hardwar it is about 100 cm/km but it changes to less than 6 cm/km in Bihar. This explains several characteristics of the region such as very high erosion and sediment loads in the Himalayan region, sedimentation in the plains, and severe flooding.

2.5.2 Brahmaputra

The Brahmaputra rises in the great glacier in the northernmost chain of the Himalayas in the Kailash range just south of a lake called Konggyu Tsho. The river under its Tibetan name of Tsangpo flows eastward through southern Tibet for about 1700 km at a fairly low gradient along the bottom of a flat tectonic valley parallel to and about 160 km north of the Himalayas. In this long journey, it meets a number of tributaries. At the extreme eastern end of its course in Tibet, the Tsangpo suddenly takes a turn to the south and cuts a deep and narrow gorge upon crossing the high Himalayan range. The gradient of the river in the gorge section is extremely steep. After turning south or southeast, it enters Arunachal Pradesh in India as the Dihang River. The river then traverses mountainous terrain before debouching into the Assam Plain near Pasighat at an elevation of only 155 m. At the town of Sadiya, India, the Dihang turns to the southwest and is joined by the two mountain streams of Luhit and Dihang. After the confluence, about 1500 km from the Bay of Bengal, the river is known as the Brahmaputra (the son of Brahma, the creator).

The river then rolls down the Assam valley from east to west for about 720 km, with its channels meandering from side to side, forming many islands. In Assam, the river is mighty even in the dry season, and during the rains, its banks are more than 10 km apart. The largest island Majuli covers an area of 1250 km². As the river follows its braided 700 km course through the valley, it receives several rapid Himalayan streams, including the Subansiri, the Kaneng, the Dhansiri, the Manas, the Champamati, the Saralbhanga, and the Sonkosh. On the south bank, the main tributaries are the Noa Dihing, the Buri Dihing, the Disang, the Dikhu, and the Kopili. The Brahmaputra also has some important tributaries flowing through the North Bengal. They are the Tista, the Jaldhaka, the Torsa, the Kaljani, and the Raidok. Emerging swiftly from narrow gorges with steep slopes, these rivers widen out considerably in the plains.

Swinging round the western spurs of the Garo Hills, near Goalpara, the river enters the alluvial plains of Bangladesh, through which it flows southward for another 270 km until it joins the Ganga at Goalundo. Below the confluence of the Tista, the old channel of Brahmaputra branches off the left bank. From here to Goalunda, the river is called Jamuna.

In Tibet, the precipitation is in the form of snow, which begins to melt in March following the intrusion of warm air. Coming down the central basin in India, the annual rainfall, brought by the southwest monsoon, is about 250 cm. Further down in Bangladesh, the rainfall over the basin averages about 240 cm. The river drains a total catchment area of 23.41 mha (excluding

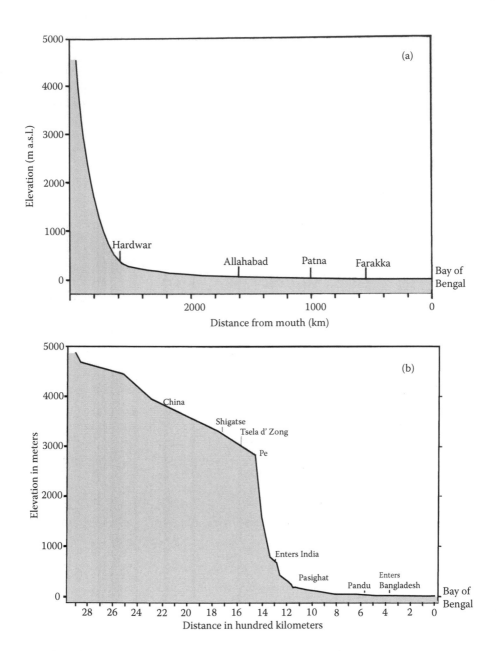

FIGURE 2.10
Approximate longitudinal profiles of Ganga (a) and Brahmaputra (b). (From Bruijnzeel, L.A., C.N. Bremmer, "Highland–Lowland Interactions in the Ganges–Brahmaputra River Basin: A Review of Published Literature," ICIMOD Occasional Paper No. II, Kathmandu, 1989. With permission.)

Tibet and Bhutan) at Bahadurabad (Bangladesh) and the average annual run-off at this site is 61.5 m ha m. The hydrograph of River Brahmaputra is shown in Figure 2.9 along with that of River Ganga. The approximate longitudinal profiles of Ganga and Brahmaputra are shown in Figure 2.10a and b. It leads to development of a novel technology, which may change the development of water resources of the GBM basin as discussed later.

2.5.3 Barak/Meghna

The Barak River, the headstream of the Meghna rises in the hills in Manipur (India) at an elevation of about 2900 m. It flows southwesterly, winding its way along hill ranges for about 250 km. Thereafter, it takes a sharp reflex bend and flows north. It changes direction to the west when emerging from the hills and follows a meandering course until it enters Bangladesh. Many tributaries join the Barak. Near the border, the Barak bifurcates into two rivers, the Surma and the Krishiyara, which again join and flow in a southerly direction under the name of Kali. It takes the name Meghna before it meets the Padma, which carries the combined flow of the Ganga and the Brahmaputra, at Chandpur.

Below Chandpur, the combined river, also known as the Lower Meghna, becomes a very wide and deep estuary. It enters the Bay of Bengal through four principal channels. The distance between Chandpur and the sea is about 130 km. The Meghna is thus about 950 km long of which 550 km including the estuary from Chandpur to the sea lies in Bangladesh.

The river has a steep slope while flowing in the hills in India. After entering Bangladesh, it travels through the low-lying alluvial basin. When the Meghna flood meets that of the Padma at Chandpur, its floods are backed and its basin acts as a gigantic detention reservoir, without which the area between Goalundo and Chandpur would be under much deeper water. At flood stages, the slope of the Meghna downstream of Bhairab Bazar, where it meets the old Brahmaputra, is only about 1:88,000 which reflects the influence of backwater from the Ganga–Padma.

There is copious rainfall in the watershed of the Meghna and the Barak rivers. Around the foothills in Assam, the rainfall reaches 500 cm on the windward side. At and near Cherrapunji, where rainfall is amongst the highest in the world, the average annual precipitation is about 1100 cm. In the plains of Bangladesh, the rainfall varies from 200 to 300 cm. The drainage area of the Meghna at Bhairab Bazar is about 8.02 m ha and average annual runoff is 11.1 m ha m.

2.5.4 Delta

The GBM basin has one of the world's biggest deltas, a reflection of the phys-iographic–hydrologic characteristics of the basin. The delta, the seaward pro-longation of silt deposits from the Ganges and Brahmaputra river valleys, covers an area of about 56,500 km^2 and is composed of repeated alternations of clays, sands, and marls, with recurring layers of peat, lignite, and beds of what were once forests. The new deposits of the delta, known in Hindi and Urdu as the Khadar, naturally occur in the vicinity of the present channels.

The southern surface of the Ganges Delta has been formed by the rapid and comparatively recent deposition of enormous loads of silt. To the east, the seaward side of the delta is being changed at a rapid rate by the formation of new lands, known as Chars, and new islands. So much silt is deposited here that the 100-fathom line lies much farther out to sea than it does, for example, off the mouth of the Indus in the Arabian Sea. The western coastline of the delta has, however, remained practically unchanged since the eighteenth century.

The rivers in the West Bengal area, being sluggish, have been described as dead or dying. Little water passes down them to the sea. In the Bangladesh delta region, the rivers are broad and active, carrying plentiful water. They are connected by innumerable creeks. During the rains, from June to October, the greater part of the region is flooded to a depth of several meters, leaving the villages and homesteads, which are built on artificially raised land, isolated above the floodwaters. Communication between settlements during this season can be accomplished only by boat.

To the seaward side of the delta as a whole there is a vast stretch of tidal forests and swampland. The forests, which are called Sundarbans (Sanskrit meaning "beautiful forest"), are protected by India and Bangladesh. For conservation purposes, no permanent settlement is permitted in them. In certain parts of the delta, there occur layers of peat, composed of forest vegetation and rice plants. In many natural depressions, known as bil, peat, still in the process of formation, is used as a fertilizer by local farmers. In recent years, it also has been dried and used as a domestic and industrial fuel.

The Ganga, as well as its tributaries and distributaries, are constantly vulnerable to changes in its course in the delta region. Such changes have occurred in comparatively recent times, especially since 1750. In 1785, the Brahmaputra flowed past the city of Mymensingh. It now flows more than 40 miles west of it before joining the Ganges.

2.5.5 Erosion and Sedimentation

In his reconnaissance survey of amounts of sediment transported in suspension (as opposed to rolling along the river bed) to the seas by the major rivers of the world, Holeman (1968) concluded that Asia's rivers were by far the greatest contributors, possibly supplying up to 80% of the world total.

Of the Asian rivers, the combined Ganges–Brahmaputra basin ranked as the first, with an estimated annual sediment yield of 2.4 billion tons (15 t/ha), of which the Ganges alone contributed about two thirds (Holeman, 1968). A more modest estimate (1.67 billon tons, or 11.3 t/ha) was published more recently by Milliman and Meade (1983). A 1968 marine seismic expedition in the Bay of Bengal estimated that the undersea fan of sediment deposited in the bay by the Ganges and Brahmaputra rivers, originating principally from the Himalayas, was 1000 km wide, perhaps over 12 km in depth,

300 km long, and is extending far south of Sri Lanka. A calculation back from this would indicate that the Himalayan area gives up 0.7 m in eroded depth every 1000 years (Curray and Moore, 1971). Although sediment loads transported by such very large rivers can never be determined with great precision, it will be clear that the amounts of sediment carried by the two rivers must be enormous. It is further to be noted that the drainage basin of Ganges and Brahmaputra is quite small compared to other international rivers.

The high sediment load and its deposition later will readily be appreciated in the context of the geology, physiography, and hydrology of these rivers. Major contrasts in sediment yield may be expected between "Himalayan" and "Peninsular" rivers within the Ganges–Brahmaputra basin by contrasting the sediment load of the Himalayan and Peninsular rivers in India (Figure 2.11). Contrasts will also be seen between various Himalayan

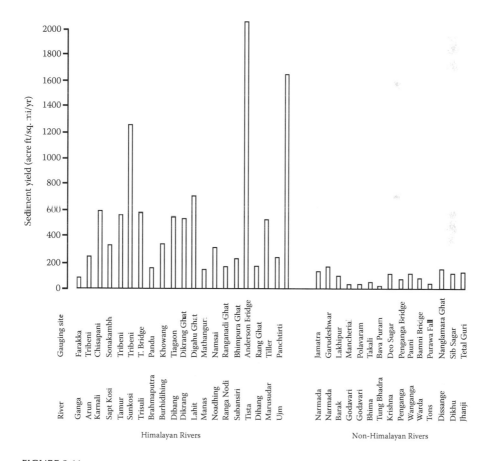

FIGURE 2.11

Sediment yields for Peninsular and Himalayan rivers in India. (From Bruijnzeel, L.A., C.N. Bremmer, "Highland–Lowland Interactions in the Ganges–Brahmaputra River Basin: A Review of Published Literature," ICIMOD Occasional Paper No. II, Kathmandu, 1989. With permission.)

tributaries depending upon the characteristics of the catchment's size and hydrologic condition. For instance, the larger the catchments, the lesser the yield, as larger basins have opportunities to store sediment. Again the larger the portion of the catchment in the dry Trans-Himalaya the lower will be its sediment load. Finally, certain rock types (e.g., phyllites, slates, or unconsolidated sandstones) are much more susceptible to erosion and/or mass movement than others, resulting in widely different stream sediment loads under otherwise similar conditions. As such, each "mesoscale" basin naturally represents a more or less unique combination of these variables.

Not all of a river's sediment load is transported in suspension. A significant but often unmeasured part moves along the channel as bed load. Although separate measurement of bed load transport in Himalayan rivers are rarely available (Galay, 1987), there is a growing body of information regarding total (i.e., suspended plus bed) loads deposited behind dams or several major rivers in the basin.

The very high amounts of sediment carried by most major Himalayan rivers causes them to adopt a braided pattern upon reaching the piedmont zone. A good example is the Kosi River in eastern Nepal, which has not been able to cut a deep and stable bed for itself after leaving the Chatra gorge. During times of high flow, the river easily overtops its shallow banks and spreads over a vast expanse that may in places be up to 30–40 km wide. Due to the continued deposition of sediment, the riverbed has risen to several meters above the surrounding plain, thereby creating a highly dangerous situation.

As shown in Figure 2.12, the Kosi has been notoriously unstable for more than 250 years (Gole and Chitale, 1966; Galay, 1987). As reported by Sharma (1985), if we trace its history, Major Rennel in 1779 and James Ferguison in 1503 reported that the oldest Kosi flowed eastward and joined the Mahananda in the middle and finally with Brahmaputra. The second old Kosi flowed just west of Purnea and joined the Ganges on the north of Rajmahal. Later on, it was established by Shilling in 1883 that (i) the bed of Kosi oscillated in the region between the Brahmaputra and the mouth of Ganges, and the event was repeated after a long period of time, and (ii) the westward movement was slow and in a series of steps, whereas the eastward movement was in a great swing. From the record, it was seen that the Kosi had moved a distance of 114 km, i.e., 0.54 km/year (Sharma, 1985). Interestingly, the shift has currently been progressively westward (Pal and Bagchi, 1975).

Also, according to Sharma (1977), the rivers lying east of Butwal (Nepal's Western Development Region), are shifting toward the west, and those west of Butwal toward the east, suggesting a tectonic origin for such movements. Several other cases of major river shifts in this region have been described (Morgan and McIntire, 1959; Coleman, 1969; Klaassen and Vermeer, 1988).

In view of the strongly seasonal character of most of the rivers in the region, the bulk of the annual sediment transport takes place during the summer monsoon. Also, major variations between years are to be expected as a result of interannual variations in rainfall distribution. In addition,

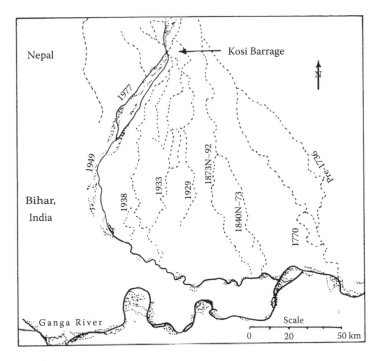

FIGURE 2.12
Meandering of river Kosi. (From Bruijnzeel, L.A., C.N. Bremmer, "Highland–Lowland Interactions in the Ganges–Brahmaputra River Basin: A Review of Published Literature," ICIMOD Occasional Paper No. II, Kathmandu, 1989. With permission.)

within the rainy season, the contribution made by a few days of abnormally high flows can be disproportionally large. Secondly, unusually heavy sediment loads could be generated by landslides. Considerable data and analysis is yet required to establish meaningful quantitative analysis of the erosion and sediment dynamics and environmental impacts.

2.6 Geohydrology

The GBM basin occupies the shallow downwarp of ancient crystalline rocks of the Indian subcontinent that are slowly sliding under the Himalayan mountain front. Himalayan orogeny contributed to the formation of a vast depression or foredeep. Each phase of Himalayan upheaval rejuvenated streams and accelerated the sedimentation of the depression, which, through periods of geological time, has been filled with a considerable pile of detritus. Rapid deposition continued uplift, and glacial and interglacial intervention created a rhythm of sedimentation, aggradation, reworking, stratification,

and partial consolidation that led to an extensive, thick sequence of unconsolidated to semiconsolidated beds. The thickness was, however, limited
by the bedrock configuration and the influence of a number of subsurface
ridges, shelves, and depressions (Dutt, 1992).

The rivers emanating from the Himalayas are youthful and torrential, have
a high gradient, and, by and large, carry a considerable amount of mixed
sediments ranging from big boulders to fine clastics. On the other hand,
the north-flowing rivers emanating from the peninsular region and joining
the Ganga system bring gravel- and clay-grade sediments of considerably
smaller dimensions. These diverse aspects of the deposited environment
offered by the peninsular and extra peninsular provenance have given rise
to conditions that have built up from hydrogeological units within the basin.

The foothills of the Himalayas are underlain by a cobble–boulder formation known as the Bhabar zone, with a thickness of the order of 1500 m. Most
rivers and streams traversing this zone are known to lose a considerable
amount of their flows, sometimes all of it. Exploratory wells drilled to depths
of between 200 and 250 m are reported to have yielded 0.6 to 0.85 m^3/s of
water in artesian conditions. It forms a prominent geological formation with
great recharge potentialities.

Next to the Bhabar zone is another distinctive hydrogeological characteristic
known as Terai. This tract begins where the groundwater table and the land
surface intersect. The river flows slows, and often, marshy conditions occur.
The Terai has a width of about 15 km but with a much greater lateral extension. The Terai is contemporaneous with the Bhabar zone, but the boulder–
gravel content diminishes considerably. In the Terai belt, alternating clay and
sediment deposits have effectively succeeded in confining the sand horizons
(aquifers), so much that artesian conditions are produced. Wells in the UP
Terai region at depths of 130 to 200 m have exhibited prolific free flowing characteristics with hydrostatic heads several meters above the land surface.

South of the Terai lie the Great Plains. The central tract of this region has
accumulated an enormous pile of unconsolidated sediments comprising
sand, gravel, silt and clay admixed with "Kankar." Aquitards and aquifers
alternate to form four to six regionally confined, deep, fresh-water aquifers,
each hydraulically separate from the other, which may underlie this plain at
depths of 100 to 250 m. Both in the Terai and in the central plains, the alluvial
aquifers exhibit wide variations in both their vertical and lateral extensions
building up an interconnected and regionally extensive aquifer system.
Typical fence diagram showing aquifer groups in the upper Yamuna basin is
shown in Figure 2.13 (Chaturvedi, 1987).

Further south, fringing the peninsula, lies an alluvial tract that is made up
of sediments from the shield provenance. The sediments deposited over the
undulating bedrock are mostly fine clastics having layers of granular material
that form the chief aquifers. The granular beds have a limited vertical thickness and areal extent, displaying a discontinuous aquifer system. The fringe
sediments in the western and southwestern parts of UP have aquifer horizons

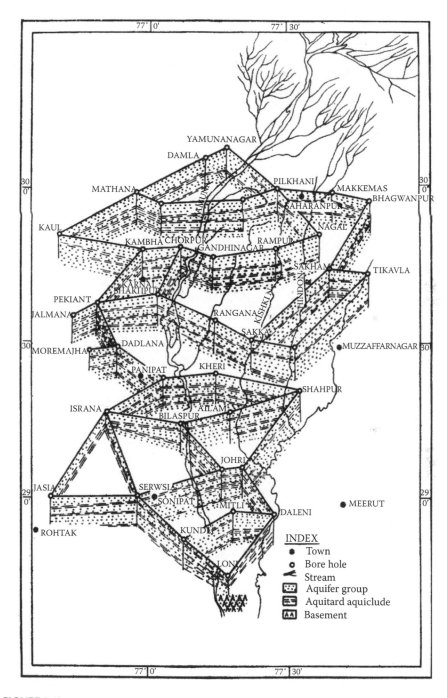

FIGURE 2.13
Typical fence diagram showing aquifer groups in the upper Yamuna basin. (From Chaturvedi, M.C., *Water Resources Systems Planning*, Tata McGraw-Hill, New Delhi, 1987. With permission.)

that often display salinity hazards, but in the rest of the basin, particularly east-ward, the sediments in the fringe zone exhibit an increase in granular materials and a near absence of salinity hazards. These changes in the sub-surface struc-ture lend dependability and ease to groundwater development.

This fourfold hydrogeological system is displayed all over the GBM basin, although the extent and details vary. For instance as one moves east, the system gets compressed laterally, so to say, with the northward thrust of the peninsular shield. On the north, the Bhabar zone is almost nonexistence in Bihar, while on the south, occupied by the Archean crystallines, ground-water occurs along buried and abandoned river channels and fractured hori-zons. The Chota Nagpur plateau has excellent hot springs.

In the Brahmaputra and Barak region, the basin is much narrower but the four hydrogeological units, the bhabar, Teria, plains and fringe zones are again present. Naturally, the characteristics vary.

The delta is another characteristic zone. Vast alluvial flats have been built up over geological periods with the deposition of the sediment load as the numerous streams gradually disappeared in the ocean. The flood plain in the upper reaches gradually merges southward with mangrove swamps and an estuarine and marine environment. The deltaic tract may be consid-ered in two parts: one is west of River Hooghly in India, and the other, east of Hooghly in Bangladesh.

The Indian side has been formed by sediment deposits brought by River Ganges and the rivers joining it from the plateau on the west. The northern parts, the plains of Murshidabad and Nadia, have productive aquifers and support a number of shallow and deep tubewells. In the central region, a clay capping appears to have confined aquifer horizons from 25 to 65 m below the surface from Calcutta to sea. The deltaic tract is characterized by a thick clay blanket that varies in thickness from 15 to 76 m. Underlying the top clay layer is a succession of sand, gravel, and clay horizons. The thick top clay and minor sand horizon is marked at its base by a gravel bed which is a marker horizon in as much as the upper aquifer above it is saline while that below it to a depth of 300 m has fresh water of potable quality. The lower aquifer is not a single aquifer but a group of three to five aquifers that are interconnected and also regionally extensive. All these aquifers dip gently toward the sea.

On the eastern side in the Assam valley the foredeep narrows and is bounded on the south by the Shillong Plateau, a highland composed mainly of the same ancient crystalline rock. The Shillong Plateau rises 1000 to 1500 m above sea level and is separated from the crystalline rock of the Indian Plate by a low saddle, now blanketed by alluvial deposits of the Ganges and Brahmaputra rivers. It is bordered on the south by the Bengal foredeep, a structural trough that extends southward beneath the Bay of Bengal. Lying to the east of Shillong Plateau and trending southward to form the Tripur and Chittagong hills that flank the deltaic plain on the east is the Aracon Yoma Fold belt. In this eastern fold belt, uplifted Pliocene and Pleistocene beds of sand and clay have cumulative thickness greater than 6000 m and lie at altitudes ranging from 80 to 500 m above sea level. Beneath the deltaic plain, sediments in the Bengal foredeep range perhaps from 13,000 to 22,000 m in thickness, according to geophysical data obtained for oil and gas exploration. Sediments in the Bengal foredeep appear to be mainly the deposits of an ancient Brahmaputra river, which during most of the Neogene time entered the Bengal basin east of the Shillong Plateau.

The Ganga–Brahmaputra system is the largest in the world in terms of sediment transport carrying some 2.4 billion tons of sediment almost half of which is carried to the Bay of Bengal. And yet, the estuary of the Meghna persists, probably because structural downwarp in the Bengal foredeep is very active. This inference is confirmed by the topography of the Surma. The Meghna River Basin's land surface contour 3 m above sea level can be traced northward to the foot of the Shillong Plateau. The Danki fault, which marks the southern edge of the Shillong Plateau, has been mapped in the subsurface to a depth of 10,000 m or more.

2.7 Land and Soils

Land is a nonrenewable resource base that supports all primary production systems as well as provides the essential social environment in terms of habitat, roads, and other infrastructural facilities. A critical input to all primary production systems as well as social facilities is water along with land. The land–water system supports the biotic environmental vectors and is itself dynamically transformed into a characteristic soil in which climate has also an important role. The land replenishes the hydro system much after the input from the natural processes of precipitation. The physical, chemical, and biological health of soil profiles determines the ability of land to serve the socioeconomic needs on a sustained basis.

The land use is currently classified under nine heads, (i) forests, (ii) land put to nonagricultural uses, (iii) barren and uncultivated land, (iv) permanent pastures, (v) land under miscellaneous tree crops and groves not included in the net shown area, (vi) culturable waste, (vii) fallow other than current fallows, (viii) current fallows, and (ix) net area sown. The last is the difference between the total cropped area and area sown more than once.

As an interaction of the land–water vegetation system, a wide variety of soils is met in the Ganges basin. The soils in the northern mountainous part, which is mostly in Nepal and the Himalayan states of India, are generally shallow and are (i) red loams along the slopes of the lower hills or along ridges, (ii) brown forest soils, (iii) podzol on mild slopes and in pockets of hills and ridges, and (iv) meadow soils near streams. The submountainous tracts have porous and pebbly soils varying from clay loams to sandy loams. This basic trend continues at the eastern edge. The northern sub-Himalayan region has brown podzolic soils, with porous structure and poor soil retaining characteristics. This has been found good for tea plantation.

The soils in the Gangetic plains fall into two broad categories, i.e., (i) older alluvium (Bangar) and (ii) newer alluvium (Khadar). The soil composition varies over the region. The Tarai area consists mostly of shallow dark grey soils varying from loam to sandy loam. The western regions have similar soils but of greater depth and are very fertile. Further east, the soils are heavy loams and are generally somewhat alkaline. The soils of the central part of the UP plains are slightly acidic loams and sandy loams. In drier, south central and western parts, saline and alkaline effloresces (known as Usar or Reh) occur. The main soils in the eastern parts are (i) Bhat soils, generally low lying and sandy loam in texture with high lime content, (ii) Banjar soils, varying from loam to sandy loam with low lime content and slightly alkaline to slightly acidic reaction, and (iii) some Dhub soils near riverbanks. The southern part of the state consists of mixed red and black soils. The black soils (Mar and Kabar) are clayey, calcareous, expensive, and fertile. The red soils (Parwa and Rankar) generally occur on top of plateaus and on the upper slopes of hills.

In Bihar, which has higher precipitation and silt load in the rivers, the sandy alluvial soils of the north region of the plains are among the most fertile in India. They are rich in lime and contain a high proportion of clay. The soils of the south Bihar plains consist of alluvium of heavy textures, with a heavy subsoil. Highly ferruginous red soils cover almost the entire southern portion.

Mangrove soil occurs in the estuarine swamps of the deltaic region. It has a high saline content and is of very limited use for cultivation, though substantial areas have been reclaimed and made use for cultivation by leaching out salts.

The Brahmaputra basin has a very rugged physiography, and the soils in the mountainous regions are generally red loam alluvial, laterite, or brown hill alluvial depending on the slopes and land formations. In the plains, soil is new alluvium. In the deltaic region, soil characteristics are similar to those in the adjoining Ganga basin. The position in Meghna is similar to corresponding physiographic regions of Brahmaputra.

2.8 Fauna and Flora

2.8.1 General Features

The GBM basin has a diversity of climate as well as rich land and water availability. It was densely forested even in recent historical times (Habib, 1982). Anthropogenic activities have severely deforested the region and degraded the environment and the landscape.

As brought out by Kaul (1992), out of a total basin area of 141.24 million ha (excluding Tibet), 27.14% of the total land area of the basin is classified as forest with or without any tree growth. With 496 million people living in the Basin (1986–1987) per capita effective forest area works out to about 0.07 ha which compares very poorly with the world average of one ha. The spatial distribution of forests in the basin is also unequal and unbalanced. A major portion of forests exists in undeveloped areas. The outer and middle Himalayas constitute the main forest belts in the basin followed by large areas in the northeastern hills and the mangrove forests of the Sunderbands in the south. Based on the percentage of recorded forest area to geographical area, the Indian basin states of the southeast like Arunachal, Manipur, Mizoram, Nagaland, and Tripura have over 50% of their land area under forests, while it is less than 10% in Rajasthan, Haryana, and Delhi. The recorded forest area in other basin states ranges between 13 to 39%. In Bangladesh forests are extremely limited and unevenly distributed. More than 94% of government forests (2.18 million ha) are concentrated in the eastern hilly region (over 63%), comprising Sylhet, Chittagong, Chitagong Hill Tract, and Cox's Bazaar districts, and the southwest (over 30%) with its large mangrove ecosystem. The central and northern regions, where the population is the highest, have the least forest resources. In fact, out of 64 districts, 28 districts have no public forests at all.

Over 70% of the geographical area of Bhutan is under forests, the bulk of which (68.6%) is located in the central zone between elevations of 2000 and 4000 m, comprising temperate and sub-Alpine species. Nearly 22% of the forest area is in the southern belt from the foothills to an altitude of 2000 m. The remaining 9.4% is in the Alpine zone above 4000 m. Of the 5.50 million ha of forested land in Nepal, nearly 32.4% is in the midmountains followed by the high mountains (30.1%) and the Siwaliks (26.5%). There is thus a fairly even distribution of forests in these three zones. Only 11% of the forest area is in the Terai (8.2%) and High Himal (2.8%) zones, although major wood supplies for urban centers like Kathmandu come from the Terai because of its accessibility.

Whereas the Himalayan and northeast evergreen forests are fairly well stocked, large areas in central India, the Indo-Gangetic plain, and other plain and valley systems, are either partially stocked or entirely devoid of forest. However, in the more humid parts of these plains and valleys, the lack of forests is made up by tree lands, homestead woodlots, and groves. A major proportion of fuel wood in India is met from these "unrecorded sources." Approximately 70% of the timber and 90% of the fuel wood and bamboo supplies in Bangladesh are obtained from village woodlots, which constitute about 11% of the total forest area of 2.456 million ha.

Of the actual forest cover of 33.56 million ha in the basin, only 20.75 million ha are estimated to be dense forest with a crown cover of 40% or more, while the rest is open forest with a crown cover varying from 10% to 49%. The growing stock of the actual forest cover per hectare in Indian forests ranges from as low as 10 m³ in Rajasthan to as high as 277 m³ in certain Himalayan forests. The average actual forest cover works out to 0.81 m³ per ha, which is much less than the world average of 2.1 m³. The potential productivity is, however, much higher than the actual production. The figures of annual production per hectare for the basin area are 2.21 m³ (western Himalayas), 2.03 m³ (eastern Himalayas), 1.66 m³ (Northeast), 1.05 m³ (central India), and 0.80 m³ (Gangetic Plain). Assuming an average growing stock of 65 m³ per ha, the total growing stock in the effective forest area of the Indian basin works out to 1.570 million m³. Allowing a net annual increment of 1.24%, the annual production from the basin forest could be of the order of 19.5 million m³.

Only 61% of the forests controlled by the Forest Department in Bangladesh are productive, the rest being encroached, barren, scrub, or grassland. These forests carry a growing stock of 46.3 million m³ or 31.7 m³ per ha. Removal rates in the past have far exceeded sustainable yield limits, causing a steady decline of growing stock over the years. There was a 48% depletion of tree resources in the Sunderbans between 1960 and 1985.

The total growing stock in the operable forests of Bhutan is of the order of 529 million m³. The gross annual allowable cut is 21.2 million m³, and the net annual allowable cut is 13.9 million m³ or 2.6% of the growing stock, which is rather high. The current consumption of fuel wood is estimated at about 1.38 million m³ per year, of which over 90% originates from natural forests and the rest from private lands.

The average growing stock in Nepal (5.5 million ha) is less than 100 m³ per ha, with heavy exploitation particularly in the middle mountains. The production of timber and fuel wood in 1985–1986 was 0.88 and 11 million m³, respectively, or little more than 2% of the growing stock, which appears to be on the higher side.

Natural grasslands and cultivated fodder crops play a key role in any improved and more intensive system of land. Under population pressure, most of the cultivable land in the basin has been brought under food crops, leaving mainly unproductive and eroded lands to serve as natural grasslands. If the condition of forests in the basin is bad, that of the grasslands is worse.

Grass and grazing lands constitute the major source of fodder in the basin. While grasses are not cultivated as fodder on any scale in the basin, most grass is obtained from forest areas. Other grass-producing areas include permanent pastures and other grazing lands, land under miscellaneous tree crops and groves, culturable wastelands, fallow lands other than current fallows, and areas under the nonagricultural use category. Most of the grasslands in India are degraded, overgrazed, and in very poor condition. Their productivity is therefore very low and varies widely. The productivity of forest grass is generally higher than in other areas. The production of dry grass generally varies from 0.5 to 6.0 tonnes/ha/year; the average grass yields from forest areas and other grass-producing areas are taken to be about 3 and 1.5 tonnes/ha/year, respectively.

While grasslands in Nepal extend over 1.75 million ha, Alpine pastures/meadows cover an area of 0.27 million ha in Bhutan. These grasslands are in poor condition and suffer from erosion due to overgrazing and fires.

2.8.2 Fauna and Flora Characteristics

Himalayan vegetation is broadly classified into four groups, with respect to altitudinal zonation: (i) below 1000 m: tropical zone; (ii) 1000–2000 m: subtropical zone; (iii) 2000–3000 m: temperate zone; (iv) 3000–4000 m: sub-Alpine zone; and (v) above 4000 m: Alpine zone, with the elevation being approximate. While the classification is valid all over the Himalayas, vegetation distribution is influenced by slope, soil type, and rainfall region besides elevation (i.e., temperature). The vegetational zonation in the Nepalese Himalayas, which is typical of the Himalayas in the GBM basin, and a cross-sectional representation of vegetation zones are shown in Figures 2.14 and 2.15, respectively.

As far as the subtropical zone, leaved forests are dominated by Schima Wallchin and the evergreen dipterocarps—a group of timber- and resin-producing trees, and are common. *Mesua ferrea* (rose chestnut), bamboos, oaks, chestnuts, and alder trees are some of the varieties. At higher elevation are found fountain forests in which the typical evergreen is *Pandanus*

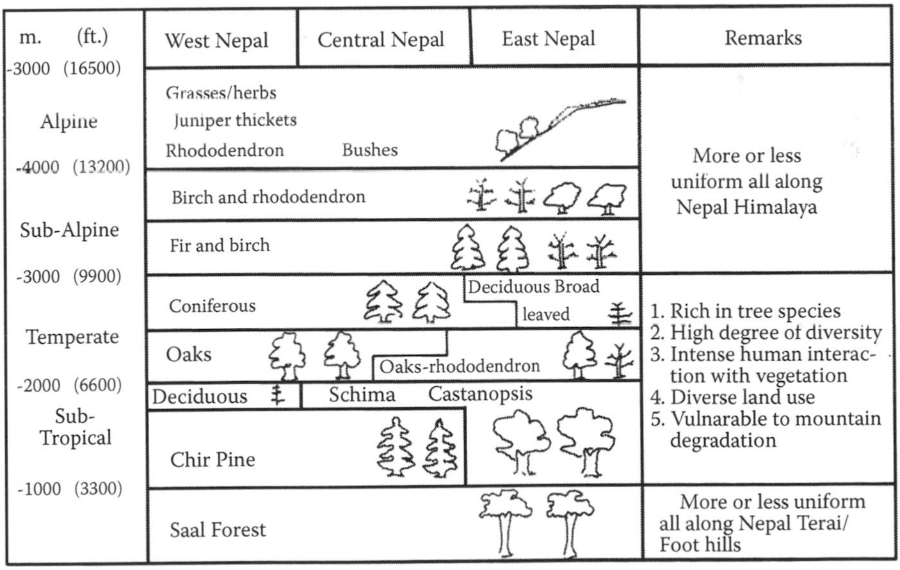

m. (ft.)	West Nepal	Central Nepal	East Nepal	Remarks
-3000 (16500)				
Alpine	Grasses/herbs / Juniper thickets / Rhododendron	Bushes		More or less uniform all along Nepal Himalaya
-4000 (13200)				
Sub-Alpine	Birch and rhododendron			
	Fir and birch			
-3000 (9900)				
Temperate	Coniferous		Deciduous Broad leaved	1. Rich in tree species
	Oaks	Oaks-rhododendron		2. High degree of diversity 3. Intense human interaction with vegetation
-2000 (6600)	Deciduous	Schima Castanopsis		4. Diverse land use
Sub-Tropical	Chir Pine			5. Vulnarable to mountain degradation
-1000 (3300)				
	Saal Forest			More or less uniform all along Nepal Terai/Foot hills

FIGURE 2.14

Vegetational zonation in the Nepalese Himalayas. (From Bruijnzeel, L.A., C.N. Bremmer, "Highland–Lowland Interactions in the Ganges–Brahmaputra River Basin: A Review of Published Literature," ICIMOD Occasional Paper No. II, Kathmandu, 1989. With permission.)

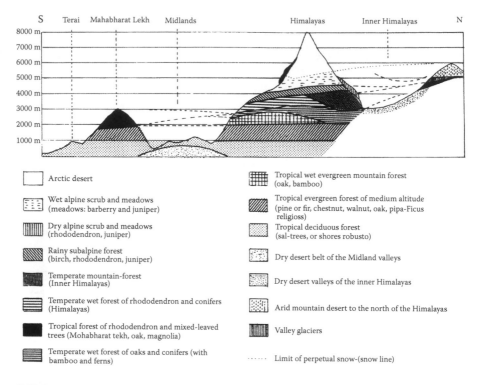

FIGURE 2.15
Cross-sectional representation of vegetation zones. (From Bruijnzeel, L.A., C.N. Bremmer, "Highland–Lowland Interactions in the Ganges–Brahmaputra River Basin: A Review of Published Literature," ICIMOD Occasional Paper No. II, Kathmandu, 1989. With permission.)

furcatus, a type of screw pine. Tropical evergreen rain forest is confined to the humid foothills of the Eastern and Central Himalayas. *Shorea* is the dominant tree species all over the tropical zone. Besides these trees, some 4000 species of flowering plants, of which 20 are palm, are estimated to occur in the Eastern Himalayas. The temperate zone (2000–3000 m) is situated above the upper limit of widespread agriculture, the major limiting factors for cropping being high cloud incidence during the summer and low temperatures in winter. The oaks and conifers are the dominant trees. Lopping for fodder and grazing are common practices in oak forests. Above 2400–2700 m, where a snow cover is present for at least two months, the main canopy is still dominated by oaks. Deciduous species, especially maple, come to the fore as well. On the wettest sites, pure stands of rhododendrons replace the oaks. Grazing pressure is high throughout the upper temperate zone, as high-altitude livestock move down in search of warmth and low-altitude cattle migrate upward in search of fodder.

In the sub-Alpine zone (3000–4000 m), winter becomes much more severe, and rooting opportunities are often restricted. This zone is dominated by a relatively open stand of silver fir with oak and especially junipers on the drier (western) side and rhododendrons on the wetter (eastern) side. At higher elevations, birch is quite common as well.

The vegetation becomes increasingly stunted with elevation above 4000 m (roughly coinciding with the timberline) in the Alpine zone, it is reduced to shrub size (junipers, rhododendons, and finally consists of herbs and grasses only). Finally, a steppe zone can be distinguished by open low vegetation, and its flora is strongly related to that of Tibet.

The animal life is also abundant and varied. Elephants, bison, rhinoceroses, musk deer, black bear, the clouded leopard, the langur monkey, and the cat are some of the denizens of the Himalayan forests. The streams are rich in fishes and aquatic life of wide variety. Bird life is equally rich but is more evident in the east than in the west. Among some of the common Himalayan birds are different species of magpie, titmouse, chough, whistling thrush, and redstart. However, the fauna and flora are severely under pressure to the point of extinction in many cases.

The Vindhyan region in the south also had very rich fauna and flora. The climate was not so varied, and there may not be as much variety, but it was tremendously rich in forests to the point of constituting an insurmountable barrier between the northern and southern parts of this region. It was the natural domain of various deciduous tropical forest types (containing *Shorea*, *Tectoha*, *Dalbergia*, etc.).

The alluvial plains were also very rich in fauna and flora although they have suffered devastating human onslaught since prehistoric times. Historical writings indicate that even until the sixteenth and seventeenth centuries, the scarce resource for agriculture was not land but people. The land, however, is now intensely cultivated, and most of the original natural vegetation and wildlife has disappeared from the Gangetic basin. The Brahmaputra valley is still rich in natural vegetation and wildlife, but it is under intense pressure.

The Sunderban area of the delta provides a distinct ecosystem. Bengal tigers, crocodiles, and marsh deer are still found.

2.8.3 Biodiversity

Biological diversity is the sum total of species richness and encompasses all species of plants, animals and microorganisms, ecosystems, and the ecological processes of which they are parts. The enormous biological diversity of the basin reflects its highly diverse physiography, wide range of climates, latitudes, and biogeographical history. About 185 plant species are endemic to the Himalayas, rhododendrons and orchids included. The maximum plant diversity is to be found in Northeast India, which has witnessed a high rate of evolution.

Bangladesh, an important transition zone between Indo-China, the Himalayas, and the rest of the Indian subcontinent, once abounded in a variety of wild flora and fauna. Its moist tropical forests were botanically among the richest in the subcontinent and supported the greatest diversity of mammals and a high diversity of birds. The Chittagong area has many Indo-Chinese plant species. Although species richness is relatively large for the small area of Bangladesh, endemism is low, and the population of most of the species has declined drastically. Eighteen species of wildlife are now extinct, nearly 15 of them having disappeared within the last two decades.

Because of its unique setting and relatively unexploited environment, Bhutan probably possesses the greatest biological diversity of any country of its size in Asia. The country still has much of its original forest cover and contains some of the best remaining representatives of habitat types found in the Himalayas. However, Bhutan's forests and wildlife are increasingly threatened by poor land management. Organized poaching is depleting fauna, particularly in the south.

Biogeographically, Nepal lies in the transitional zone between the Indo-Malayan and Palearctic realms. While few endemic animal species have evolved in the Nepal Himalaya, a large number of species of mammals, birds, crocodile, fish, and butterfly have been recorded.

3

Socioeconomic Scene

3.1 Introduction

Geography has been said to be the matrix of history. Environment, in turn, is affected by people, increasingly so in the modern times. We briefly review the current sociopolitical scene. A brief historical perspective is given, as the Ganges–Brahmaputra–Meghna (GBM) area provides the foundations of India's sociopolitical economic history. We briefly describe the people and identify the population dynamics and projections. The political scene after Independence, which is going to shape the environment of the future and vice versa, is next reviewed for each of the riparian countries. Then, the economic scene is also reviewed briefly. Finally, an overview on the infrastructure and services, including the energy, agriculture, and industrial scenes, is given. The historical sociocultural settings still strongly influence the management of the economy and the environment.

3.2 Political Geography

The GBM basin has been the scene of intense political activity. Although several empires have risen and crumbled over history, considerable political and administrative unity was achieved beginning from about the sixth century BC. The current scene evolved from the British period essentially building on the crumbling Mogul empire. The area was one of the first in India to be conquered by the British, with their base in Calcutta. By 1836, considerable political integration over the region had been established, as shown in Figure 3.1, with British control spreading almost up to the western end of the basin, which was called North Western Province. Over a period of time, the region was organized under the provinces of Punjab, United Provinces, Bihar, Bengal, and Assam, with large areas in the south and west under princely Indian states. After Independence, these acceded to India and were organized into new states or amalgamated with the earlier ones, which

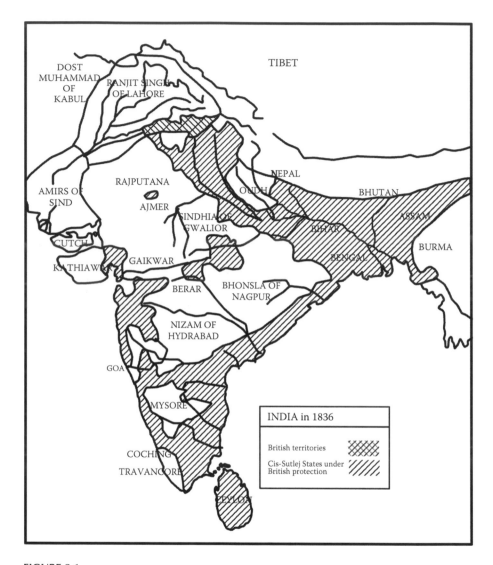

FIGURE 3.1
British India in 1836. (From Spear, P., *A History of India*, Vol. 2, Penguin, London, 1978. With permission.)

were reorganized. The recent (as of 1990) political map of India is shown in Figure 3.2. In India, the states have primary jurisdiction over environmental management.

The basin area, countrywise and, for India, by constituent states, has been given in Table 2.2. Bangladesh, Bhutan, and Nepal are completely in the GBM basin. The basin constitutes about a quarter of India, with some states like Uttar Pradesh and Uttarakhand completely in the Ganges basin and the states in the northeast completely in the Brahmaputra–Meghna basin. The

FIGURE 3.2
Political map of India (1990).

Ganga riparian states in India, the percentage of the basin in the state, and the percentage of states in the basin are as follows: Rajasthan (13.3/33), Haryana (4.0/78), Delhi U.T. (0.2/100), Himachal Pradesh (0.5/8), Uttar Pradesh and Uttarakhand (34.2/100), Madhya Pradesh and Chattisgarh (23.1/45), Bihar and Jharkhand (16.7/83), and West Bengal (8.3/81).

The Himalayan region in the Ganges basin, except for a small portion in the west that was under British control as part of Uttar Pradesh, was an independent country, Nepal. The Himalayan region of Brahmaputra was in Sikkim, Bhutan, and Assam. Nepal and Bhutan, though independent, were under British suzerainty.

3.3 Historical Perspective

As Nehru (1947) wrote, "The Ganga is above all the 'River of India' which has held India's heart captive and drawn unaccounted millions to her banks since the dawn of history. The story of the Ganga from the source to the sea, from old times to new, is the story of India's civilization and culture, of the rise and fall of empires, of great and proud cities, of the adventure of man and the quest of the mind which has so occupied India's thinkers, of the richness and fulfillment of life as well as its denial and veneration, of the ups and down of growth and decay of life and death."

Indian civilization is noted for its historical continuity and for the fact that elements of Indian tradition are so firmly embedded in the country's culture that they persist in influencing contemporary social and political behavior. A bit of historical perspective may therefore be in order. This will be up to the period of Independence.

The Indo-Gangetic basin had an outstanding natural resource base and environment, leading to one of the earliest establishments of civilization. Archeological findings have revealed some of the earliest human settlements and highly urbanized Indus Valley civilizations (2500–1700 BC) in the western part of the Indus basin extending south up to the middle of western India. Earliest archeological findings were at Mohenjodaro–Harappa and were later found to be more extensive. These societies, however, came to an end between 2000 BC and 1500 BC. The reasons are not yet understood, but it may be on account of environmental changes. It is only in the last few decades that these archeological findings have helped scholars deduce that the Indus Valley civilization made an important contribution to the formation of Hinduism. The script used by the Indus Valley people (which has yet to be deciphered) was lost along with the disappearance of the Indus Valley people.

At about the same time, the so-called Aryan or Vedic tribes were moving into northwestern India. They had superior aggressive powers due to their possession of horse chariots and iron weapons. Apparently, this enabled them to clear the forests in this region, and further extension of settled agriculture took place. Perhaps it was easier to clear and inhabit the western parts of the GBM basin, such as the Ganga–Yamuna region, in the first instance.

Indeed, the cultural history of India starts from this period, which is estimated from about 1100 BC. According to legends, and the legendary epic Mahabharata, which is the world's longest single poem, the founding kingdom was established in the region between Ganges and Yamuna around present Delhi by Emperor Bharat, after whom India is again named as Bharat in the constitution after Independence.

Over the course of time, the wetter and more densely afforested areas of the central and eastern GBM basin were also brought under settled agriculture. With the increasing availability of fertile land, more prosperous and settled communities developed. Ganges became a natural highway of trade. Settlements developed along its banks.

Increasing levels of surplus in agriculture resulted, allowing an increasing fraction of the population to take to crafts, trade, and warfare, resulting in the development of kingdoms and empires for the first time in the Indian subcontinent. The center of gravity of this society lay in present day Bihar, which provided an excellent combination of fertile land, plentiful water, navigable watercourses, and iron and coalfields. The completion of agricultural colonization is considered to have taken place by sixth century BC. The next 1000 years that followed was a period of empires and prosperous trade, of many advances in the so-called high culture and in science and technology (Thapar, 1977). Although many of the early kings were elected, the institution finally became hereditary.

Over the long relatively undisturbed period of 1000 years that allowed the Aryans to penetrate the subcontinent and evolve from seminomadic tribes to principalities and kingdoms, a characteristic socioreligious tradition developed which has given the Indian society historical continuity. It has been pointed out that, in contrast, the Chinese traditional ideology, popularly associated with Confucianism, which has given it a strong historical identity, is sociopolitical in nature (Vohra, 1997, p. 11). This has strongly impeded the sociopolitical integration of society and its development.

Aryan society was divided into four classes (varnas): the Brahmin priests, the aristocratic Khatriya warriors, and then followed by the Vaishya. Commoners, in that order and at the bottom of the scale, were the Shudras, who were mostly of indigenous extraction and considered only fit for menial tasks. There was also a fifth, though unrecognized, class entrusted with the lowliest and most loathsome duties such as scavenging and handling of dead bodies; these unfortunates, the so-called untouchables, were severely ostracized and discriminated. Even in the four classes, only those belonging to the first three classes were entitled to be initiated into the Vedic ritual and gain education. Unlike Confucian China, where theoretically the peasant masses were not barred from education, Hindu India kept over 50% of its population illiterate by design (Vohra, 1997, p. 24). Since the term also means color, the class system may indicate the arrogance of "white" Aryans. By the end of fourth century BC, division based on occupation, lineage, and so on arose within the varna, and these divisions (jatti) are the ones that can truly be described as castes (the term *caste*, which is derived from the Portuguese word for race, has been popularly used to describe the four varnas; in recent years, the word has been more appropriately applied to the subgroups [jattis] that exist within the varna).

The origin of castes is extremely complex and is not well understood as of yet. Briefly speaking, jattis were and are endogamous groups representing functional distribution within a varna (e.g., barbers, carpenters, and weavers in the Shudra class). New jattis were formed when members of different varnas intermarried, when a group was degraded within a varna for not performing the religious rites expected of it, when a local or a foreign tribe was absorbed into the Hindu fold, or when members of a caste moved away from their original habitat or changed their profession. In brief, while the four varnas provided a broad class division of society, it is the over 3000 hereditary, endogamous, craft-exclusive jattis or castes, each with its own name and norms of conduct, that govern everyday family life in traditional India. Social relations are further regulated by rigid concepts of purity and pollution. The caste system, by delineating the occupation and daily life of an individual and his behavior pattern within the larger community in the village, provided stability and continuity to the social system. The caste system in the narrower framework of the village thus became a world in itself, and politics was irrelevant. Kings and empires came and went, without any impact except on the authority of the dominant local castes, the local power holders.

Hinduism also perpetuated the caste system. Hinduism is a complex composite of disparate beliefs. It is not a religion of revelation. It has no church or hierarchy of priests that can make authoritative pronouncements on doctrinal disputes. It has no set doctrines. It is not congregational. It does not aspire to establish a theocratic state. What makes Hinduism the most tolerant of religions is its syncretic nature and its acceptance of the fact that there are many paths that lead to God and that God accepts all of them. To sum up, Hinduism attempts to be all things to all people. Unfortunately, this liberalism of Hinduism and the caste system was based on the ghettoization of society. Intercaste communication was limited to business dealings, and there was no intercaste socializing. Thus, neither political nor social unity nor purpose ever developed, leading to repeated conquests by foreign invaders.

The Indian economy was characterized by what would be called today an "appropriate technology" geared to small-scale production in family units of peasants and artisans. Land and labor as factors of production were abundant, and therefore, the third factor, "capital," which is substituted for the other factors whenever they are scarce, was not required. There was, of course, some capital formation in trade and also in terms of construction of wells and tanks for local irrigation, but there was no capital accumulation that would lead to a concentration of the ownership of the means of production. The tools and implements were simple and could be made locally and cheaply. Manufacturers, sponsored by some rulers for the making of arms or for luxury goods, were very rare indeed. Successful merchants or victorious warlords sometimes amassed wealth, but such wealth was usually lost as quickly as it was gained.

In agriculture, the small family farm was predominant. Land was not yet a commodity that was freely bought and sold. The peasant family who tilled the soil was in greater demand than the land, "which belonged to him who first cleared it," as the ancient laws proclaimed. Landlords were not landowners. They had only the right or privilege to collect taxes from the peasants. They kept some of these dues for themselves and handed over the rest to the ruler or another privileged person in the hierarchy of those who lived on the work of the peasantry. If these dues were too high, peasants would flee and look to the protection of a less rapacious lord. Nobody ever thought of organizing agricultural production on a large scale with hired labor. The vagaries of the monsoon would have made such an operation very risky. It was much better to let the peasant families bear the risk and, usually, the entire cost of production, and get hold of their surplus when the harvest was in.

A strong ruler would tend to eliminate middlemen and concentrate the collection of the surplus in his capital city, but there were several problems that he would face when doing this. Such concentration was impossible without coercion, but this would lead to growth of overheads and taxation, which would finally be met with resistance. Moreover, the collection of surplus in kind was difficult because of transportation, and the high degree of monetization, which was required for the large-scale collection of revenue in cash, could not be attained in many periods of Indian history.

The large republics established in the early history were gradually broken up under foreign invasions, and although attempts to establish centralized republics were made, decentralization prevailed most of the time. Medieval Indian kings controlled only a circle of about 100-km radius around their capital in terms of direct taxation: from more distant parts, they could at most exact some tributes. These kings were used to granting a great deal of local autonomy to their subjects. A new kind of politicoeconomic structure termed feudalism emerged.

The self-sufficiency and integrity of the village communities have been a striking feature of the Indian socioeconomic system. Production was approximated to local requirements, with little attempt at producing surplus to be used specifically for trade or exchange. Surplus production would hardly have benefited the peasant, since it would have led to a demand from the landowner for a larger share. The existing system led to the acceptance of the standard of minimum production, since incentive to improve production was absent. Subsistence economy was the characteristic feature of the society, and it also inhibited the development of towns, which was significantly less than that in China at the same time.

A caste system had developed based on professional activities, which created strong social rigidity and hierarchy. Kings and empires developed and disappeared without almost any perceivable impact or consciousness on the village and villagers. What was immutable in the Indian society was not freedom or slavery, but caste (Thapar, 1977, p. 77). There was little change in knowledge, perception, values, or technology. The society had almost stabilized into a set pattern shaken only by pestilence, famines, floods, and so-called acts of God.

The traditional structure of the medieval kingdoms was rudely shaken when Northern India was conquered by Islamic horsemen at the end of the twelfth century whose new strategy of swift cavalry soon spread throughout India, as no ruler could hope to survive unless he adopted the new style of warfare and military organization. This required increased taxation, and the man on horseback was a more formidable tax collector than his pedestrian predecessors. Local autonomy was crushed by the man on horseback. New urban centers arose in the countryside, but they were not for trade. They housed the garrison of the cavalry and the treasury of the tax collector, serving as marketplaces as well. Increased international trade brought precious metals to India, and this contributed to the spread of monetization, which enabled the tax collectors to get their revenue in cash.

The military commander and tax collector, who ruled the countryside from his garrison town, was usually a stranger with no local roots, often a foreigner, Turk or Afghan, speaking a foreign language. Whenever central control was reduced, such a commander would turn into an independent warlord and start a dynasty of his own. In order to guard against such a challenge, the central rulers tried to concentrate a larger military force in their capital. The type of urbanization bred by this system of military feudalism was a strange one: there were no patricians and no municipal autonomy in these towns. The military elite were in complete control, and civil administration was already subject to the control of military officers. Thus, the rule was essentially military and the regime something of an armed camp.

The universal spread of cavalry warfare and the distribution of the cavalry units in various garrisons made the assertion of central control more and more difficult. Only the introduction of an even more powerful force could help to support a new central dynasty. The Mughal field artillery was such a force. Starting from the year 1526, the Mughal Empire lasted until it crumbled in disorder by the middle of the eighteenth century. At its peak of glory during Akbar's regime until his death in AD 1605, it covered most of north India, covering an area larger than that controlled by the British later. There were several notable features that shaped the later colonial rule, and even modern India. One was the re-creation of the imperial idea in India. Another was the organization of a bureaucratic administration and imperial service. The imperial service consisted of a graded hierarchy of commanders. The titles were not hereditary. Appointment and promotion were by imperial favor, and rank did not in itself confer office. The high officers were rotated, with one office not being held for more than 3 or 4 years. Secondly, the assignments were for life only, with resumption of their property at death. The Mughal nobility was thus an official aristocracy, which was hereditary as a class but not as individuals and was land holding but not feudal.

This class was spread over the country to work the administrative machinery. Akbar divided the empire into 12 subas or provinces, which grew to 18. These were in turn divided into sirkars, the ancestor of the British district, and further into parganas, the ancestor of the subdistrict. Throughout this system, the principle of division of authority prevailed. From the suba downward, there were two sets of officers, the magisterial (Subedar) and the revenue (Diwan). The former controlled the armed forces and was responsible for law and order, while the latter controlled the land revenue and was responsible for the land assessment. In general, the proportion of the gross produce or its value taken was one third. There was no elaborate system of judicial courts such as those later introduced by the British.

The Mughals were successful in conquering most of India with their guns, but in this process, they finally exhausted the land revenue resources on which their empire was based. The enormous overheads of innumerable soldiers, courtiers, and administrators were too much for the revenue-paying peasant, who had to sell an ever-increasing proportion of his produce in order to pay his revenue in cash. This monetization under a military feudal regime did not lead to a genuine commercialization, just as the new pattern of feudal urbanization did not lead to the rise of a bourgeoisie. The Indian traders who bought and sold grain and provided credit to the ruling class could never hope to achieve a corporate autonomy. They were very astute in going about their business, but the conditions under which they had to operate were not conducive to the growth of genuine capitalism. The sluggishness of the seasonal circulation of money in the vast agrarian economy also handicapped the financial operations of the traders. Sometimes, they emerged as large-scale revenue collectors, but more often, they were cautious and did not want to get involved in such risky business. Knowing the rapacity of the military elites, they hid their wealth and adapted themselves to the prevailing conditions. In addition to the appropriate technology, there was an appropriate capitalism of merchants, as Rothermund (1988) has called it, which was used for dissimulation rather than for an active assertion of their influence.

The picture of India was a vast area studded with villages in dire poverty, mainly concerned with sowing and reaping, engaged in subsistence agriculture and economy, and diversified with a few large centers of political power, where display, luxury, and a teeming life were the rule. Of course, before the agricultural revolution in Europe, there was dire poverty in the European countryside also, but the big advantage that Europe possessed over India was capital accumulation and investment by the merchants, more productive spending by the government, and development of independent institutions and local autonomy.

The monetization of the land revenue was very important for the rulers of the great agrarian state because only cash could be easily transferred to the central level to support a powerful government and a large army. However, spurts of inflation took place on account of international interlinkages or the so-called acts of God in the form of famines and floods, and the rational system of land revenue assessment and the carefully graded salary scales for the hierarchy of imperial officers developed by the founding Mughal rulers were ruined toward the later period of the rule. There were increasing uprisings and wars, which severely strained the economy and administration. Military feudalism was at the end of its tether. By 1707, with the death of Aurangzeb, the last of the powerful Mughal emperors, decline started to set in. The succeeding Mughal kings became a shadow, keeping a facade of the Mughal Empire, as several regional warlords broke it up. Finally, the bourgeois, as servants of the East India Company, stepped in where the Mughals had failed.

From ancient times, India had active maritime trade relations with many countries around the Indian Ocean. In the medieval period, South Indian states were particularly involved in this trade. Powerful corporate empires rose in several parts of Asia in the medieval period. The political order of these corporate empires differed from the land revenue–based agrarian states that were developing in Northern India. There was not that much of territorial sovereignty, but there was a corporate network of rulers, merchants, temple priests, and/or royal officers. It is considered that in the eleventh century, Asia was definitely ahead of Europe in most aspects. However, as the land-based conquests, which introduced a military kind of feudalism, were increasingly established, these corporate empires were run over, and they started to decline. With it, their navies also disappeared, and the Indian Ocean emerged as an enormous free-trade zone uncontrolled by any sea power whatsoever. Gradually, the maritime trade was taken over by Western traders, pirates, and later trading companies who established coastal bridgeheads. Although initially, Portuguese, starting from AD 1498, and later Dutch and French companies were more important, the British East India Company, established on the last day of AD 1600, gradually held the sway. There were three bridgeheads, Madras, Bombay, and Calcutta. While the British stuck to their coastal bridgeheads in other parts of India, there was considerable textile trading in Bengal, which increasingly took them further and further up the River Ganges, into the interior of the country.

As the Mogul Empire was crumbling, local administrators became independent warlords. The military power of the British traders was also increasing. They were extracting trading concessions and were even obtaining the responsibility to collect the land revenues and other taxes, plundering the country in the process. Anarchy and, correspondingly, the political power of the British increased day by day. As Rothermund (1988) has observed, the

East India Company as a modern capitalist corporation of an advanced bourgeois nation entrenched itself like a parasite in the agrarian state dominated by a decaying military feudal regime. The parasite adjusted to the system of the host and benefited from it without changing it very much. The company was well geared to function in this way. It had developed a modern bureaucracy in the course of its trading operations, and through better organization and discipline, it soon acquired control of the entire India, although direct political and administrative control was limited to a small region, with the larger parts being placed under the reign of native rulers, who were, however, only titular heads. The lower Ganges basin, including Bengal, which has been considered to be the most prosperous region of the world at that time, bore the brunt of this plunder and was thoroughly ravished (Spear, 1978). By AD 1818, the hegemony of the British was established, and by AD 1836, the entire India, up to the banks of River Sutlej, was under the control of the British. The Company had tacitly become the maritime auxiliary of the empire. After a feeble mutiny by the Indian princes in AD 1857, which was the last convulsive movement of protest by the traditional India, the country came under the direct political control of the British Government and became part of the British Empire.

India became a dependent agrarian state. The Company, and later the British Government, did nothing to stimulate the economy and simply collected tribute as its military feudal predecessors had done, but whereas these predecessors had spent the tribute in India, the Company transferred most of it abroad. Indeed, in the Indian land revenue system, the British contractual law was injected, making it even more exacting to the detriment of the peasantry. As the British regime continued, the objective was governance for the benefit of the ruling country and, correspondingly, maintenance of law and order. There was little development of industry, as it was not in the interest of the ruling country. There was some infrastructure activity such as construction of railways, roads, and bridges, but the consideration was maintenance of law and order rather than economic development. Irrigation was developed with the objectives of increasing revenues and maintaining law and order through mitigation of famines, and the policy became stabilization of the sustenance agriculture rather than development of productive agriculture. The per capita food availability declined as population increased. Even education was not for introducing mass literacy or development of technological capability but for the production of a subordinate staff for administrative requirements.

Even worse, the social development was derailed by the very fact of a foreign colonial government. While in the Western world dynamic indigenous development of the social system was taking place through political, educational, economic, and technological revolution, India was becoming increasingly impoverished socially and culturally. In contrast to the cumulative causative process of self-reliant dynamism, sociopolitical integration with resultant structural changes, institutions and attitudes, that was taking place in the Western countries, in the developing countries, a corresponding process of dominance, disarticulation, and poverty developed (Sagisti, 1979; Chaturvedi and Chaturvedi, 1985). Domination implies that the underdeveloped country does not have the capacity for autonomous decision making and that it exercises little control over its destiny, thus being an extreme form of dependency. The primary form of dominance is political and economic, although it is closely related to cultural and technical domination, with one of them leading to or implying the other. Disarticulation means that the underdeveloped country does not constitute a homogeneous unit from the cultural, social, and economic point of view. It is a highly stratified society with little or no interaction among the various strata. There is a comparatively small high-income group alienated from the rest of the society, which tries to adopt the values and the lifestyle of the colonial masters. In the legacy of the military and administrative garrison centers of the Mughals, civil lines were established by the British rulers in the towns, housing the bureaucracy apart from the local people. Racial discrimination was practiced. Foreign language completed the schism of the soul. A small official class, mostly foreign at senior levels, ran the Raj, maintaining law and order for sustaining the parasitical relationship. While bureaucracy is a modern world phenomenon, the historical socioeconomic–political environment in India was entirely different, giving the bureaucratic institutions a unique culture. In the West, despite national differences and departures from Weber's ideal type, modern bureaucracy developed as an antithesis of feudalism, a result of economic progress and intellectual development, as a middle-class bourgeoisie concept functioning with the elected elements of government for public administration. In India, the bureaucracy was a colonial product and tended to be alien and despotic, particularly as there was no development of alternative independent institutions of government and management. The government ruled rather than governed.

As Rothermund (1988) has put it, India's economic history, which is dialectically related to the societal system, is like a fascinating drama: an ancient peasant culture was subjected to a regime of military feudalism, which achieved its greatest success under the Great Mughals in the seventeenth and eighteenth centuries. The Mughal agrarian state and its revenue base were then captured by the East India Company. This capitalist organization was engrafted on the agrarian state. In this way, a "parasitical symbiosis" was established that benefited the alien usurpers and paralyzed the host, which survived under conditions of a low-level equilibrium. In the late twentieth century, India came under the influence of the world market to an increasing extent. The First World War, the Great Depression, and the Second World War all made a strong impact on India. As there was no national government, these external influences remained unchecked by any national economic policy, and India was fully subjected to them. Throughout this period, the vagaries of nature also played their fateful role in the socioeconomic conditions of the country, as India depends on rain-fed agriculture to a large extent.

3.4 Population

The region has been attracting people for a long time. In Toynbee's (1947) words, the region provided an environment of "appropriate challenge and response" to foster human growth. People of different ethnic compositions came at different times; the last to come were the people commonly referred to as the Aryans. Aryan is in fact a linguistic term indicating a speech group of Indo-European origin, as is not an ethnic term, but this inaccuracy has become an accepted norm (Thapar, 1977). Local population was known as Hindus, which basically denotes not a religion but people living in a certain region with a certain way of life.

People kept on coming from the west and east. Muslim religion had been established in the west, and the neoentrants were violently proselytizing even though in small numbers. Since they were the new kings, considerable religious conversions, particularly from the lower castes, took place. The last to come were Europeans, but they did not settle or intermarry to any extent.

Historically, the estimate of population of the subcontinent has been 181 million at the end of fourth century BC. An estimate for the early seventeenth century is 100 million (Thapar, 1977). The first census was carried out in AD 1891, and the population was 235.9 million, with 238.4 million in AD 1901, when the first dependable census was conducted. Assuming that about half of the population of Indian subcontinent lives in the GBM basin, the basin population at the beginning of the twentieth century was about 120 million.

From the early period of checkered growth, a period of moderately increasing growth of about 1% started from 1921. However, a period of rapid population growth started from 1951, shortly after Independence. The population of India was still 361.1 million at that time. With increasing emphasis on social welfare and economic growth after Independence, the population started exploding. Lately, emphasis has been placed on population control, but the impact has been marginal, and the population continues to grow almost unchecked at about 2.5%. About 534.80 million people were living in the basin in 1990–1991, about 76% of whom were in India, 20% in Bangladesh, and 3.5% and 0.3% in Nepal and Bhutan, respectively, as detailed in Table 2.1, which also gives some other salient details. The population of the basin is expected to stabilize at about 1 billion by about 2050. About 47% of India's population of about 1 billion resides in the basin, the bulk of this in the extensive Gangetic plains.[1]

Density of the population in the Basin was the highest in West Bengal (827 per km^2), followed by Bangladesh (736 per km^2). The population in the plains of the Gangetic basin, Bihar, and eastern Uttar Pradesh is comparable. The population in Nepal works out to 135.7 per km^2, but if it is considered on the basis of cultivable area, it works out to about 600 persons per km^2, which is comparable to the plains. Population density in the outer states is significantly lower, 137 per km^2, and those in the hills of Northeast and the Himalayan belt are even lower at 73 and 40 per km^2, respectively. But these figures do not bring out the reality of resource scarcity as brought out in the case of Nepal.

A peculiarity of many parts of India is that of having so-called "tribals." At some ancient time, some people driven from the plains took refuge in the jungles on the hill slopes and thereafter remained isolated from the plains to become noncompeting societies in their own enclaves. Later, they were to be called tribals and described, as one eminent anthropologist did, as "people on the ledge of history." They have been exploited by the forest contractors, but for the northeast, including the hill tracts in Bangladesh, they were traditionally more militant and aggressive. The resulting law and order problems were later aimed to be resolved by the British Government in India by installing an inner line to segregate the tribals from the plainsmen. The conflicts have continued in several forms, leading to the granting of statehood for each major tribal group in the northeastern region of India. Tribals constitute 8.5% of the GBM basin. They are heavily concentrated (52.0%) in the hilly tracts of the northeast and to a lesser extent in the outer states (16.3%) and the Himalayan Belt (13.6%). It is in this area where the major forest wealth of the region is found as well as most of the sites for multipurpose dam projects.

3.5 Independence and Present Scene—India

National stirrings had started in British India in the nineteenth century, and by mid-twentieth century, independence was achieved. The social system is being recreated in the countries of the basin. The historical legacies, however, have a crucial role in shaping the future development.

Winds of change blowing over the modernizing Western world were bound to affect India. Beneath the burnished cover of the British administration, the mind of India was actually in ferment. Indian nationalism or, better yet, Indian transformation was taking place. In the beginning, Indian nationalism was only a movement rather than a force, an aspiration rather than a general dynamic. However, indignation at foreign rule was growing, and it found a new leader in Mahatma Gandhi. The driving force of the Independence movement was the identification of the leaders with the masses personalized in the image of Mahatma Gandhi and the trust of the people in the leaders. The foreign government had to bow to this force, and the Independence of India was achieved in August 1947. Along with Independence came the partitioning of British India into India and Pakistan.

While there had been an ardent freedom struggle, India attained independence not because of a revolution, and paradoxically, a transfer of power took place from the Raj. Herein lie the enigma and seeds of many of the failures of the new society (Chaturvedi and Chaturvedi, 1994).

It was obvious that the task facing India was to start the process of modernization and bring about a social revolution, but the definition of this and the means to this end were not clear. Even before it could be undertaken, there were the immediate tasks of continuing the process of government, maintaining law and order, providing the basic needs of sustenance, coping with the trauma of partition, and establishment of the national government. All the new leadership had were well-institutionalized administrative machinery, the so-called "permanent government" of the colonial government, the legal and educational systems, and the entire infrastructure as a growing concern. There were no alternative societal institutions such as local self-government and commercial, urban, scientific, and intellectual groups. There was not even a sizable literate group. The society was already deeply divided with a miniscule elite group, principally in the Government or its ancillary functions such as law and other minimal services. The society was thus involved in the contradiction of construction of a civic society from the top in the context of the legacy of a long colonial rule, backwardness, poverty, and illiteracy. This posed serious problems, which had many strands.

First of all, a modernizing revolution from above is, by definition, ill suited to provide for the ever-increasing complexity entailed by modernization. Whether in the sphere of institutions or of social groups or even in theory, state-promoted modernization tends to work toward models of simplicity and predictability that make for easy administration; it shuns complexity and even rejects it as some theories of nationalism have. In a general and noncommittal level, need for radical social and economic change is freely and passionately proclaimed, but in planning and policies, utmost care is taken so that traditional social order is not disrupted.

Thus, in the sphere of governance and sharing of power, democracy was constitutionally proclaimed. There were ready-made democratic ideology and even a framework of ideological apparatus of governance at the central and state levels, but there was neither democratic reality nor even a facade at the grassroots level. On the contrary, the administrative machinery deriving its institutions and expectations created by about 350 years of Mughal feudalistic and British colonial heritage had the perception and tradition of ruling and not governing. Even society had internalized the long experience of being ruled. The newly elected leaders, most of whom did not have the commitment of the freedom fighters, soon started identifying themselves in the image of the departed colonial rulers, as there were no restraints from alternative power centers, nor were the political party institutions strong enough to keep constraints on them. Over the years, even the roots of the party, which led India to independence, starved. Thus, over the years, the political system and environment have increasingly become corrupt, undemocratic, and an exploiter, indifferent to the people whom they are supposed to represent and serve. A sizable number of the political leaders have well-known long jail experiences under proven criminal charges and embezzlements, yet they continue to be elected and achieve important positions of power.

Economic development was a central objective, but development was considered in linear evolutionary terms. It was not realized that developed and developing societies represent different quality, structure, and dynamics of individuals, institutions, and organizations. In the social system, there is a complex set of heterogeneous motivational and regulatory mechanisms. With increasing development maturity, these mechanisms develop for the creation and development of "positive causation process" (Myrdal, 1971). On the other scale of development, these processes and structures have not sufficiently developed. Development is not merely a matter of investing in capital and technology as hardware projects. It is the whole social system that has to be moved upward, characterized by growth capacity at all levels of the socioeconomic hierarchy to create new production environments for the producer to respond adequately and respond to the new production environment. Interventions are required, of which technology is an important one, aimed at improving the cybernetic structure of the coordinating level and of the production level in order to elicit from the structurally modified system a new variety that would be expressed in advancing and upgrading the production environment, in improving response patterns, in strengthening coordinating control, and in vertical integration. In developing societies, there are strong political, psychological, social, and institutional inertial forces resisting change and progress from a low ultra stable to higher dynamic equilibrium. Considerable planned and determined effort is required to break the vicious cycle of backwardness, which is often not displayed.

Development has also led to increasing internal disarticulation, as benefits have not gone proportionately to the poor. As commitment of the Independence movement or Indian transformation increasingly becomes a past event, consumerism increases, and the elite urban group tries to ape the lifestyle and trappings of the Western world, creating a dual society, a parallel culture, which further exacerbates stratification and fragmentation of society and also imposes unsustainable strain on the environment. Paradoxically, in the modern world of shrinking time and space, with increasing power and prosperity of the Western world, the developing countries face an increasing threat of distortion of their developmental effort, disarticulation, and dependency. As linkages in the two worlds increase with increasing knowledge, technology, trade, and better international transportation and communication systems, while the developing countries are helped to modernize, sometimes unattainable expectations, distorted attitudes and lifestyles, and consumerism are also generated, which are totally inimical to the developmental policy for equitable, sustainable, and rapid economic growth of these societies. Unless there is commitment to self-reliance and eradication of poverty, which implies involvement with the poor and a sense of creative cultural identity, dependency and disarticulation can be exacerbated in the modern world.

This is not a matter of nationalist chauvinism. The developing countries have much to learn from the advanced countries, and collaborative development can be to their considerable advantage, but this has to be pursued with extreme care on terms designed to generate self-reliance and modernization rather than stifling creativity and generating dependency. For instance, as the modern technology is skilled labor intensive, collaborative research and production, if properly planned, can be mutually advantageous. The central issue is that in view of the dichotomies introduced over colonial heritage and problems introduced by neocolonialism, inherent in view of the extreme social disparities, utmost care has to be exercised in bridging the disparities rather than letting them be exacerbated, as will be the natural outcome if due care is not exercised.

3.6 Sociopolitical Perspective—Bangladesh

The region, currently called Bangladesh, was in the mainstream of Indian religion and culture. The population was overwhelmingly Hindu, with some Buddhists and animists. The region was one of the most prosperous parts of India and the world until the mid-eighteenth century (Spear, 1978).

The advent of a handful of Muslims at the beginning of the thirteenth century and rapid expansion of their rule permanently changed the character and culture of the area. Even as late as AD 1872, there were more than 18 million Hindus in Bengal compared to about 16 million Muslims (Encyclopaedia Britannica, 1984). By 1947, the majority of the population of the region was Muslims, leading to the partition of Bengal, with West Bengal remaining as part of India and East Bengal becoming part of Pakistan. It was a most anomalous arrangement, with two parts of a country separated by more than 3000 km and having nothing common except religion. The language, ethnic background, cultural background, and social practices were all different. The breakup of Pakistan was natural, but the exploitation of East Pakistan by West Pakistan administratively and economically hastened the process. A regional party, Awami Muslim League, in contrast to the national party Muslim League, which led the formation of Pakistan, was constituted as early as 1949. It became the dominant party. By 1955, it became Awami League and was later led by Sheikh Mujibur Rahman to contribute significantly to the formation of Bangladesh.

The Government constituted after the 1954 elections demanded autonomy for East Bengal. The popular government was dismissed, and rule of the Central Government was imposed through Governor's rule in May 1954.

The Pakistan Constitution, promulgated in 1956, turned the two wings of Pakistan into two provinces. The official name was changed from East Bengal to East Pakistan while its demand for autonomy was ignored. Martial law was clamped in Pakistan in 1958, which lasted 11 years.

Sheikh Mujibur Rahman crystallized the demand for autonomy in 1966. He was arrested but had to be released. East Bengal had a long history of nationalism as well as a tradition of resistance to alien domination. It was fermenting under the economic and political domination. The ingrained urge for independent national identity personified in Sheikh Mujibur Rahman got its fullest scope with the surrender of the Pakistan army in Bangladesh in the Indo-Pakistan war in 1971. The administration of Bangladesh began in a *de facto* manner. Bangladesh constitution came in force on December 16, 1972, and Sheikh Mujibur Rahman formed the Bangladesh Government.

Bangladesh rose out of a bloody war in which the Pakistan army before it left and which had destroyed the intellectual elite of the country. The administrative and economic structure was in chaos. According to a United Nations (UN) estimate, the damage caused to Bangladesh was more than $1.2 billion. In addition, most of the industries had been in the hands of West Pakistan magnates, who, of course, left the country.

Mujibur Rahman had enormous charisma, but he faced an enormously difficult task of building a nation and a state, following a devastating war. Mujibur Rahman failed to transform Bangladesh nationalism into a workable state. This was due first to heavy reliance on charisma and second to the lack of a coherent ideological framework and strategy for nation building. He had little practical experience of government, whether administrative or economic, and he failed on the latter front as well.

Mujibur Rahman was assassinated on August 15, 1975, by a conspiring group of junior army officers and some politicians. Four senior Awami League leaders were also murdered to ensure that Mujib's camp was left with no leadership. There was a period of uncertainty for a year. Ultimately, Major General Ziaur Rahman and his junta installed themselves.

Ziaur Rahman's tenure of about 5 years underlined that Bangladesh, with Mujib's assassination, had entered a period of destabilization. There were two coup threats in 1 year alone. He tried to gain legitimacy by forming a Bangladesh Nationalist Party (BNP) and holding a general election followed by a presidential election, but his efforts did not succeed in a country as highly politicized as Bangladesh. Parties without mass contact, organization, and ideology cut little ice.

After his assassination in 1981, as a result of faction fighting within the army, there came a few months of disequilibrium. General Ershad, as Commander-in-Chief, held the reins of power *de facto* until early 1982, when through a painless coup he formally installed himself in power by re-imposing martial law. General Mohammed Ershad also sought to prove his acceptability to the people through a referendum and another political party, but he too was not very successful.

Military rulers in Bangladesh have tried, without success, to reinforce their regime through a contrived campaign for Islamic revivalism. The ban of 1972 on purely religious parties has been scrapped, and reference to secularism from the constitution has been removed. Since 1983, the number of Islamic political parties increased from 15 to 65 in 1985.

General Ershad was faced by two powerful political alliances: the old Awami League and the BNP, one led by Mujib's daughter, Hasina Wajed, and the other by Ziaur Rahman's widow, Begum Khalida Zia. Beneath the struggle for political power between General Ershad on the one hand and the political parties on the other, there were several contending forces. The two fundamental ones were the conflict between rural Bangladesh and a rising urban middle class, and the growing forces of Islamic fundamentalism against a secular thrust. The struggle continues. Begum Khalida Zia came to power in the first instance, through elections. These were, however, considered to be rigged, and there were violent protests. Elections had to be held again, and Hasina Wajeed and her party came to power. Basically, a continuing state of uncertainty has emerged. The administrative and sociopolitical background of East Pakistan, now Bangladesh, was akin to other parts of undivided British India. However, over almost 50 years, political stability and social unity have eluded the country so far.

3.7 Sociopolitical Perspective—Nepal

Nepal's name is derived from the Valley of Nepal or Kathmandu. It has been closely and culturally related to the Gangetic Plains at least for the last 2500 years. A coherent dynastic history for the Nepal valley becomes possible, though with large gaps with the rise of the Lichhivi dynasty in fourth or fifth century AD. They were of Indian (i.e., plains) origin, and this set a precedent for what became the normal pattern thereafter—Hindu kings claiming high-caste Indian origin, ruling over a population that was neither Indo-Aryan nor Hindu in ethnic origin. Several dynasties and rulers followed, ruling various hill principalities of the region. By the sixteenth century, virtually all of these principalities were ruled by dynasties claiming high-caste Indian origin whose members had fled to the hills in the wake of Muslim invasions of northern India.

In the early eighteenth century, one of these principalities—Gorkhas, ruled by the Shah family—began to assert a predominant role in the hills. The great Gorkha ruler Prithvi Narayan Shah conquered the Nepal valley in 1769 and moved his capital to Kathmandu shortly thereafter, providing the foundation for the modern state of Nepal. Prithvi Narayan Shah (1742–1775) and his successors established a unified state in the central Himalayas and set the present boundaries of the kingdom.

Throughout most of the modern period, Nepali politics has been characterized by a confrontation between the royal Shah family and several noble families. The regents, as three minor kings sat on the throne in succession, and nobility competed for political power using the young rulers as puppets. The rivals were exterminated, exiled to India, or placed in a subordinate status. The Thapa family (1806–1837) and, on a more extensive basis, the Rana family (1846–1951) obtained control. The Shah ruler was relegated to an honorary but powerless position, while effective authority was concentrated in the hands of the leading members of the dominant family. Although interfamilial arrangements on the distribution of responsibilities and spoils were achieved, no viable national political institutions were created. The court conspiracies and the family feuds of yester years have gradually expanded to assume the present-day complex forms of broad-based political competition and conflicts at the national level.

The emergence and consolidation of the British power in India, which found in the Ranas a dependable ally, sought to perpetuate the eclipse of the monarchy. An agreement was reached under which Nepal "accepted" British "guidance" on foreign policy. In exchange, the British guaranteed the Rana regime against foreign and domestic enemies and granted it virtual autonomy in domestic affairs.

The British withdrawal from India in 1947 created an altogether new situation. The restoration of the Nepalese monarchy from eclipse took place in 1950–1951. India and the democratic forces led by the Nepali Congress, which had its inspiration from the India National Congress, played a key role in this restoration. The change of 1950–1951 restored full status and authority to the king and released a tremendous upsurge of democratic aspirations and forces.

The process of introducing a democratic political system in a country accustomed to autocracy with almost no democratic traditions and experience proves to be a formidable task. King Tribhuvan, who was the symbol of this restoration, readily endorsed the concept of popular legitimacy of heredity monarchy. His successor, King Mahendra, was, however, uncomfortable with representative institutions and elected elites. Although a constitution for parliamentary government was drawn up and a government was formed in 1959, it was dismissed in 1960. Then Prime Minister B.P. Koirala and his colleagues were imprisoned. The constitution was suspended, and direct rule was imposed. After a 2-year period of gestation, the king came out with a cover of legitimacy in the name of a Panchayat system. Under it, the political parties were outlawed. The king's wish became the law of the land, and the king also proclaimed himself to be an incarnation not only of Lord Vishnu but also of the "popular will."

To sustain, aggregate, and legitimize power, King Mahendra painstakingly and vigorously pursued four sets of strategies. They were (1) the consolidation of a state apparatus under his control, (2) direct contact with the masses, (3) repression and fragmentation of the opposition led by banned political parties, and (4) mobilization of international diplomatic and economic support for his domestic policies and programs. Relations with India, which had expressed its sympathies and support for the democratic forces in Nepal, became the worst casualty of this strategy.

Considerable confusion, political uncertainty, and unrest followed. King Birendra came to power in 1972. Notwithstanding his liberal education in the democratic environments of Eton and Harvard, he inherited the ethos of his father and the institution of monarchy. Political stability still eludes Nepal. Indeed, on the contrary, there is increasing political confusion.

3.8 Sociopolitical Perspective—Bhutan

Bhutan is a small mountainous country with a population of about 1.3 million. It is one of the least developed countries (LDCs) and is almost totally insulated from the world. For three centuries, Bhutan has faced a problem, internally and externally, of establishing its own personality as a nation. This becomes entangled with the problem of economic development, modernization, and integration with the emerging global village faced by all developing countries.

Prosecuted Buddhist sects in Tibet formed the core of Bhutan's identity in the seventeenth century. The intersectoral conflicts from the eighth century until the seventeenth, with frequent invasions from Tibet, gave little chance for the consolidation of a separate Bhutanese personality. Although the Drupas of the Kagjupa section ultimately prevailed over other sects under the leadership of Nawang Namgyal and formed the geographical and political identity of the country, these migrating sects still looked toward Tibet for their religious and cultural roots. After the death of Namgyal, until the establishment of the monarchy in 1907, there was no let up in the inner struggle for power. The theocracy, consisting of reincarnate Shabdungs, subsequent to Nawang Namgyal, was inherently unstable and was eliminated one by one by the civilian power that had emerged through the Druk Desi (the secular head), the Penlops (Governors), and other powerful local officials. The Druk himself became a weak personality, dependent on the goodwill of the Penlops, who were fighting among themselves for supremacy. It was ultimately the Governor of Tangsa, Ugyen Wangchuk, who emerged as the most powerful to become the first king in 1907, with the support of the British. Monarchy, therefore, is only 105 years old in Bhutan. The institution of Shabdung was never revived although incarnates were born thereafter. The religious establishment was passed onto the Head Lama, the Jey Khenpo, who headed a monastic council and, for all practical purposes, is now subordinate to the monarchy.

Bhutan came under the suzerainty of the British in 1865, with the treaty of Punaka of 1910 confirming the status. Bhutan could have no external relations independent of the British government. In return, Bhutan was assured of internal autonomy.

The first two kings up to 1952 were involved in the centralization of their power, integrating the various factions. It was only under the third king, Jigme Dorji, that Bhutan moved toward a new direction of a more liberal and modern administrative system, seeking recognition as a sovereign independent country with its own distinct identity and culture. The setting up of the national assembly called Tshogdu was the first step toward liberalization and people's participation in governance. Continued steps are being taken toward modernizing the executive, polity, and judiciary.

Initially, the departure of the British from India in 1947 did not change the pattern of the Indo–Bhutan relationship in any significant way. However, with the emergence of China as a powerful neighbor and its occupation of Tibet, conditions for change in this relationship were created. By 1958, Prime Minister Nehru had moved away from the suzerainty concept and toward Bhutan's independence and made a commitment to sponsor it for membership in the UN, which it joined in 1971.

Bhutan has completed few development plans. The first four plans were funded entirely by India. It was only in 1972 that international aid apart from India's started coming in. India, however, still remains to be the major donor.

Bhutan's foreign policy and external relations have a limited perspective. Sandwiched between the giants China and India, its relation with both would be the key factor in the formulation of its policy. For historical and other reasons, India has been and will continue to be, for better or for worse, the dominant element, but China cannot be ignored.

3.9 Economic Scene

3.9.1 General Background

The generous land and abundant water provided the foundation for a developed civilization based on agriculture. However, while the Western world changed, these societies' farming remained traditional and at a subsistence level. Waves of foreign domination and, later, colonial rule had little concern for the people. As a result, the population remained overwhelmingly rural, unorganized, illiterate, and desperately poor. Deeply embedded cultural, social, and economic practices inhibit the modernization of the society, economy, and agriculture. Attempts are being made, since Independence, for economic development, which might break the chains of tradition that now bind the people in misery and abysmal poverty, but there are strong cultural obstacles.

To get a relative idea of the socioeconomic state, reference may be made to the international statistics developed by the World Bank and the UN for this purpose. National progress had been measured on the basis of the gross national product (GNP/per capita). Nepal, Bangladesh, and Bhutan (until recently) are classified by the World Bank as LDCs, with a per capita GNP of barely a dollar a day. The eastern quadrant of India, which broadly comprises the GBM region in India, is the equivalent of LDCs on every count. This part of India has the lowest land–man ratio and a low level of industrialization and urbanization, resulting in high out-migration. Much of it is caught in a hunger–poverty trap aggravated by harsh feudal agrarian relations, particularly in Bihar. There is a prescription of socioeconomic sickness and political instability. Parts of the southern fringe of the Ganga basin in India, marked by a plateau, lie in a rain shadow area and are drought prone. This tract is home to a large concentration of tribal people of which India has the largest number in the world.

Of late, the human development index (HDI) has been developed as a more comprehensive socioeconomic measure. As explained in the Human Development Report (1994), the HDI is a composite of three basic components of human development: longevity, knowledge, and standard of living. Longevity is measured by life expectancy. Knowledge is measured by a combination of adult literacy (two-thirds weight) and mean years of schooling (one-third weight). Standard of living is measured by purchasing power, based on real GDP per capita adjusted for the local cost of living (purchasing power parity or PPP).

To find a common measuring rod for the socioeconomic road traveled, the HDI sets a minimum and a maximum for each dimension and then shows where each country stands in relation to these scales—expressed as a value between 0 and 1. So, since the minimum adult literacy rate is 3% and the maximum is 100%, the literacy component of knowledge for a country where the literacy rate is 75% would be 0.75. Similarly, the minimum life expectancy is 25 years and the maximum 85 years, so the longevity component for a country where life expectancy is 55 years would be 0.5. For income, the minimum is $200 (PPP) and the maximum is $40,000 (PPP). Income above the average world income is adjusted using a progressively higher discount rate. The scores for the three dimensions are then averaged in an overall index.

The conditions in the GBM region on the human development indicators are given in Table 2.1. Conditions deteriorate as one moves from west to east. Review of the figures brings out, obviously, the backwardness of the region, which stands at rank 135–162 in the total range of 1–173.

Another depressing fact is that of missed opportunities. To take two examples, Korea and China started at the same time at almost the same, or even worse, level of poverty and backwardness. Korea is far ahead, at 32 levels in this group and with an HDI value of .859, ranking well in industrialized countries and comparing well in economic and human development while the GBM region is one of the poorest and most backward even among the least developing countries.

The statistics do not convey the backwardness, poverty, and misery of the people. About 75% of the people live in villages, mostly in decrepit huts, with little to eat, hardly anything to wear, and without any running water facility or sanitation. Women may have to go long distances to collect water from ponds, wells, or rivers, which is stored in a few pitchers. There is hardly any furniture, some broken cots, if any at all. Mostly they are uneducated and largely illiterate. There are hardly any lighting facilities, if at all a lantern. The food is cooked in makeshift stoves (chulahs), with the fire tended by twigs collected by the women and the children. A family is fortunate if it owns land, which only about 50% has, with the rest being landless laborers. The meager average land holding of about 0.4 ha is not enough to sustain the increasing population, which is continuously moving out to exploding slums in the towns and cities.

The condition of the slum dwellers in towns is not much better than what it was in the villages. The urban conditions, in general, are very poor. Housing, sanitation, transportation, medical, and educational facilities, in short, all that constitutes the urban environment, are shockingly primitive. A microscopic community in the urban centers flaunts Western advanced lifestyle, but the overwhelming filth and disorder cannot be masked.

The Indian economic scene at the time of Independence was of unimaginably abject backwardness and poverty. Subsistence agriculture provided employment to the people. There was virtual stagnation in agricultural production in the British regime, with output growing at 0.4% per annum between 1891 and 1946. The growth of agriculture could not keep pace with the growing population after the first quarter of the twentieth century, and even the per capita food availability continued to fall as shown in Figure 3.3. There was hardly any industry or infrastructural development, except for a tweaky railway system. Habitat was hardly developed except for the colonial officers and their supporting Indian population.

Organized effort at economic development was undertaken after Independence. The first 5-year plan (1951–1956) was a relatively modest venture, but increasingly larger and larger investments have been made. The focus on the first three plans was on industry. From the fourth plan onward, agriculture also got its due share. Economic development was upset by the great drought of 1965–1967, internal political problems, and external wars, but a long period of planned growth has continued. The achievements may not be as spectacular or as good as they could have been when one notes the achievements of the South Asian countries, but the foundation for economic development in all the sectors has been laid. Structural changes have taken place as shown in the share of GDP by different sectors. But poverty has proved to be a hardy perennial and has almost stayed put regardless of the sophisticated debates about it (Rothermund, 1988). Each sector is overviewed specifically in the context of interaction with natural resources and environment.

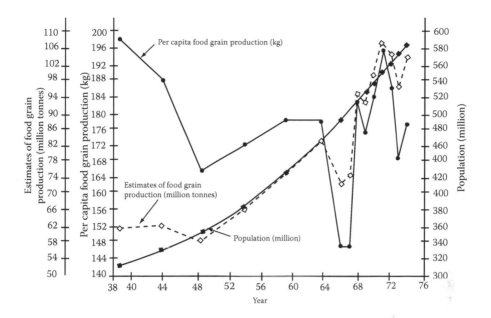

FIGURE 3.3

Per capita food availability decline during British regime. (From Chaturvedi, V., and M.C. Chaturvedi, *Development, Technology and Education*, Proceedings of the 12th Comparative Education Society of Europe, Antwerp, July, 1985. With permission.)

3.9.2 Agriculture

Agriculture is a complex issue and has been studied in detail by several scholars. We follow a study by the World Bank (1991).

For the majority of people in India, what happens in agriculture directly shapes their daily lives and their hopes and prospects for a better life. This has been true for centuries and is likely to remain so in the future, even with the progress in industrialization. What happens in agriculture also influences how the region will tackle the problems of poverty, employment, and environment. Yet at the very time when agriculture is called upon to make a greater contribution to economic growth, there is growing concern that this challenge cannot be met in these countries. This applies with even greater concern in the GBM region, which is the poorest in South Asia despite its richest resources.

We focus particularly on eastern India. Conditions in the Bangladesh and Nepal portion of the region are even worse. An overall all-India agricultural GDP growth rate of 2.6% has been achieved since 1965. This is modest compared to the performance in other Asian countries and to India's population growth rate of 2.1% per annum in the same period. Although famines have largely been eliminated, hunger remains a persistent problem in 20% of the Indian households.

The conditions are worst in the GBM basin. Agriculture is the predominant activity in the region—agriculture accounted for 42% of the state domestic product in eastern India compared to an all-India average of 35% in 1985–1987. Taking two states, Bihar and West Bengal, the growth rate of total factor productivity from 1956 to 1983 was only 1.0 compared to the rest of country's figure of 1.4 (World Bank, 1991).

The roots of slow growth in eastern India lie in the relative neglect of the factors that provide higher growth: technologies, infrastructure, institutional arrangements, and investment and private policies. These are poor all over the country but worst in the GBM basin. First, suitable technologies for the regions' major crop, rice, have not been developed. Existing rice varieties are not suited to conditions of flooding and deep water, which characterize nearly 50% of rice lands in this area. Significant resources have not been devoted to resolving this problem. Thus, the yield per hectare of rice has grown much more slowly than the rest of India (0.8% versus 1.7%).

Second, while basic infrastructure—roads, irrigation, markets, communications, and power infrastructure—is very poorly developed all over India, it is even worse in eastern India. Eastern India agriculture is predominantly rain-fed and, as such, has poor yields and subject to greater variation and risk. Thus, only 29% of the land is irrigated, compared to 91% in Punjab. Even where irrigation exists, it is neither assured, adequate, nor timely. The continuation of lack of HYVs and limited and poorly performing irrigation and flood control measures explain much of the poor performance of the agricultural sectors.

Levels of public investment per capita of the rural population in GBM states have been less than half of those in comparatively wealthier states. This is partly due to the fact that these are poor states, but the eastern Indian states also tend to spend relatively less on irrigation as a proportion of public expenditure.

Institutional factors figure predominantly in lagging investment rates. Eastern India has a much higher percentage of small farms, tenant farmers, and fragmented land holdings, which discourage investment. Sociocultural factors, primarily the conception of agriculture as a residual sector, have worked against the basis for growth in this sector. Finally, institutions in the public sector in eastern India are among the weakest and most politicized in India.

Eastern India needs more investment in research, human capital, and infrastructure to increase agricultural productivity and reduce rural poverty. There is at least implicit concern that such investments would not have the high payoffs that they did elsewhere in India because of the daunting physical, institutional, and socioeconomic conditions. However, analysis of productivity indicates that these have been as productive or more productive in eastern India than in the rest of India, despite the floods, drought, and institutional constraints. Slow growth rates in eastern India have been caused by the slow growth of these inputs rather than slow growth of productivity. This is in contrast to the common perception of eastern India being a technically backward and inherently low-productivity region. In particular, irrigation has a larger impact on total factor productivity in eastern India than in the rest of India, together with higher untapped potential in the region. This argues for more investment in irrigation in eastern India, particularly in groundwater potential. The large impact of low growth in cultivated land suggests that investments in land reforms, despite their political difficulty, would have high payoff (World Bank, 1991).

Agriculture has a prominent role in addressing several key development issues—poverty and employment, women development, rural nonagricultural growth, and environment. There are, however, growing concerns that agriculture will not be able to increase its contribution to overall GDP growth or sustain its role in resolving these problems high on the agenda of development without shifting it to a higher growth path.

The framework for sectoral policy reform has been considered to include four main elements: (i) improving the structure of incentives, (ii) increasing the efficiency of public investment and promoting greater investment, (iii) encouraging technology adoption and generation, and (iv) improving the performance of irrigation. The last relates particularly to water resources development and environmental management and has been considered in detail in later chapters.

3.9.3 Industry and Manufacturing

From the primitive state at the time of Independence, considerable advances in industry and manufacturing have been made, but the conditions even now remain very poor. The structure of production has improved, with industry and manufacturing sharing percentage ranks 27/15, 12/4, and 12/10 in India, Nepal, and Bangladesh, respectively, being almost nonexistent in Bhutan. This sectoral composition is even lower than the average of the low-income economies, 35/15, and much lower and inefficient than the 35/25 of the industrial market economies. However, even with this elementary level of industrialization, serious problems of environmental pollution are increasingly emerging, as will be discussed later.

3.9.4 Services, Infrastructure, and Energy

What differentiates the developing and the advanced countries most poignantly and their developmental impact are the level and conditions of services, infrastructure, and energy. Indeed, to lump the sector under the same literal or sectoral terms is a distortion. Infrastructure, habitat, and energy use provide a measure of man-made environment and are very poor. The commercial energy consumption in Bangladesh, which will represent the eastern part of the region, and in Nepal, which will represent the mountain regions, is only about 10%/4% of the current level of developing countries and about 1%/0.5% of the industrial economies. For a majority of people, houses are hovels, the roads are just dirt tracks, and energy needs are only twigs and shrubs for cooking food. Little effort has been made to make advancements in this region.

Note

1. The population of India is estimated to stabilize at about 1.640 or 1.581 billion by 2050, according to the current high or low development scenarios. Assuming about half the population in the GBM basin and increase in other riparian countries of the GBM basin in about the same proportion yields the figure of about 1 billion. It is just an indicative estimate.

Part II

Current Development Policy and State

4

Environmental Management—
Historical Perspective

4.1 Introduction

Civilizations are intimately related to the natural resources and the environment. People develop practices and technology for natural resources management which are dialectically and circularly related to society and environment. Technology is a societal activity, and policies, practices, and impacts have historicity. A brief historical overview is therefore undertaken. The historical perspective is divided into two parts, pre British and, more importantly, the British period, on which the foundations of the current development were laid. We look at each vector, water, land, and fauna and flora separately, as there was no integrated policy of natural resources and environmental management. In the British India, division of the country had not taken place, and the overview related to India and Bangladesh integrally. There was hardly any activity in Nepal and Bhutan, and no separate mention is made of these areas at this stage.

Water resources development was given increasing importance, just prior to Independence, from the consideration of rehabilitation of the returning armed personnel as they went back to farming and in view of some new technological developments. It became the foundation of the continuing and increasing activities in Independent India. It will therefore be considered under Independent India, in Chapter 6.

4.2 Historical Perspective—Pre-British Scene

The subcontinent of India is a region rich in natural resources. Indigenous people developed modes of management of the resources, which were modified as the region was colonized by wave after wave of human populations beginning more than a hundred thousand years ago. A variety of practices of ecological prudence also evolved.

The mass extinction of large mammalian prey at the end of the ice age in many parts of the world seems to have ushered a revolution in resource use through the cultivation of plants and domestication of animals (Ucko and Dimbleby, 1969). It is considered that there were a number of centers of plant and animal domestication. Although India is not considered to be one of these centers, many of them were close by, and diffusion of these technologies took place as they had a definite comparative edge over the nomadic hunter gathering technologies. Increasingly, agricultural practices developed in India in view of the hospitable environment. While shifting cultivation was the practice often used for maintaining the fertility of soils, in areas where fertility was maintained by flooding and deposition of nutrient-rich silt by rivers, such as the Indo-Gangetic basin, more stable social systems developed as evinced in the Harappan civilization. The archeological findings of the cities of this period display remarkable water management in terms of baths, drains, and sewers. These societies also display their sense of ecological prudence in their veneration of trees, particularly pipal (genus *Fecus*) trees (Gadgil, 1985).

These societies, however, came to an end between 2000 BC and 1500 BC. The reasons are not yet understood, but it may be on account of environmental changes. At about the same time, so-called Aryan or Vedic tribes were moving into northwestern India. They had superior aggressive powers in view of possession of horse chariots and iron weapons. Apparently, this enabled them to clear the forests in this region, and further extension of the settled agriculture took place. Perhaps it was easier to clear and inhabit the western parts of the Indus and Indo-Gangetic basin such as the Ganga–Yamuna region in the first instance, but over the course of time, the wetter and more densely afforested lower regions were also brought under settled agriculture. With the increasing availability of fertile land, more settled and prosperous communities developed. Since cattle provided the motive power needed for agricultural operation, careful use of cattle would assume importance. It is considered that earlier animal sacrifices were gradually given up and the principles of animal care, as brought out in Buddhism, evolved (Koshambi, 1970).

The completion of agricultural colonization is considered to have taken place by sixth century BC. Large levels of surplus in agriculture resulted, enabling an increasing fraction of population to take to crafts, trade, and warfare, leading to the development of kingdoms and empires for the first time on the Indian continent. The center of gravity of this society lay in present day Bihar, which provided a combination of fertile land, plentiful water, and iron and coal fields. The following thousand years was a period of empires and prosperous trade, of many advances in the so-called high culture and in science and technology (Thapar, 1977). Such a society thriving on high levels of surplus would attempt to enhance this further and stabilize it against the vagaries of nature.

While agriculture is reported to be well developed, as brought out by observations of Megasthenes in 300 BC, famines are also recorded in documents like Puranas, Mahabharata, and Ramayana and are understandable in view of the climatic conditions. Efforts have therefore been made since prehistoric times to provide irrigation, through ponds, tanks, wells, canals, and dams, as mentioned in Vedas, Puranas, and Smritis. Kautiliya's Arthashashtra gives rules for land management, irrigation, establishment of sanctuaries, and hunting in forests (Kangle, 1969; CBIP, 1954). References are made to construction of dams, canals, and groundwater development in ancient literature, as brought out by Paranjpye (1988) and other references (e.g., ICAR, 1964). Mention is also made of different types of rain gauges and other water measuring devices (NIH, 1990). However, the activities appear to be on a limited scale and were derived on the basis of practices rather than science.

Political divisions of the Ganga–Brahmaputra–Meghna region under different kings continued to take place historically, but it was one integrated socioeconomic–cultural unit. For instance, Arjuna is shown as marrying and living in the easternmost kingdoms. Lord Buddha's homeland was in the Nepal region. Regional divisions accentuated with the intrusion of Muslim invaders and even more with that of the British intruders.

There is an increasing record of water resources development over time (CBIP, 1954). Differing geological conditions in north and south India led to different irrigation technologies. In south India, emphasis was on tanks in view of the undulating terrain and hard rock foundations. In north India, in view of the flat alluvial plains, tanks were not feasible. Instead, wells and canals were developed. These were undertaken by individuals as well as by kings and emperors. Ferozshah Tughlak (1351–1388) built the Western Yamuna Canal near Delhi in 1355 (GOI, 1972). Akbar (1556–1605) renovated the canal in 1568 to irrigate lands in the Hissar district. Later, Shahjahan (1592–1666), the builder of the Taj Mahal, improved it, taking a branch canal to supply water to Delhi. The Eastern Yamuna Canal was constructed by Mohammed Shah (1719–1748) and was later partially restored by the Rohilla Chiefs in 1784 (Stone, 1984). As Cautley (1860) reported, diversion canals from perennial rivers were widely seen in northwestern India as he traveled over the region during the Sikh wars. The objective of the canals, besides the irrigation of some lands, was also to serve other purposes, as use of WJC to supply water to the emperors' hunting grounds or to Delhi. The technology was elementary, with the result that they often got damaged by floods and deposition of silt (Stone, 1984; Deakin, 1893). Wells were even more extensively used for irrigation. It has been estimated that before the construction of the pioneering Upper Ganga Canal (UGC) in Ganga–Yamuna Doab in the mid-nineteenth century under the British regime, there were 70,000 pucca (i.e., brick lined) and 280,000 kuccha (i.e., unlined) wells in the command area itself (Stone, 1984).

Navigation was the main mode of transportation. The navigation technology was of course just that of country boats. Yamuna is reported to be navigable up to Delhi, and similarly, other perennial tributaries were navigable. Navigation was the main transportation means in the lower reaches of Ganga–Brahmaputra–Meghna and was the predominant means of transportation in that region. The comparative prosperity of Bengal is considered to be due to better transport facilities on account of navigation (Spear, 1978). It has even been suggested that on account of good navigational facilities, large regions could be administered, which could be one of the factors in the development of large empires and oriental despotism (Wittogel, 1956, 1957).

4.3 British Period

With the British rule, water resource development and environmental management entered a new phase. Although the British's role in India's sociopolitical sphere started from the mid-eighteenth century, the British hegemony was established only in 1818 (Spear, 1978). It was a strange phenomenon, a foreign trading company gradually emerging as a successor of the decaying Mughal Empire. The British India, however, consisted mainly of three regions called "presidencies." One was the Bengal Presidency spread over Bengal, Bihar, Orison, and a tract to the northwest end of Ganges and running up to River Sutlej, embracing the capital of the Mogul Empire, Delhi. The second was the Madras Presidency in the south, consisting of coastal Carnatic. The third was on the west coast, called the Bombay Presidency. British India in 1836 is shown in Figure 3.1. The India that British usurped was very different from the India of the British Moghuls and in a large measure as it was down to 1748. The British found a country in ruins. They encountered not only dismantled fortresses and deserted palaces but also canals that run dry, tanks and reservoirs that were broken, roads neglecting towns in decay, and the whole region depopulated. The lengthy wars and famines in between had ravaged the country and have completely broken down the social system (Spear, 1978). They themselves had a dominant role in this social and environmental devastation.

In the ensuing reconstruction, natural resources management was of primary importance, as the land revenues were the primary source of the company's income. Environmental management, as yet and for a long time to come, meant only natural resources development. We will therefore consider them in three parts: water resources, land management, and forestry.

4.4 Water Resources Development

4.4.1 Historical Perspective

Water resources development under the British rule started in a dramatic fashion. With the establishment of the British hegemony, the East India

Company became the virtual ruler of the country. One of their earliest public works was renovation of the Western Yamuna Canal, constructed in the fourteenth century, which was constructed to supply water to the capital of Delhi and irrigate lands en route. It consisted of diverting the low season flows of River Yamuna after monsoons, as it debouched from Himalayas near Dhakpather. Irrigation was undertaken by inundating the agricultural lands, and these canals have been called "inundation canals." The canal had fallen under disuse over the long period of political vagaries. It was taken up for reconstruction in 1817 by G.R. Blance of the Bengal Engineers (GOI, 1972). The work of reconstruction was completed in 3 years but without permanent headworks at Hathnikund.

There were several reasons for undertaking the work. One was to demonstrate the emerging royal role of the East India Company, which came to be called Company Bahadur (valiant in local language). The work was appreciated by the people. It was found that the work was also very remunerative, yielding a return of about 14%.

Rehabilitation of the Eastern Yamuna Canal (EYC) was soon undertaken. It is believed to have been originally excavated in the reign of Mohammad Shah (1718–1748) but fell into disuse later. It was renovated by the British in 1830 (GOI, 1972). Proby T. Cautley was a young lieutenant in charge of restoration of these canals.

In view of the climatic conditions and the agrarian economy, irrigation was getting attention in the emerging British governance, as it contributed to their political interests. Famines were endemic in the country on account of the vagaries of the monsoons (Bhatia, 1967). Irrigation could mitigate the hardships on account of the famines, thereby also contributing to political stability and acceptability of the new regime. The region that was the westernmost part of the then British India had importance in view of proximity to Delhi, and irrigation was thus important based on climatic as well as political considerations.

The British had no experience of irrigation engineering, and these were just measures to restore the earlier works. The nature of work carried out was very much experimental. Funds were severely restricted, and the original alignment of these works was, in consequence, largely adhered to for reasons of economy. Depressions were crossed on earthen embankments that intercepted drainage and were prone to collapse and cause damage. These works eventually required the extensive remodeling of main and distributing lines.

Developments in other parts of British India were also being undertaken. The Grand Anicut, a barrage on River Cauvery, built in second century AD, was by far the greatest engineering feat of India, irrigating considerable land in the fertile delta of the river. The system had fallen into disuse during the long period of political anarchy. Its repair commenced in 1834, followed later by the repair of the Godavari system in 1846, all by Major General and later Sir Arthur Cotton, who acquired great eminence in the field.

Inspired by the technological and financial achievements of the two Yamuna Canals, Colonel Cautley of the Bengal Artillery, later Sir Cautley, next began the construction of the UGC taking off River Ganges at Hardwar at the foothills of the Himalayas, to irrigate the doab (tract between two rivers) of Rivers Ganges and Yamuna. The construction of the canal is a fascinating story (Cautley, 1860). There were small diversion canals in this region, which gave the inspiration for these works, the alignment following a canal already existing at the location. However, the new work was an entirely different order of activity and marks the beginning of large-scale canal irrigations in the world.

Preparatory works for the project started in 1836. The work had to be suspended for some time, as the Sikh Wars started shortly afterward. As Cautley (1860) reports, his confidence in building the canal was strengthened as he saw many small diversion canals in the even more arid lands in the west in the Indus basin. The canal was completed in 1854, at a cost of Rs.2.15 million, about half the cost of constructing Taj Mahal (Smith, 1856; Rogers, 1989). This was the largest irrigation canal in the world at that time, incorporating 650 mi. (1000 km) of main and branch lines designed to carry 191 m³/s and irrigate 1.5 million acres. It still is one of the largest in the world. It is an outstanding pioneering engineering achievement, as even the science of hydraulics was not developed at that time. It established certain technological and policy practices that continue to haunt Indian irrigation until today.

Momentous changes were taking place in the Indian history at this period. The British annexation of India was being steadily undertaken. The annexation of the Indian state of Oudh in the heart of the Ganga basin was carried out in 1837, thereby bringing the entire Ganga–Brahmaputra–Meghna basin under British control, the region representing the centerpiece of British India. By 1849, the Indus basin was also annexed. This region received even greater importance in irrigation on account of its arid climate and need for ensuring political stability in the frontier province, as earlier western GBM basin had received.

As the efforts at development were being undertaken with increasing confidence and pace, the Indian Mutiny occurred in 1857–1858. As Spear (1978) notes, it came as a profound shock to the British, but it was a shock to their complacency rather than to their self-confidence. Several major changes in administration and policy took place, the most important being that the administration of India was taken over directly by the British Parliament and Queen Victoria was declared the Empress of India. The development works continued in which irrigation had a prominent place and continued with greater vigor and control, even though there was a bit of slackness in the beginning.

Until 1858, before India came under direct control of the British Government, irrigation development was carried out by the engineering department of the army, under the supervision of the military board in each of the three presidencies. All expenditure was treated as ordinary expenditure and charged against the revenues of the year. Lord Dalhousie, the development-oriented governor general, proposed that expenditure on the construction and maintenance of works necessary for the administration of the colony should be classified as "ordinary expenditure" and set against the revenues of the year. Outlays on projects that were "calculated to increase the wealth and promote the prosperity of the country," such as irrigation, harbors, and railways, were to be met mainly with borrowed money.

Dalhousie's retirement and the Mutiny of 1857 postponed the adoption of these financing proposals. In the postmutiny period, there was considerable pressure on the Government of India to promote irrigation by the same indirect agency as that by which they were then extending railways, by guaranteeing a return to private companies, and the period 1858–1866 was marked by what turned out to be a remarkable experiment in irrigation development through private companies. Two private companies, the Madras Irrigation Company and the East India Irrigation Company, undertook construction of irrigation works. The Government of India guaranteed assured return to the companies at 5% and, as for the railways, supervised the plans and sanctioned their expenditure. These companies started the work on a grandiose scale, aiming to link Karachi with Calcutta via Kanpur and Cuttack to Madras and across to Poona and the west coast, creating an all-India navigational link (GOI, 1922; Deakin, 1893). But all they were able to achieve was a series of disconnected waterways, like the Midnapur Canal, the Orissa High-Level Canal, and the Kurnool–Cudappah Canal (IC, 1972). The venture ended in a failure and had to be taken over by the Government. This episode led to a policy review, and it was decided that, in the future, irrigation development would be departmentally undertaken by the Government itself, the reason being that there was no concurrent development of the Indian industry even in the construction sector. This practice continued even after Independence for a considerable time.

Financial and economic criteria were established gradually. The works were divided into "productive," "unproductive," and "protective" works. A productive work is one in which the net revenue derived from the project within 10 years after the date of completion is more than a definite percentage of the total capital outlay. The percentage was fixed from time to time by the central government as follows: (i) before April 1919, 4%; (ii) between April 1, 1919, and August 1, 1920, 5%; (iii) between August 1, 1921, and March 31, 1937, 6%; (iv) after April 1, 1937, provinces set the rates within 4%–6%; (v) from April 1, 1949, to March 31, 1954 (Central Government), 3.75%; (vi) from April 1, 1954, to February 1960, 4%–5%; (vii) from March 1960, 5%; and so on.

An unproductive work was one that did not meet the above criteria of productive work. The protective works were essentially famine relief works, as discussed later. The productive works had to depend on public loan to raise capital. Certain stipulations were, however, also laid down besides the financial gains. For instance, equity and averting the risk of famine already introduced in the planning of the UGC were formally laid down. It was stipulated that "the first limitation of the supply of water would be with reference to the proportion of cultivable area within the reach of each canal, for the irrigation of which provision should be made." The proportion might either be fixed at one uniform rate for the whole canal, say "one-third or one-fourth of the cultivable area." Interestingly, these principles were followed unwittingly even in the design of canals after Independence even when they were backed by storage dams (Chaturvedi, 1967a, 1967b).

A number of projects were taken up shortly after the administration was organized. These included the Agra Canal, which takes off from the right bank of the River Yamuna about 11 km downstream of Delhi, which was undertaken in 1868 and completed in 1873, and the lower Ganges Canal, which was completed during the period 1870–1873. Both these canals were constructed downstream of the upper canals and utilized the regenerated flows in River Yamuna and Ganga, from groundwater or return flows from irrigated lands by the Upper Yamuna Canal and the UGC. Similarly, activities were undertaken in Bihar and Bengal, with the distinct difference that as one moves east in the basin, emphasis on irrigation decreased.

Following the great famine of 1876–1878, a Famine Commission was appointed in 1880, which emphasized the need for direct state initiative in the development of irrigation, particularly in the vulnerable areas. A special fund of Rs.15.0 million, known as the Famine Relief and Insurance Fund, was set apart every year from 1982. Half of it was earmarked for the development of railways and irrigation if it was not spent on famine relief. Betwa Canal in Uttar Pradesh was one of the earliest irrigation works undertaken under this scheme.

As Rogers (1983) has stated, "the scale of these developments and subsequent developments in the second half of the nineteenth century did not escape the notice of the rest of the world. In 1890–1891, Herbert M. Wilson of the U.S. Geological Survey was sent to India by the U.S. Department of the Interior to see how the Indian experience could be translated to the U.S. where large-scale western irrigation was just beginning. The report (Wilson, 1892) is a fascinating document." There were, according to Wilson, 35 major productive works expected to cover more than 10 million acres. Of the 6000 mi. of main canals, over 2300 mi. was used for navigation (the figure is close to zero now) and there was 18,000 mi. of distributaries. By 1900, only 12 of these 35 projects were returning more than 4% profit—but the profits from these 12 were sufficient to cover the losses of the other 23. For the first 20 years, the works were not profitable, but for the next 10 years, the work as a whole averaged more than 4% return on the gross capital.

As opposed to these productive works, there were five major protective works costing Rs.6,866,000 and designed to irrigate 723,720 acres. These major protective works had 400 mi. of canals and 800 mi. of main distributaries. There were also 80 "minor" works in operation in 1900 costing Rs.15,040,000. These works were composed of 6500 mi. of canals and 2600 mi. of distributaries and irrigated an area of about 7,201,000 acres. These works averaged

5% net return on the total capital outlay. Even 90 years ago, these minor irrigation works were economically more attractive than the larger ones.

The gross area irrigated by the productive, the protective, and the minor works amounted in 1900 to 18,611,106 acres. Wilson claimed that 50% of the gross value of the crops is due to irrigation—a very handsome return indeed. The average water rate charged was less than Rs.1.40 per acre (1890), and the return was between 3.3% and 8.25% of the value of the crops.

Following two great famines in 1897–1898 and 1899–1900, the First Irrigation Commission was appointed in 1901 to report on irrigation as a means of protection against famines in India. The Commission drew up a 20-year program of public works and suggested a number of measures to stimulate the construction of private works, which accounted for 44% of the area irrigated.

Following the First Irrigation Commission's report, an accelerated construction of public irrigation works was undertaken. There was some slackening during the First World War, but works were again taken up after its cessation. A significant activity was undertaken for irrigation works in the Indus basin for the rehabilitation of the soldiers returning from the war.

Some of the important works after the recommendations of the First Irrigation Commission are the construction of the Sharda Canal, irrigating the tract between Ganges and Sharda, developments in the famine-prone Bundelkhand and Mirzapur–Allahabad area of Uttar Pradesh and Tribeni and Dhaka Canals in the Champaran district of Bihar under the famine relief works. Sharda Canal was another big canal system like the UGC. The technology and management of canal irrigation had sufficiently been tested over the long period of time since the construction of the UGC, but the socio-economic milieu continued to be the same. The policy and technology continued to be that of providing an apology of extensive irrigation.

In 1921, the constitutional reforms brought about a transfer of power to the provinces. In the next 10 years, a number of works that had already been undertaken or new ones were completed. This included Damodar Canal in 1926–1927. The earlier concern for protective works, however, slackened.

The Government of India Act of 1935 placed the Irrigation Department under the popular ministers. It was expected to speed up irrigation development, but the increasing tensions in the context of freedom movement and the outbreak of the Second World War placed severe restrictions on expenditure, and as a result, there was a virtual stoppage of all works not related to the war effort.

Despite the various incentives recommended by the First Irrigation Commission for the promotion of private works, particularly wells, the Royal Commission on Agriculture did not find any significant expansion in well irrigation. Indeed, with steady expansion of canal irrigation, there was a steady decrease in well irrigation.

One of the early developments was the development of tube wells for irrigation. They were very popular with the farmers because water availability was more assured, coming from a neighboring source, but the development was very limited. Similarly, some small storage projects were also developed in the Vindhayan region of Uttar Pradesh, but again, the development was very limited.

Bengal suffered a most severe famine in 1944, and the Famine Enquiry Commission of 1944, which was appointed to investigate it, recommended that it will not be enough to irrigate through public works so private works had to be greatly expanded. The Grow More Food Campaign was launched in 1943 and subsequently integrated with the First 5-Year Plan after independence, placing for the first time greater reliance on minor works.

Statistics for development of water resources according to river basins is difficult to collect. On the all-India basis, the net sown area in undivided India was 116.8 million ha (mha) of which 28.2 mha or 24.1% was irrigated. A much higher percentage was in Pakistan, as the arid Indus basin received the highest emphasis. In India, the GBM basin and the coastal tract of eastern India were the primary beneficiaries.

4.4.2 Review

The canal irrigation technology and management in British India were an outstanding development. There was no experience, technological or managerial, for these works in India and abroad. Its development may be briefly reviewed, as it has implications even now.

Construction of the large canals on the major Indian rivers posed formidable challenges, particularly because construction technology was poor and the science of hydraulics was almost undeveloped. Some details are given by the original builders (Norton, 1853; Cautley, 1860; Buckley, 1880; Nicolls, 1888; Deakin, 1893) and by Brown (1978) and Stone (1984), which make fascinating reading. It was a unique achievement, yet the main canal was only part of the system. Water had to be supplied to numerous small fields owned by illiterate peasants belonging to a different culture through a complex hierarchical distribution system of main canals, branches, distributaries, minors, and, finally, a fixed outlet or a Kolaba leading to the fields themselves through guls on which the peasants took turns to obtain their supplies. These had to be managed to provide a scarce variable resource according to the variable crop demands of the numerous cultivators. Appropriate prices had to be fixed to provide incentives to the farmers; revenues had to be collected, and the complaints and appeals had to be judiciously attended to. A finely tuned system did evolve, but it grew with time.

In the early stages, headworks were temporary structures, which had to be constructed every year on Ganga Canal. Permanent headworks consisting of a weir was put up only in 1913–1920. Placement of headworks at Hardwar involved the construction of four major cross-drainage works; one of them, the Solani aqueduct, cost a quarter of the total cost of the canal at the time of opening. (This was one of the severe points of criticism against the canal design raised by Cotton in the controversy with Cautley. It has bearing on some modern interbasin transfer proposals being made currently, as shall be discussed later.) In the earliest stages, water was supplied to cultivators' water courses through cuts in the canal banks. Water distribution in the early years depended almost entirely upon adapting natural drainage lines (nullas) to the purpose through the construction of simple earthen dams. After a number of years, provision was made for a basic distributary network on EJC—constructed at the cultivator's expense—and such a network was allowed as a matter of course on the plans of UGC, which was often poorly aligned, ill maintained, and environmentally disruptive. Introduction of Kolabas came much later, but they were sanctioned on the basis of demand and supply. The water supplies, naturally, remained an apology of irrigation.

Cautley described the northern plains of India as a "region designed by nature a great field for artificial irrigation" (Cautley, 1860). Numerous wells had been constructed for irrigation. Canal irrigation was a new technology in a peasant economy governed by a colonial rule. Detailed studies about its economic, social, and environmental impact are very limited. It has been considered as the "greatest monuments to British Rule" (Macpherson, 1972, pp. 144–145). A section of a book, published in 1972, on the Uttar Pradesh agrarian economy in the closing decades of the last century contained some analysis (Whitcombe, 1972). According to her, the canals proved a costly failure (Whitcombe, 1980, p. 91). It was argued that due to the way irrigation was applied in the Doab, the policies over its use, and the responses of the cultivators themselves, the canals caused pronounced environmental and economic disruption (Guha, 1994). In a similar vein, it has been argued that canals assisted in the creation of structures within the rural sector that were exploitative and obstructive to efforts to promote growth. A detailed study has been carried out by Stone (1984), and it has been shown that "canal irrigation was substantially absorbed by the society rather than bringing about fundamental changes in the underlying structure. Canal irrigation could play a key role in providing the appropriate technical environment for the release of the expansionary forces, though it was, by itself, limited in its potential to increase output" (Stone, 1984, pp. 10–11). "For all its advantages, the canal provided more nearly for the needs of most peasant family producing units. The canal was an innovation which met their requirements, and it did so because it slotted into the productive aspects of the peasant system in a way which made it generally more advantageous than even the most favorable well irrigation" (Stone, 1984, p. 70). Similar views are expressed by some other observers (Rogers, 1983; Sengupta, 1985).

The disadvantages of canal irrigation as currently practiced, as often pointed out, are water logging, salinity, and unhealthy conditions leading to the possibility of malaria (Klein, 1993). These, however, are simple matters of poor design that can be easily eradicated. However, a serious shortcoming of the design is that it provides only extensive irrigation. Basic revolutionary changes in water resource development and management are needed to change the design to provide intensive irrigation.

Reviewing the policy of water resources development in British India, it can be concluded that it emerged from a constellation of sociopolitical–technological considerations in an ad hoc manner. The overall objective of irrigation was to stabilize sustenance agriculture from considerations of increasing the land revenues and mitigating famines. Extensive irrigation was therefore the logical corollary. Extensive irrigation would also be indicated from several technological considerations. Irrigation tends to cause water logging and must be provided with drainage facilities concurrently, but this would raise costs substantially. Extensive irrigation would lead to minimum water logging, which was hoped to be mitigated by the natural drainage conditions. Furthermore, these were diversion canals, diverting the low flows without any storage, and even the construction of diversion arrangements were temporary in the beginning, as the diversion work on these big rivers was a difficult job. Initially, the diversion was through the construction of temporary obstructions, and this arrangement continued for nearly 50 years, with the obstruction being constructed every year at the end of the monsoons. Therefore, large areas were designed to be irrigated, providing water to only about 30% of the cultivated area in the canal command called the cultivated command area (CCA), so that agriculture over large areas could be stabilized and that, simultaneously, the benefits could spread over a large region. Only three waterings were scheduled. The first one was critical and was available to the entire designated irrigated area. The second and third depended on the availability of water and often could not be provided to the entire area. Irrigation thus meant only one or two waterings, except for the head reaches, even though full irrigation charges were levied even if one watering was taken. The concept of irrigation was limited to the construction of canals and delivery at the canal head of a 0.028 m^3/s capacity pipe or outlet for further use by a number of farmers as arranged among themselves. Field level developments were not considered due to the difficulty of administrating these activities, as it would require considerable supervision and involvement in local rural affairs. Even with the current arrangements, considerable management was required. During the period of low supplies, the distributaries and branch canals had to be run in rosters, generally of 1:3 to 1:5, which means that they are open for one week in a period of three weeks to five weeks. During this period, the irrigation was done according to a period of roster called "Osrabandi." According to this, each cultivator got irrigation water during his turn for a certain period in the week in proportion to the area he was cultivating. The cultivator was free to choose his cropping pattern and area to be cultivated. Considering the large number of small holdings, it was a tremendously bothersome and troublesome practice (Gustafson and Reidinger, 1971).

It is also significant to note that the entire irrigation activity was focused principally on winter (Rabi) irrigation due to technological considerations. Irrigation during the monsoons, to assure availability of water, as it was highly unreliable and the famines were a result of the failure of the monsoon, would be a logical activity. But providing irrigation in the monsoons was, technologically, very difficult at that stage of technological capability, and secondly, it would not be remunerative as the demand would often not be there in periods of good monsoon as it was still subsistence agriculture.

Thus, a policy to stabilize the sustenance agriculture by providing extensive irrigation during Rabi evolved from a constellation of technological, policy, environmental, management, and economic considerations. Even this had serious environmental impacts. Water logging and salinity were introduced over large tracts, as drainage was not provided. Diversion of low flows completely cut off the water supplies to the downstream communities, converting the free-flowing rivers to open drains. Perception of environmental impacts was, however, still a distant issue, and with rapid regeneration, low urbanization, and poor industrial development, pollution was still limited.

Water supply for drinking purposes and habitats and sanitation is the first priority in water resources development, yet these were not attended to in view of the socioeconomic considerations. The role of the Government was to rule and not to govern. Water supply for drinking purposes was not its concern. Moreover, the local self-government was yet to develop, which could have the responsibility for these activities. This came only as late as 1921, and even after it, since urban development was slow, the facilities were not developed. The rural areas, of course, were totally neglected.

Groundwater, through pucca (brick lined) and kuccha (unlined) wells worked through a pair of bullocks, was the dominant technology for irrigation in the region in the precanal era. Canal technology completely eliminated it as canal water; though meager and unreliable, it was, on the whole, cheap and convenient. Ground water development through modern technology of energized pumps could not develop even though the region is abundantly rich. Some state tube wells were installed in the western region of Meerut from 1931. They were a great success, but the development was limited. By mid 1946, a few years before Independence, only 1847 tube wells had been installed in the region. They were all of 0.04 m³/s (1.5 cusec) capacity. Again, in accordance with the policy of stabilizing the sustenance agriculture, extensive irrigation policy was also followed on the tube wells, and for the same reasons, they were located outside the area commanded by the surface canals. There was hardly any development of private tube wells by that time.

Water resources development in British India is a good example of technology as a facet of the socioeconomic milieu. A colonial government ruling a peasant economy could only provide extensive irrigation. It cannot be said that the technological policies and practices were harmful environmentally and socially, as has been suggested by Whitcombe (1972), but as has been demonstrated by Stone (1984), these were merely an outcome of an attempt under a paradoxical socioeconomic milieu of a colonial government interfacing with a peasant economy and a political objective to maintain peace and order with economic benefits to the governing country. The technology of canal irrigation, with the system of management that the colonial government developed, was an innovation that met their requirements, and it did so because it slotted into the productive aspects of the peasant system in a way that made it generally more advantageous than even the most favorable well irrigation (Stone, 1984, p. 70). To conclude, it must be said that the canal irrigation developed in British India was an outstanding and pioneering achievement.

4.5 Land

Agriculture was the main source of revenue, and attention was paid to its organization and collection of revenue from the very beginning of the British rule. Indeed, the administration revolved around the collection of revenue and maintenance of law and order. The principal officer in a district was known as the "collector," who also had overall responsibility for the administration of the district. A system of permanent settlement was introduced through which land revenues were fixed on a long-term basis irrespective of yearly yields. However, in keeping with the basic characteristic of a foreign rule and in not being interested in the welfare of the people, there was no deliberate policy for the development of the resources of the national economy. A separate Department of Agriculture was created in the Government of India only in 1871, and subsequently, similar departments were set up at provincial levels. Centuries-old subsistence agriculture continued. Management of land and soils was almost totally neglected, as it only constituted expenditure.

A quasi-stable environment had developed in the basin, but as large-scale irrigation was introduced from the mid-eighteenth century, problems of land management increasingly emerged. With the commissioning of the Western Yamuna Canal, problems of water logging were noted as early as 1850, as drainage was not provided in the colonial irrigation that was being developed. As further large-scale irrigation works were undertaken increasingly, complaints from the farmers of the deterioration of land on account of canal irrigation increasingly started pouring in with the revenue authorities (NCA, 1976). As early as 1876, the Reh Commission was established to investigate the causes of deterioration of the soils of Uttar Pradesh, which had been fertile. Some corrective steps were undertaken, but on the whole, there was neglect of land management.

After the Great Famine of 1876–1878, a Famine Commission was appointed in 1880. Some piecemeal activities for development of agriculture were undertaken. In 1926, the Government decided to appoint the Royal Commission on Agriculture in India, but the recommendations could hardly be implemented, as soon afterward, the Great Economic Depression took place. With the increasing activities of nationalistic movement, the political system was under strain. Soon, the Second World War started, and all attention was diverted toward it.

4.6 Forests

The GBM basin was thickly forested when the British reign started, as the population pressure was not yet high and requirements of forests were only for local use. The forests and grazing lands were common property resources and were managed from community resource and ecological considerations. As time went on, land was increasingly brought under cultivation, with increasing population and commercialization of agriculture. Deforestation was primarily on land that could be cultivated. The alluvial plains thus bore the brunt. This, however, became complicated with the introduction of commercial forestry, which seriously disrupted existing patterns of resource utilization. The landmark in the history of Indian forestry is undoubtedly the building of the railway network. The large-scale destruction of accessible forests in the early years of railway expansion led to the hasty creation of a forest department set up with the help of German experts in 1864.

Successful forest administration required checking the deforestation of the past decades, and for this, the state monopoly right was considered essential. An initial attempt at asserting state monopoly through the Forest Act of 1865 was found wanting, and a comprehensive All-India Act was drafted in 1878. An elaborate procedure of forest settlement to deal with all claims of users was established.

However, management of forests, either from a conservation point of view or from an economic point of view, was very limited. For instance, the gross revenue from forests all over India was only about Rs.300 million, and the expenditure on management was about Rs.200 million, giving a surplus of about Rs.100 million (NCA, 1976).

A National Forest Policy was constituted as early as 1894, which recommended the following:

1. The sole object with which state forests are administered is public benefit. In general, the constitution and preservation of a forest involve the regulations of rights and the restriction of the privileges of the user in the forest by the neighboring population.

2. Forests situated on hill slopes should be maintained as protection forests to preserve the climatic and physical conditions of the country and to protect the cultivated plains that lie below them from the devastating action of hill torrents.

3. Forests, which are the reservoirs of valuable timbers, should be managed on commercial lines as a resource of revenue to the states.

4. Wherever an effective demand for culturable land exists and can only be supplied from forest area, the land should ordinarily be relinquished without hesitation, subject to the following conditions:

   a. Honeycombing of a valuable forest by patches of cultivation should not be allowed.

   b. Cultivation must be permanent and must not be allowed to an extent as to encroach upon minimum area of forest requirements, present and prospective.

   c. Forests that yield only inferior timber, fuel wood, or fodder, or are used for grazing should be managed mainly in the interest of the local population, care being taken to see that the user is not exercised so as to annihilate its subject and the people are protected against their own improvidence.

The policy was based on the desire that forests serve agricultural interest more directly than before (NCA, 1976).

Conflicts over forest and grazing rights, particularly in the hills, where forests are an integral part of village life, arose as state monopoly rights were established with the introduction of the 1878 Act. These were a feature of precapitalist and early-capitalist Europe as well. However, the nature of social conflict in the transition to industrial capitalism in Europe inevitably differed from that of the endemic conflict over forest rights, which were germane to the artificially induced capitalism under colonial rule (Guha, 1994). These conflicts continue.

5

Developments in Nepal and Bhutan

5.1 Introduction

Nepal and Bhutan are entirely in the Himalayas. The developments there are governed by the Himalayan environmental characteristics, interlinked with the plains in view of land–water ecodynamics. The Himalayan region presents challenging opportunities but demands very high societal and technological capabilities. The developments are governed, as usual, by the socioeconomic milieu, which are similar except that Nepal is even further away from the glimmer of modernity compared to India, and Bhutan is even further removed.

Nepal is in the Ganges Basin while Bhutan is in the Brahmaputra Basin. Nepal straddles the Himalayas on the north of India and is flanked by India, with Uttranchal on the west and Sikkim on the east. Bhutan is also on the north of India, flanked by Sikkim on the west and Arunachal Pradesh on the east in India. Nepal's area is about three times that of Bhutan, and its population (1999) of 24 million is about 14 times Bhutan's 1.7 million. We, therefore, concentrate on Nepal and briefly describe Bhutan later, as the environmental challenges are almost the same. A detailed National Water Vision has been formulated for the development and management of the water resources of Nepal to which reference may be made (Malla et al., 2001).

5.2 Nepal

5.2.1 Environment

The physical environmental characteristics and dynamics have been discussed in Chapter 2 and the socioeconomic features in Chapter 3. Some further specific details, however, may be brought out.

Covering a geographic area of 147,181 km², Nepal extends 885 km along the east–west with an average width of 193 km along the north–south. It has five almost parallel zones: (i) High Himalayas, (ii) High Mountain, (iii) Middle Mountain, (iv) Siwaliks, and (v) Terai, which show visible differences in topography, climate, geology, soils, and vegetation. Nepal may accordingly be divided into five regions from south to north (Figures 5.1 and 5.2). Although the divisions are mainly on the basis of altitude, they are equally different in the physical terrain as well.

All across is the Terai, a tropical zone of river plains that stretches throughout the length of Nepal with an average width of 30 km. Next comes the Churia (or Siwalik) Hills that rises to around 13,000 m and generally stands bereft of any vegetation. Beyond it lies the Mahabharata, a magnificent cluster of mountains serving as an effective barrier to the hill country up north, standing as tall as 2700 m, with deep, broad valleys and dense populations encased within. Given its wide difference in altitude, the vegetation ranges from the tropical to the alpine in the higher reaches. After the Mahabharata is the Inner Terai or Bhitri Madesh, a hilly region where rivers crisscross in every direction and where almost the entire stretch of the region is inhabited. Terraces along the slopes have made them cultivable, where crops such as paddy, barley, buckwheat, maize, and temperate fruits and vegetables are grown. Running parallel to the Mahabharata range on the northern side of this zone of hills and valleys is 885 km of the mighty Himalayas.

The southern flank of the Himalayas drains into the Ganga, or Ganges, River system while the northern side forms part of the Tsang Po/Brahmaputra

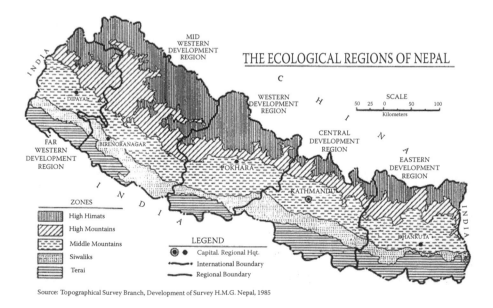

Source: Topographical Survey Branch, Development of Survey H.M.G. Nepal, 1985

FIGURE 5.1
Ecological regions. (From Thapa, B.D. and B.B. Pradhan, *Water Resources Development–Nepalese Perspective*, Konark Publishers Private Ltd., New Delhi, 1995. With permission.)

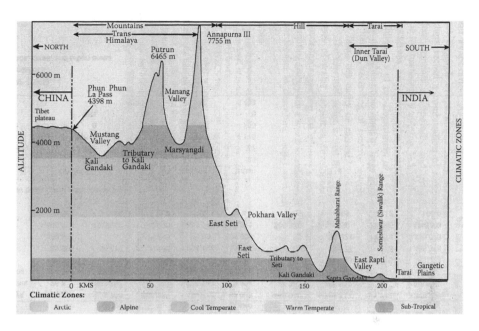

FIGURE 5.2
Physiographic profile at 84. (From Gyawali, D. *Water in Nepal*. Kathmandu. Himal Books, 2001. With permission.)

watershed. As the Himalayas are not one unbroken chain of mountains, it only lends a semblance of being a continuous line. The entire range is a series of mountain clusters, which have been divided by the rivers flowing through them. The alpine meadows constitute the resource to be utilized for excellent growth of livestock. The higher zone is often the zone of shifting cultivation whereas the lower undulating belt forms the marginal areas of traditional farming with low yield primarily for local consumption. The magnitude of noncultivated land, degraded watershed forest, and grasslands calls for management interlinkages between agriculture, forest, and pasture. The Terai does have the potential fertile lowland for intensive cultivation and is, in fact, the granary of Nepal.

The High Himalaya zone located between the altitudes of 4000 to 5000 m above sea level occupies 23% of the total area of the country. The climate is arctic with perpetual frost and low precipitation.

The High Mountain zone (transition zone) located between 2300–3000 and 4000 m occupies 20% of the area. The climate is humid continental, with high precipitation as intense orographic rainstorm. The valley slopes are usually forested.

The Middle Mountain zone comprises mainly of a network of ridges and valleys located between the altitudes of 200 to 3000 m. It occupies 30% of the area and supports about two-thirds of the population. Most of the middle mountain zone has a temperate climate, with cool and dry winters and warm and wet summers. Annual rainfall varies from 1000 to 3500 mm. The area has an extensive and exemplary terracing system. About one-third of the area is forested. Some of the most acute environmental problems in Nepal are in this area.

Sivalik zone is a chain of low ridges, with an elevation of 120 to 2000 m. The area covers about 13% of the country's area. The climate is subtropical, with mild and dry winters and hot and wet summers. Geologically, Siwaliks are of the Paleocene and Quaternary era, consisting of shale, sandstone, and conglomerate. The soils are predominantly coarse loamy to sandy loamy and fine loamy, shallow and susceptible to erosion. Siwaliks are the most fragile zone for conservation and environmental management.

The Terai zone is a strip of the Gangetic plain south of Siwaliks and north of the Indian border. This area extends to about 14% of the country. The altitude ranges from 100 to 300 m. The climate is subtropical, with hot and wet summers and mild and dry winters. Terai is composed mainly of quaternary consolidated fine alluvial material. Soils are deep, coarse textured and have a good water holding capacity.

There is another classification from east to west, created by the four river systems of Nepal. Into the Kosi, the largest of the four, all the rivers of eastern Nepal flow. In the same manner, rivers from central and western Nepal flow into the Gandaki, while those from midwestern Nepal flow into Karnali. The Mahakali, which forms the boundary with India, flows in the far-western region of the country. All these rivers have the potential for extensive harnessing.

The snowfed rivers richly supplied by the monsoon precipitation nourish the land. The heavy and torrential monsoon that is of four months duration is often destructive, resulting in intensified soil erosion and landslides in the hilly terrain.

River system of Nepal has been brought out in Figure 2.2 but may be brought out with more specificity in Figure 5.3. The rivers of Nepal, as in Himalayas, have been classified into three groups: rivers originating from the Himalayas and Tibet plateau, the four major rivers (from west to east, Mahakali, Karnali, Gandaki, and Kosi); rivers originating in the middle mountain zone or Mahabharata (Bagmati, Rapti, Meohi, Kanki, Kamla and Babai); and rivers originating from the southern slopes of Mahabharata and Siwaliks. The two later groups are tributaries of the four major rivers.

All the rivers of eastern Nepal flow into Kosi, the largest of the four major rivers. Rivers from central and western Nepal flow into Gandaki, while those from midwestern Nepal flow into the Karnali, called Ghagra in India. The Mahakali (called Sarda in India), which forms the boundary with India, flows in the far-western region of the country.

The rivers originating from the Himalayas are perennially snowfed rivers. These carry heavy silt load in the rainy season. Rivers originating in the middle mountains carry less water during winters. The rivers of the third category are numerous small streams, which are often dry during the

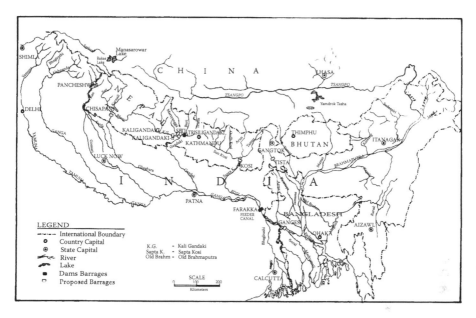

FIGURE 5.3

River systems of Nepal. (From Thapa, B.D. and B.B. Pradhan, *Water Resources Development–Nepalese Perspective*, Konark Publishers Private Ltd., New Delhi, 1995. With permission.)

winter season. There are about 6000 large and small rivers, totaling about 45,000 km in length. The total area of surface water is estimated at 0.4 million ha out of which 97% is occupied by big rivers and the rest by natural lakes and reservoirs (MPFS, 1988). The drainage density of 0.3 km/km² of lateral drainage reflect the closeness of the drainage channels.

5.2.2 Socioeconomic Scene

The socioeconomic indicators were brought out in Table 2.1. Nepal remains one of the poorest countries in the world with a per capita income of about $220 in 1997. Social indicators are well below the average for the South Asian region.

Traditional agriculture, nascent industrialization, unemployment, and underemployment are the chronic problems of the region and, even more so, of Nepal. Agriculture accounts for about half of the GDP, providing employment to 91% of the economically active population. The yields are very low, and employment in agriculture has a very low direct productivity. In 1988–1989, the 8.8 million people who worked in agriculture generated an average value added of about Rs.4734 per person. Those who worked in other sectors of the economy generated a value added that averaged about Rs.39, 729 per person (Agarwal, 1991).

Nepal's other important resources with good potential are hydropower and tourism. While the tourism sector contributed about 20% of Nepal's export earning, significant development of its hydropower potential is yet to begin.

Nepal's development efforts in recent decades have been hindered by a number of fundamental constraints, including its poor resource endowment, a weak administrative and political system, and high population growth. The domestic savings rate, averaging about 10% of GDP in recent years, is low, and Nepal relies heavily on external assistance for financing much of its development activity and exports.

5.2.3 Environmental Conservation

Nepal represents a serious problem of environmental conservation in the Himalayan ecosystem, shared by India and Bhutan. Fragile geologic formation, steep topography, shallow soils, and high intensity monsoon rainfall are the natural factors that make the Himalayan ecosystem very delicate. This is compounded by the activities of the poor people whose survival is dependent on the natural resources. With the exploding population and increasing resource scarcity, a series of interlinked vicious circles operates inexorably to drive Nepal into a downward spiral. There is a perceived progressive and accelerating shift from potential instability to massive actual instability (Ives and Messerli, 1988). The key issues are deforestation, overgrazing, soil erosion, mass wasting, and inappropriate land use practices. All are interlinked though considered separately. Finally, we consider the policy context.

5.2.3.1 Deforestation

The forest characteristics have been covered in Chapter 2. We briefly review the deforestation aspect. All the figures are from Achnet (1991), based on the Master Plan for the Forestry Sector (MPFS, 1988).

Nepal's forest area in 1985–1986 was about 5.50 million ha, declining at an average annual rate of 0.4%. The decline was much higher—3.9% annually in the Terai region. The overall annual rate of percentage change was only about half of this rate in 1964–1979. The forest area shrunk by 473,000 ha in 1964–1985. The situation of the crown cover also reflects the nature of forest degradation. The percentage of crown cover in the Hills, Siwaliks, and Terai in 1978/1979 was estimated at 41%, 54%, and 64%, respectively. The plantation target of 47,000 ha in the two periodic plans (sixth and seventh plan) is just about half of the forest devastated in a year.

As high as 81% of the total domestic household energy needs are met from forest resources. Annual fuel consumption is estimated at 11 million m^3, equivalent to 50,000 ha of well-stocked forests (MPFS, 1988). Forests also provide 50 million tons of green fodder per annum. All these pressures, including the timber demand, are growing exponentially. It has been projected that effective management of 1,250,000 ha of forest, including 410,000 ha of new plantation together with the use of 300,000 improved stoves, will not be able to meet the present challenge. But this is only a part of the forest degradation scenario.

The higher density class forest decreased considerably in mountains and hills. In Terai, it was more a case of clearance. The situation worsens as one goes east from west. More than 25% of forests have less than 40% crown cover. Two-thirds of the forests have mainly small timber. Natural regeneration is inadequate. In addition, forest area distribution in relation to densely populated Terai and hills is only 0.11 and 0.26 ha, respectively. Sustainable yields from the natural forest can meet only part of the total requirement, and the gap in demand and supply of forest resources is widening, creating a downward spiral. It has been estimated that under current trends, 0.6 million ha of natural forest would decline in another 25 years (MPFS, 1988). In Terai and midmountains where 25% of the population lives, fuel wood deficit will go from the present 0.25 million m^3 to 2.5 million by AD 2000. Fodder deficit is also likely to worsen in these areas where 80% of the livestock are raised. It is estimated that every hectare of farmland requires 2.8 ha of unmanaged forest to provide fodder, 0.36 ha for fuel wood, and 0.32 ha for timber or, in other words, 3.48 ha each year. Thus, unsustainability of farmland ensues with deforestation. Concurrently, the associated adverse environmental impacts, such as erosion change in the hydrology of the area, a reduction in sources of natural nutrients, unfavorable impacts in genetic resources and biodiversity also follow.

5.2.3.2 Pasture Land Degradation

Livestock are an important part of the Himalayan rural socioeconomic setting. The composition of livestock is 43% cow and ox, 25% buffalo, 30% sheep and goat, and 2% other (Achnet, 1991). Most of the animals are unproductive grazing and browsing animals. With the growth of human population, the population of cattle is also increasing.

Overgrazing is reflected in the growth of sinigrass (Imperata SP). The heaviest pressure of overgrazing is present in the midhills, where livestock population is highest but pastures up to the higher region are covered more and more by sinigrass, showing the widespread and intense pasture degradation (Pandey, 1982). Introduction of fodder trees as a part of the agroforestry system has been continuing, but the degradation of pasturelands is worsening. This is reflected in the enormously increased erosion rate of 34 tons/ha/year in a pilot study as compared to the rate of 9.4 tons/ha/year in the well-maintained pasture (Achnet, 1991).

5.2.3.3 Land and Soil Erosion

In view of the geologic and hydrologic characteristics, the Himalayan region is seriously susceptible to soil and land erosion. Human interference further exacerbates it. The annual sediment yield from Nepali rivers has been estimated to

TABLE 5.1

Annual Sediment Load of Nepali Rivers

River	Estimated Load (million tonnes)
Mahakali	18
Karnali	170
Narayni	169
Kosi	198
Babai	12
West Rapti	24
Bagmati	41
Kamla	7
Other rivers	80
Total	726

Source: Thapa, B.D. and B.B. Pradhan, *Water Resources Development–Nepalese Perspective*, Konark Publishers Private Ltd., New Delhi, 1995. With permission.

be 726 million tonnes, as given in Table 5.1, which gives a clue to the problem of land and soil erosion (Thapa and Pradhan, 1995). Several studies have been carried out in this context as overviewed by Bruijzeel and Bremmer (1989).

The erosion has been considered primarily under two factors: (i) due to natural factors and (ii) due to human interventions. Under the latter, two categories have been identified: (i) effects of changes in vegetation and land use pattern and (ii) effects of construction.

It has been considered that there is a high level of mass wasting prevailing in the Himalaya mainly due to a combination of geologic and climatic factors and, to a lesser extent, due to land use factors, steep dip slopes, unstable nature of rocks due to their structural disposition (e.g., degree of fracturing), depth and degree of weathering, high seismicity, and oversteepening of slopes through undercutting by rivers ranked among the most important geologic factors. The impacts thus vary considerably spatially in view of these factors.

In many parts of the Himalaya, mass wasting in the riparian zone via low level undercutting by incising streams is considered to be the dominant mechanism of sediment supply to the streams. In addition, mass movements on the hillsides are dominated volumetrically by a limited number of very large failures that are usually found in the heavily fractured zones with concentrated groundwater flow. These features are purely geologic in nature and are not in any way caused by the presence or absence of forest cover. In contrast to these deep-seated slides, the occurrence of shallow (less than 3 m) landslides is strongly influenced by the presence of deep-rooting vegetation cover.

The second important cause of mass movements and sediment production is related to construction activities, particularly roads. About 5% to 7% of all landslides have been ascribed to road construction. Naturally, sound engineering practice could reduce it significantly.

The third factor, resulting particularly in soil erosion and gully formation, is due to changes in vegetation and land use patterns (we have avoided the term deforestation, as it is used so ambiguously that it has become meaningless as a descriptor of land use change). Rates of surface ("on-site") erosion are strongly controlled by the degree to which the soil surface is exposed to rainfall or disturbed otherwise. Therefore, soil losses from nonterraced cropped fields or from overgrazed grass and scrubland rank among the highest recorded in this area.

The question generally raised is "what is the role of forests and land use in the uplands with respect to flooding, dry season flows, and sedimentation in the plains, and what downstream benefits can be expected in this regard to upland rehabilitation program?" First, it is necessary that the scale of interest should be defined. As stated by Hamilton (1987), "at a local level, sediment load is strongly influenced by human activity, stream discharge characteristics much less and on the whole, human activity has less impact than natural factors on flooding and siltation within the Himalayas. At the medium level, downstream of the catchment being impacted, we are still uncertain of the quantitative effects of the human activity but the high variability of the natural factors dominates both stream discharges and sediment load. At the macro level in large basins, human impacts at the upper watershed are insignificant on lowland floods, low flows, and sediment, but these effects can be significantly influenced by human activity in the lower reaches of the Himalayas."

5.2.3.4 Watershed Condition

Watershed condition is the present state of erosion relative to the erosion estimated under natural or well-managed conditions. About 13% of the land area of Nepal was estimated to be in a degraded condition. Among the ecological zones, Siwaliks were categorized as the area with the worst watershed condition, with the middle mountain zone following it.

5.2.3.5 Genetic Resources and Biological Diversity

In view of the complex environmental setting, including a wide range of variation in topography, climate geology, soils, and vegetation, the Himalayas are comprised of a global range of ecological region covering practically tropical to tundra region in a small area. This leads to a unique richness in genetic resources and biological diversity. In Nepal, more than 54,000 species of vascular plants, 130 species of mammals, and 800 species of birds have been reported (MPFS, 1988). The plant diversity ranges from medicinal plants to a wide variety of food and fodder plants. Many species are yet to be identified.

Deforestation and pastureland degradation have serious adverse effects on genetic resources and biodiversity. The number of endangered species is increasing. At present, 26 mammals, 8 birds, and 3 reptiles have been classified as endangered species.

5.2.3.6 Environmental Management Policy and Implementation

As brought out by Thapa and Pradhan (1995), HMG/N has endorsed the national conservation strategy (NCS, 1989) prepared with the assistance of the International Union for Conservation of Nature (IUCN). The conceptual framework adopted by NCS embodies a long-term perspective on natural resource management and conservation action agenda. The strategy not only outlines conservation problems but also recognizes the potential for increasing the productivity of agriculture and grazing grounds through the initiatives of individual users and communities. It identifies many obstacles in the attainment of NCS objectives. Because of the tendency to deal with conservation problems on a sectoral basis, a need to improve coordination is a recurring concern in each sector. Conservation action agenda includes four models of integrated resource management, one for each of Nepal's four geographic divisions, including formulated conservation awareness programs.

Strategies relating to environmental issues are explicitly stated in the following sectoral policies: (1) forestry sector, (2) power and energy sector, and (3) transportation sector. Drinking water, sanitation, and housing also include policies related to environmental protection. Although the policies relating to the environment are well formulated, action and effective coordination are yet to take place. Environmental impact assessment has been on a project-to-project basis, and the standard has varied widely.

5.2.4 Energy Sector

5.2.4.1 Introduction

Energy consumption in Nepal per capita is one of the lowest in the world. Presently, Nepal's energy consumption is small (0.5 TOE/year), and 95% of it is met from forests, agricultural residues, dung, etc., which are being over-exploited, with serious environmental consequences. Nepal has immense sources of hydropower, but less than 1% has been developed. Only about 9.2% of the population has access to electric supply, with half of the domestic connections being concentrated in the Kathmandu valley. A brief review of the energy sector follows based on Agarwal et al. (1991) and Malla (1991), which in turn is based on the Government of Nepal official studies (WECS, 1985–1991).

5.2.4.2 Indigenous Energy Resource Base

The indigenous energy resources have been estimated to be 1857.7 GJ per annum (Thapa and Pradhan, 1995). The annual sustainable forest yield accounts for 15% of the total while hydropower represents 75% of the total.

The theoretical hydroelectric potential has been estimated to be 83,290 MW (Table 5.2). The identified power potential at 60% load factor is assessed

TABLE 5.2

Theoretical Hydropower Potential of Rivers Flowing in Nepal

			Linear Potential Resource (MW)		
Sl. No.	River Basin	Area (km²)	Major Rivers	Small Rivers	Total
1.	Sapta Kosi	28,140	18,750	3600	22,350
2.	Gandaki	31,600	17,950	2700	20,650
3.	Karnali and Mahakali	47,300	32,680	3500	36,180
4.	Southern Rivers	40,141	3070	1040	4110
	Total	147,181	72,450	10,840	83,290

Source: Thapa, B.D. and B.B. Pradhan, *Water Resources Development–Nepalese Perspective*, Konark Publishers Private Ltd., New Delhi, 1995. With permission.

to be 44,582 MW (Table 5.3). As we will argue later, current estimation of the potential has two serious shortcomings that have not been taken into account in the current evaluation of the hydroelectric potential, which has serious implications for the current management and future development. These limitations may be mentioned. They will be considered in detail while discussing the future development of the GBM basin in Chapters 12 and 13.

Hydroelectric potential, particularly in the Himalayas, is developed in terms of two technologies—run-of-river schemes and storage schemes. In the upper regions of the GBM system, storages are not economically feasible, and the development is undertaken through run-of-river schemes. Hydroelectric development requires energy supply on a 90% reliability basis. Since energy cannot be stored, this means only the utilization of the low period discharge potential. This is brought out clearly by reference to the typical river hydrograph of Ganga and some other rivers of India, as shown in Figure 5.4. After storage becomes possible, the hydroelectric potential of the entire year flow potential can be utilized, though restricted to the head of this limited reach of the river only, and consistent with the proportion of the total river flow that can be reliably and safely stored. From hydrologic considerations, only

TABLE 5.3

Identified Power Potential in Nepal

Sl. No.	River Basins	No. of Sites	Identified Power Potential (MW)
1.	Sapta Kosi	40	10,860
2.	Gandaki	12	5270
3.	Karnali	7	24,000
4.	Mahakali	2	1125
5.	Other southern rivers	5	878
	Total	60	42,133

Source: Thapa, B.D. and B.B. Pradhan, *Water Resources Development–Nepalese Perspective*, Konark Publishers Private Ltd., New Delhi, 1995. With permission.

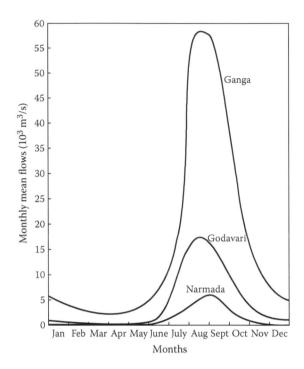

FIGURE 5.4
Hydrologic characteristics of Ganga and some other rivers of India.

the low period discharges assure dependable hydroenergy potential with a 90% reliability, which, in the Himalayan hydrologic characteristics, means limiting it to the low period, post and period monsoon flows. This also corresponds to the current technology of their development. Currently, most of the hydroelectric development has been undertaken through the development of run-of-river schemes, as discussed in Chapter 6.

5.2.4.3 Current Development and Usage

While hydroelectricity represents a sustainable 78.6% of the energy base, only 294 MW, which is less than 1% of the technoeconomic potential, has been developed. Currently, wood represents 75% of the energy consumption, agriculture residues 8.5%, dung 11%, petroleum products 4.3%, and hydroelectric energy the remaining 6.5%, with direct solar energy's contribution being negligible. Forests are being overexploited and depleted at the rate of 2.1% per annum. The domestic sector consumes nearly 95% of the energy and relies mainly on traditional energy resources, mostly in nonmonetized consumption. Commercial energy use accounts for about 5% of the total domestic energy use and perhaps more significantly for about 26% of the cost of petroleum imports considering use of kerosene for cooking and rural lighting.

The industrial sector accounts for about 1.9% of the total energy requirement and approximately 8.0% of imported petroleum products. Among all categories of industry, fuel wood still has the highest share of energy demand at about 51% of the total.

The transportation sector accounts for about 2.0% of total energy requirements. Energy demand in the transportation sector is met predominantly by petroleum products, and it accounts for 53% of the petroleum imports.

The agricultural sector accounts for only about 0.1% of the total energy requirements and 4.0% of petroleum products.

The commercial sector accounts for about 1.1% of the total energy requirements and 9.0% of petroleum products. The commercial sector accounts for approximately 21% of all power consumption and 25% of the petroleum product demand. Hotels/lodges and restaurants account for over 80% of the sector's energy requirements.

5.2.4.4 Policy, Issues, and Options

Some of the important policy issues and options that have been considered by the Nepal government policy makers are as follows (Thapa and Pradhan, 1996).

Nearly 90% of the people of Nepal live in rural areas. Although electricity has been made available to 65% out of 75% districts, it is mainly limited to the district headquarters, with only about 9.2% of the household having electricity. This signifies very little penetration of electricity in the rural areas.

Rural electrification is currently being done in four ways: (i) national grid electricity, (ii) small hydroplants, (iii) diesel generation, and (iv) private microhydroplants and solar. National grid electricity has yet to cover the rural area. Small hydropower plants have been supplying electricity to nearly 30 district headquarters in the hill/mountains. Cost per kilowatt of microhydro (less than 100 kW) is normally less than $1500, slightly higher than large hydro, which is less than $1000, but it is appropriate for rural communities not yet covered by grid. Diesel generation is costly and has problems of uncertainty of availability. It supplies electricity only to seven districts. Nearly 90 microhydroplants (totaling 920 kW) supply electricity to rural communities. Two districts have electricity supply from 50-kW solar plants.

It is realized that rural electrification is a very complex technoeconomic–social problem. Scattered poor population creates low demands. The mountainous terrain itself leads to high cost of transmission/distribution. The poor population has low paying capacity, and development of industries has several complex issues. Rural electrification has to be part of integrated rural development.

Transportation is important for development, but the transportation sector is very poorly developed. Roads are very expensive in the mountainous terrain. The petroleum bill represents requirements of scarce foreign exchange. Proposals have been made for an electrified railway between Kathmandu and Birgunj, which has the highest traffic density. Other routes are also being considered.

With Nepal's potential of hydroelectricity, electricity-intensive industries are being considered. Fertilizer projects electric arc furnaces, integrated textile mill, craft paper industry, and cement industry are being considered.

Substitution of electricity for diesel in the agricultural sector is also under consideration. Presently, there are more than 17,000 shallow tube wells, which use mainly 5-hp diesel pump sets for power. Electrification is better suitable than diesel pump sets for areas having a command area of more than 4 ha. Thus, wherever there are clusters of tube wells, electrification could be appropriately introduced, and it will have other positive linkages.

5.2.4.5 Policy Framework

Some broader energy policy issues also emerge in the context of the overall development and various sectoral policies; these could be stated as follows.

Excessive reliance on forest-based energy sources is going to pose serious environmental problems. Thus, there is urgent need for sustainable forest management and to move away from the forest-based energy sources to other renewable forms of energy sources.

Despite abundant hydroelectric resource endowment, the role played so far by this source is extremely limited. Major hydroelectric or multipurpose water resource projects could be developed based on the export market. In addition to generation of export earning, they will contribute significantly to meeting the domestic energy need at comparatively cheaper prices and will also open opportunities for nationwide development of transmission and distribution networks. However, so far, there exists only one buyer, and implementation of these projects has been considered solely in terms of the financial benefits to Nepal without taking into consideration the immense benefit to be derived by enhancing the hydroenergy potential itself besides the widespread socioeconomic benefits to both parties through collaborative development.

Imported hydrocarbon plays a predominant role in the monetized sector of energy consumption. The import of petroleum products, however, eats away 34% of the export earnings.

5.2.5 Water

5.2.5.1 Resource

The water resources of the GBM basin were described briefly in Chapter 2. The four major rivers from east to west in Nepal are Kosi, Gandaki (known as Gandak in India), Karnali (known as Ghagra in India), and Mahakali (known as Sarda in India).

The Mahakali River makes the western border between Nepal and India. The Mahakali emerges from the plains at Brahmdeomadi, and before it, the Pancheshwar dam has been proposed. It makes an easterly bend before joining the Karnali (Ghagra) River at Bahramghat. It is, therefore sometimes included with Karnali (Ghagra).

Several streams join Karnali in Nepal. It makes a deep gorge at Chisapani while crossing the Ghuria hill (the Siwalik), where the Chisapani dam is proposed. Two major streams joining Ghagra (as Karnali is called in India, after Sarda joins it in India) are the Rapti and the Little Gandak. It joins the Ganges a few kilometers downstream of Chapra in Bihar.

The Gandak is called Sapta Gandaki or Narayani in Nepal. It joins the Ganges near Patna.

Kosi, called Kaushika in Sanskrit, is the largest tributary of the Ganges, if Mahakali and Karnali are considered separately, and is next in size and discharge to the Indus and Brahmaputra. It drains a large area between Langtang and Kanchenjuma Himalayas. It has three major early tributaries, Sunkoshi, Arun, and Tamur, joining at Tribeni, which lie a few kilometers inside the hill from the plains.

Salient statistics of surface waters of the Nepal area were given in Table 2.5. The annual surface run-off of the rivers of Nepal is estimated to be 175 billion m^3. The Himalayan Rivers have considerable variation between the maximum and minimum flows. The ratios for Kosi, Gandaki, and Karnali are 14:1, 19:1, and 13:1, respectively. When the instantaneous flows are considered, the ratio is much higher, being 130:1 for Karnali.

The Terai region is also rich in groundwater, which occurs both under confined and unconfined conditions. The clayey sand and sandy formation yield groundwater under an unconfined condition down to an average depth of 50 m. Confined groundwater occurs in sand–gravel beds at depths exceeding 50 m. The wells tapping confined aquifers show both flowing and nonflowing conditions. The Terai aquifers show a strong confining condition; the piezometric head in the flowing wells ranges between 6.60 and 8.90 m above ground level (magl), and in the nonflowing wells, it ranges from 1.65 to 11.20 m below ground level (mbgl). The tube wells yield 25 to 50 l/s for a pressure head varying between 1.5 and 8.7 m in the case of flowing wells while the nonflowing wells discharge 8.86 to 37.94 l/s at drawdown from 4 to 9 m. It is considered that 250,000 ha of land is irrigable by groundwater (Agarwal, 1991).

5.2.5.2 Demand and Development

The first and most important demand of water is for habitat water supply and sanitation; the requirements are small compared to other uses. These, however, have been neglected and sadly developed, as are the basic infrastructure needs of shelter, transportation, and other civic amenities. Urbanization is only 9%, and the urbanization is only an apology of minimal facilities. In view of the physiographic characteristics, provision of these will pose serious difficulties.

5.2.5.3 Irrigation

Irrigation makes the major demand for water. Most of the irrigable area, about 78% of the total, is in the Terai. Of this, about 58% (57% of the total) has been provided with irrigation. But, as seen in the case of India, it is an apology of irrigation.

Irrigation schemes may be divided into two categories: (i) the farmer-managed irrigation schemes (FMIS) and (ii) the government schemes. Irrigation has been practiced in the hilly and mountain areas of the Himalayas for a long time by diverting the waters from the streams and conveying them through contour channels called Kuhls (guhls or rajkuls in different parts). The supplies are seasonal, as the streams are often not perennial. The structures and canals are often primitive and inefficient.

Larger and, so to say, modern systems have been developed by diversion from larger streams and rivers. The first modern irrigation system was built in conjunction with Sarda Canal in India at Banbassa, a few kilometers below the point where it debouches from the hills. At this point, the river forms the boundary between India and Nepal. Inspired by the example, the Nepal Government employed a British engineer and built the Chandra canal system at Fatehpur in the Trujuga valley in 1930 (Sarma, 1983). Later on, the Judha Canal at Manusumara in the Rauthat district was added. Several works have been constructed since the 1951 changes under several successive 5-year plans.

The Indian Government has undertaken projects on major rivers. Conjunctively, development in Nepal has also been undertaken. When the barrage on Sharda was negotiated with the British-ruled India in 1920, an agreement was reached which provided Nepal with some land area in exchange and entitlement to specified supply of water for irrigation. Development of the Mahakali Irrigation Project was undertaken in 1970 to use these waters. The 1954 Kosi agreement led to the construction of the Kosi Project and the Chatra canal with a net command area (NCA) of 25,000 and 66,000 ha, respectively. The 1959 Gandak agreement led to the construction of the Narayani and West Gandak development with a total NCA of 51,800 ha.

From 1960, the Minor Irrigation program was initiated. Under it and under several other programs, efforts have been made to improve the existing FMISs.

5.2.5.4 Hydroelectric Development

Hydroelectricity is central to Nepal's economic development and environmental conservation by reducing the pressure on forests for fuel wood.

Despite the tremendous potential, less than 1%, amounting to an installed capacity of 244 MW, has been developed.

As noted in the energy sector review, there are two strategies for hydroelectric development. One is the development of those projects, which can be developed indigenously. These are essentially run-off-river projects. However, development of hydroelectric energy can only be carried out in collaboration with India for several reasons, as discussed briefly in the energy sector review and in detail in Chapter 9.

Some hydroprojects have been studied at a feasibility level or a prefeasibility level to identify the projects to be undertaken by Nepal on its own or until agreement with India on collaborative development is reached.

Considering the urgent need for developing hydroelectricity, a least cost generation expansion plan (LCGEP), which outlines a process of selection based on available information to the available hydropower projects and the system load forecast, has been undertaken. The process takes into account two parameters viz. earliest commissioning date of project and energy cost of the project. Based on the above considerations, Kaligandki and Upper Arun have been considered to be the most attractive sequence. Negotiations on export of power to India, if fruitful, could lead to an earlier and accelerated development of Arun 3. After Arun 3, some of the priority projects chosen are upper Arun (360 MW), Budhi Gandki (600 MW), and upper Karnali (240 MW). An entirely different picture emerges if collaborative development between India and Nepal could take place, as discussed in Chapter 8.

5.2.6 Transport and Inland Water Transport

Nepalese planners lay considerable emphasis on inland water transport. These are briefly referred to following Thapa and Pradhan (1995), although in our view, a inland water transport in the Himalayan region does not appear to be feasible, as briefly discussed later, and has hardly received any attention in Indian or Pakistani Himalayan ecosystem development.

Provision of surface transport in the Himalayan region is very difficult and expensive. Ninety percent of the population of Nepal lives in a rural areas, most of which is yet to be connected to the road network. They depend on foot trails and mule track, and human portage is still the main form of transport. These trails feed into the hills from road heads. The longer-distance east–west transport is mostly taken up by roads. The length of the main trails is around 7000 km, with 70% of them being in the western regions. Travelling is by foot or horseback, and commodities are moved by mules, donkeys, yaks, etc. Bullock carts are extensively used in the Terai. Roads are gradually being developed. The road network increased from 976 km in 1951 to 7400 km in 1991. About 4304 km of these roads lies in the hills and mountains, the rest being in the Terai.

The first narrow gauge railway was established in 1928 from Raxaul to Amlekhganj. Another 53-km-long narrow gauge railway line was established in 1935 connecting the border town Jainagar to Janakpur and Byalpure.

Two ropeways have also been installed. A 2.5 km long 8-ton/h capacity ropeway connecting Dhursing and Matatirtha was established in 1927, and later extended to Teku in 1947. A 25-ton/h ropeway between Kathmandu and Hetanda was constructed in 1964.

Inland water transport is almost nonexistent. Locally used boats used to be operated in the 27-km stretch of Karnali, 58 km in Gandaki, and 40 km in Kosi, but they are mostly limited to crossing.

Reference to prospects of inland waterway transport from Calcutta to Nepal via the three main rivers, Kosi, Gandaki, and Karnali, does not display any possibility of feasibility (Thapa and Pradhan, 1995). Considerable emphasis is, however, laid on inland water transport by Nepalese planners (Thapa and Pradhan, 1995). These may be referred to following Thapa and Pradhan (1995), although in our view, inland water transport in Himalayan region does not appear to be feasible, as briefly discussed later, and has hardly received any attention in Indian or Pakistani Himalayan ecosystem development.

5.3 Bhutan

5.3.1 General Features

The "Druk Yul," popularly known as the Kingdom of Bhutan, lies in the eastern belt of Himalayas and is wedged between China in the north and India on the rest of the sides. Geometrically, it is nearly an oval-shaped country and it lies between latitude 26°45′ to 28°10′ north and longitude 88°45′ to 92°10′ east. The east–west length is approximately 320 km, and the north–south width is about 175 km. The area of Bhutan is 47,000 km². The capital of Bhutan is Thimpur, which is 180 km from Phunsolving, the Gateway of Bhutan.

Bhutan is a very underdeveloped Himalayan region. Its per capita income is $80, and the gross national product is $982 million. The population is 1.32 million (1981). The population growth rate is 2.24% per annum.

Bhutan has 4500 villages. There are a few small cities. The average population density is 28 persons per km², but the density is higher in the southern parts.

5.3.2 Geomorphology

The geomorphologic feature of Bhutan is similar to the relation between the palm and fingers of a hand, the palm being the higher Himalayas, the fingers the north–south running hills, and the gaps between two fingers being the river valleys. Bhutan is compartmentalized in different units by north–south flowing rivers, i.e., the Manas, the Sankosh, and the Raidak from east to west.

The terrain rises abruptly from the Indo–Bhutan border and reaches to different levels in different parts. Elevation-wise, Bhutan can be classified into three units: (i) southern foot hill region, (ii) mountain and valley belt, and (iii) the Himalayan chain. It is entirely in the Himalayas; the features have been described in Chapter 2.

5.3.3 Land Resources

The forest area is 32,600 km^2 or 69.37% of the geographic area. The distribution is as follows: (i) broad-leafed zone, 12,000 km^2; (ii) coniferous zone, 17,000 km^2; and (iii) alpine pasture and vacant area, 3600 km^2. Actually, the real productive forests will be much less than the above. It is believed that growing stock is 181 million m^3 in the coniferous zone and 189 million m^3 in the broad-leafed zone. The rate of harvesting is negligible at present.

Bhutan has a limited area under agriculture, and the scope of further extraction of land from the forest zone is very limited. The total cultivated land is 1152 km^2. It is divided into three categories: (i) wet land (river valley and foothill), 23 km^2; (ii) dry land (rainfed), 601 km^2; and (iii) shifting cultivation, 316 km^2. The per capita agriculture land comes to 0.085 ha (0.21 acres). The principal crops are paddy, maize, and millet. Cropping intensity for paddy works out to 123. The yields are very poor.

5.3.4 Water Resources

The average rainfall is about 1152 mm. Most of the rainfall (76%) goes off as surface runoff, about 19% percolates as groundwater, and 5% is retained as snow. The yearly low and high discharge difference is not much, i.e., 16 to 30 times, unlike in the other Himalayan Rivers, where it is about 100 times. The density of primary rivers is 0.0138, which indicates a hard topography. The density of snowfed rivers in Bhutan is from 0.01050 to 0.0305, and that of the nonsnowfed rivers is from 0.0395 to 0.075.

The important rivers, from east to west, are the Manas, the Sankosh, the Raidak or Wangchu, and the Amochu or Torsa. Only the Torsa River rises in Tibet, whereas the rest originate in the southern part of the Himalaya in Bhutan.

On the basis of very rough estimates, the annual runoff of Bhutan Rivers has been estimated to be about 43,240 million m^3 (Sharma, 1983). Again on the basis of rough estimates, a hydroelectric potential of 20,000 MW has been estimated. About 40% of this potential lies in the Manas basin, 25% in the Sunkosh basin, and 25% in the Wangchu River basin.

5.3.5 Development and Conservation

The opportunities and problems in Bhutan are the same as in other Himalayan countries and regions. Very little development has taken place so far. Land has been terraced over centuries, but the productivity is low, and environmental deterioration prospects are very high. The same applies to forests.

Water is the key to resource and economic development and environmental conservation, but again, very little development has taken place.

6

Post-Independence Development—India

6.1 Introduction

With the Independence of India, she became responsible for her future. There was tremendous euphoria. Development of natural resources received highest attention in view of the agrarian economy and the climatic–hydrologic characteristics of the region. The concepts, practices, and institutions of colonial heritage, however, shaped the activities in the initial phases. For instance, water resources management continued to be primarily in terms of stabilizing the sustenance agriculture even though multipurpose dams were attempted to be undertaken from the very beginning. Vectoral approach toward environment continued although environmental concerns have become of interest of late. Therefore, we consider the development in terms of each vector, viz. water, land, and forests in that order in view of emphasis accorded to each, and follow with an overview of environmentally oriented activities that have emerged of late.

6.2 Water

The Ganga–Brahmaputra–Meghna river basin, as briefly presented in Chapter 2, is one of the world's richest in terms of land and water, and it is predominantly in India, as shown in Figure 6.1. Development of water has been undertaken, essentially in terms of the various purposes, e.g., (i) water supply for habitat, industry, and infrastructure; (ii) irrigation and hydroelectric energy; (iii) floods; (iv) inland navigation; and (v) management of wetlands. We may therefore consider the development of each purpose separately. In each section, we consider (i) the present state, (ii) the historical perspective of activities, and (iii) an evaluation.

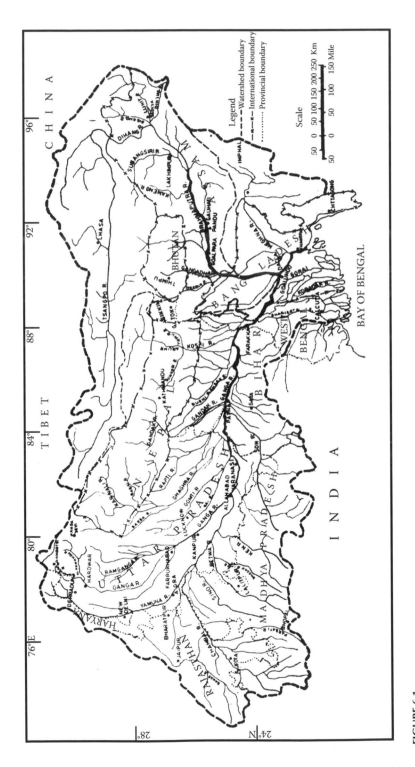

FIGURE 6.1
Ganga system.

6.2.1 Water Supply for Habitat, Industry, and Infrastructure

Supply of drinking water and provision of sanitary facilities are the most important activities in the water sector. Reviewing the position, it was noted in the Eighth Plan that "Operation and maintenance of water supply and sanitation installation in the country is badly neglected." (GOI, 1992).

The statistics are very deceptive, as the norms and quality of service make water supply and sanitation an apology of the activity. For instance, for rural water supply, the norms are 40 l per capita per day (lpcd) of safe drinking water within a walking distance of 1.6 km or elevation of difference of 100 m in hilly areas to be relaxed as per field conditions to arid, semi-arid, and hilly areas. At least one hand pump/spot source for every 250 persons is stipulated to be provided. Regarding rural sanitation, the norms are coverage of about 5% of the rural population with their full involvement with sanitation facilities, adopting a concept of total environmental sanitation, providing guidelines to the rural population in regard to the proper environmental sanitation practices, and construction of biogas plants adjacent to sanitary complexes. These standards continue. Thus, neither for water supply nor for sanitation are individual house facilities contemplated, simply because most of the people have hardly a house worth the name. Secondly, most of the statements are pious bureaucratic pronouncements, observed more in breach than achievements.

Similarly, in the urban areas, the norms are 125 lpcd for urban areas where piped water supply and underground sewerage systems are available, 70 lpcd with piped supply but without an underground sewerage system, and 40 lpcd for towns with spot sources/stand posts, with one source for 20 families within a maximum walking distance of 100 m. It may be pointed out that even in the capital city of New Delhi, even in the most affluent sections, water supply is available only for a couple of hours at pressures enabling supply only at the ground floor levels. It has been estimated that actual rates are only 60 lpcd in affluent areas and 10 lpcd in slums. Thus, the water supply and sanitary facility can be said to be only notional.

The sanitary facilities are even worse. As statistics show, most of the cities do not have sewerage systems, and even where these exist, as in the capital of the country, Delhi, it is only in the affluent sections. The large section of the population, about 60%, lives in jhuggi-jhopris (ramshackle temporary hutments), which have been neither drinking water nor sanitary facilities. The people and the environment are subject to the worst pollution hazards. The return flows from industry are even more neglected. The result, with the fact that the river flows have been almost completely diverted for irrigation requirements, has been that the rivers have been turned into open sewers. Some efforts to clean the rivers have been undertaken of late, as we shall discuss later, but they have been followed more in neglect rather than implementation.

The provision of water supply and sanitation is very complex and difficult, as it is a societal–developmental problem. The technology is well established, and the quantity of water, proportionate to the agricultural sector, is small. The provision of water supply depends upon development of the habitat and other infrastructure facilities and the economic well-being. For the rural population, it is dependent on rural development because unless the people are organized and interested in development of the basic amenities of life, it is not possible to provide this facility. With about 75% of the people living in villages in the most abominable conditions, rural development would appear to be the most important and priority task for Independent India. Yet for several reasons, rural development, in almost all facets, for example, literacy, health, population control, and infrastructure facilities, has been badly neglected. Even now, although it is a much-talked-about facet of developmental activity, this is more rhetoric than factual. Similarly, in the cities, development of housing, infrastructure, and the package of activities representing the basic amenities has been neglected. Social organization and management, again, are the first step in this context, but under the colonial heritage, they have also been sadly neglected. The result has been a most degrading condition of urban living, which is also crumbling rapidly on account of an increasing influx of people from the villages.

6.2.2 Water for Irrigation

Water resources development historically has been almost synonymous with irrigation, and it continued to receive highest emphasis, in this context, after Independence as well. This was further aided by the concern for developing stable food supplies in the context of the Bengal food famine experience recently. Multipurpose major storage projects had started to be developed all over the world, more particularly in the USA. Emphasis on development of the Bhakra dam was being laid in the adjoining Indus basin. An attempt at developing storage projects was, thus, earnestly laid all over the basin, particularly in the Himalayas. A very large complex, one of the biggest in the world, has been developed, as shown in Figure 6.2.

Comparatively easier new developments in the Bundelkhand and Mirzapur areas, where developments had been already undertaken earlier, were started, while investigation for multipurpose projects in the Himalayas was undertaken. Two major projects, Nayar dam on a tributary of River Ganges and Kishau dam on a tributary of River Yamuna, each higher than 250 m, were identified. (The author started his career before the attainment of Independence as a member of a small team designing these dams.)

However, these were given up on account of some foundation problems. Instead, storage projects in the Vindhyan region, which were easier to construct, were undertaken; Rihand dam, on the tributary of River Sone, was one of the first ones. (The author moved over to it, undertaking the planning from the first day, in the leading capacity, under the direction of the Engineer-in-Chief, Mr. A.C. Mitra.) The emphasis on high dams in the Himalayas was, however, restored, and the Ramganga Project, on River Ramganga, a tributary of River Ganga, was undertaken next. (The author moved over to undertake its design as the Director of Designs after having completed his doctoral

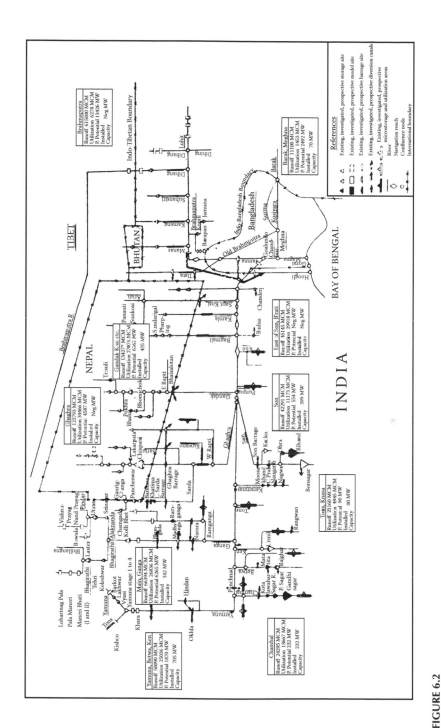

FIGURE 6.2

Schematic development of the Ganga–Brahmaputra–Meghna River system. (From Chaturvedi, M.C., *Water Resources Systems Planning and Management,* Tata McGraw Hill, New Delhi, 1987. With permission.)

studies in the USA.) With its completion, confidence was gained to undertake the very high dams, more than about 250 m on the Himalayas. Tehri dam was the first one. After its completion, more projects in the Ganga–Yamuna complex were planned to be developed. (The author was a member of the Board of Consultants since their conception.) A complex of multipurpose and hydroelectric development has gradually evolved. The Ganga–Yamuna–Ramganga, as shown in Figure 6.3, represents a typical subsystem.

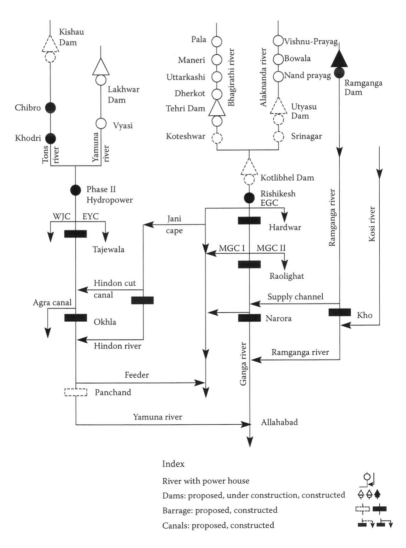

FIGURE 6.3
Ganga–Yamuna–Ramganga system.

The complex is developed on a simple principle. Irrigation works had already been developed on River Ganga and River Yamuna for about a century, by construction of diversion works at the foot of the Himalayas. Storage dams were proposed on these rivers, wherever it was feasible to store the monsoon waters for stabilized and additional irrigation and to generate hydroelectric energy. Flood mitigation is considered incidental. Run-off-river schemes are developed both upstream and downstream of the storage projects wherever feasible. The Ramganga Project was undertaken first, diverting the waters to River Ganges upstream of the lower Ganga Canal headworks, so that more adequate supplies could be made available in the command area of the latter. Development of power is considered secondary to irrigation, with releases to be made according to irrigation, with a small reserve storage called power cushion for power, for releases according to power demands. It was considered to be one of the well-designed projects by the leading engineers of that time, constituting the Ramganga Board of Consultants. However, considerable improvements in the field have been made since then, as reference to modern literature on the subject will bring out, but even now, these projects are being designed based on outdated styles and practices (Chaturvedi, 1981, 1988; Chaturvedi and Rogers, 1985). The subject requires a revolution (Chaturvedi, 2006, 2011c).

Developments were undertaken all over the basin, as water resources development is a state subject. An important activity by the Central Government was the Damodar Valley Project envisaged on the lines of the Tennessee Valley Authority. Damodar is a small tributary meeting Bhagirathi, Hooghly, in West Bengal. It suffers serious flood problems, and four multipurpose dams on River Damodar and its tributaries, each about 50 m high, were constructed.

Activities were also actively pursued in other parts of the basin, but the historical developmental policy continued to be followed. Diversion works were constructed on River Ghagra at the foothills of the Himalayas to irrigate the western lands between River Ghagra and River Sharda and to supplement the Sharda Canal by diverting 368 m³/s and linking the two.

The Gandak Canal Project was similarly undertaken on River Gandak. It comprises of a barrage built across Gandak in Nepal territory, 17.7 km north of the Uttar Pradesh–Nepal border, and two canals irrigating the eastern and western regions.

Focusing on the northern tributaries, an important activity is the multipurpose Kosi Project in Bihar on which work began in 1955. The Kosi was rendering vast tracts of fertile lands unfit for cultivation by depositing coarse silt in the process of shifting westward. Over a period of 130 years, the river has shifted westward by about 112 km (Rao, 1975). The main barrage is sited in Nepal, 5 km beyond Hanuman Nagar. It has three canals, the Eastern Kosi Canal, Western Kosi Canal, and Rajpur Canal, irrigating 0.57, 0.32, and 0.16 mha, respectively. It has a 20,000-kW capacity powerhouse, 50% of which goes to Nepal. Embankments, 270 km in length, have been constructed, spaced 12 to 16 km to serve as silt traps. Even as the project was being constructed, it was appreciated that these measures alone will not be adequate for flood control and river management (Rao, 1975).

Continuing in this fashion eastward, the Teesta Barrage Project has been constructed recently. It is to be noted that river management and control become increasingly difficult as one moves east.

On the southern side, development is comparatively easier in view of the physiographic characteristics. Starting from the west, considerable storage diversion and canal activities have been undertaken on River Chambal and the minor southern tributaries of River Ganges.

The Chambal Project is typical of the Vindhyan region. The princely states were integrated with India after Independence, and the new provinces of Rajasthan and Madhya Pradesh were created. Development of the Chambal basin in these states, prior to Independence, was by tanks across the small streams flowing into Chambal and its tributaries, irrigating small patches of land. Development of Chambal was undertaken through an integrated project consisting of three reservoirs, viz. Gandhi Sagar, Rana Pratap Sagar, and Jawahar Sagar, and a barrage downstream at Kota from where irrigation canals take off to irrigate 0.285 mha of land.

River Sone is a major southern tributary, and irrigation works had been constructed early during the British reign. Development of River Sone, except for modernization of the barrage, has lagged because of interstate disputes.

Besides the conventional storage-diversion projects, three types of activities, which are slightly different, may be mentioned. One is that of lift canals. The gravity canals, taking off from the head reaches near the Himalayas or further down, neglect the even further downstream areas close to the river. These would be and have been irrigated by lift canals, pumping water directly from the river, in contrast to the gravity flow canals. The second is the Kharif Canals. The vast canal network described so far started from the concept of the historic Upper Ganga Canal. The focus was on irrigation during Rabi (winters), diverting the low flows. Projects such as Madhya Ganga Canal, have been undertaken to provide irrigation during Kharif (monsoons).

The third type of activity is modernizing the old canals. This began through the Project of Modernization of Upper Ganga Canal, sponsored by the World Bank. The project, as the title implies, seeks to redesign the Upper Ganga Canal, developed about 150 years back. Comprehensively, the task was proposed to be undertaken over a 40-year period. The project, as developed with the help of so-called World Bank experts, has serious shortcomings. It is also significant to note that the World Bank financial aid was stopped, as the progress of the work was not satisfactory—a malady that infests almost all major projects.

Irrigation development in Brahmaputra has lagged. Irrigation was not developed in the eastern parts of the basin during the British period. The area was neglected even after Independence, and only from the latter part of the Third Plan (1961–1996) were some small major projects undertaken. Recently, several impressive multipurpose major projects have been developed. These will be discussed under the section on Hydroelectric Development, as it is the dominant role.

In addition to the major projects, considerable surface water development has been carried out through medium and minor projects. At the beginning of the first 5-year plan (1950–1951), irrigation schemes were classified into three categories: (i) major costing more than Rs.50 million each, (ii) medium costing between Rs.1 million and Rs.50 million, and (iii) minor costing less than Rs.1 million each. According to a revised clarification made in April 1978, (i) projects having CCA (cultivable command area) of more than 10,000 ha are each classified as major projects, (ii) those having CCA between 2000 and 10,000 ha as medium schemes, and (iii) those having CCA of 2000 ha or less as minor schemes.

6.2.2.1 Groundwater Development

Besides surface water development, groundwater has been developed tremendously, as the basin represents the world's richest groundwater potential. Groundwater structures may be grouped into three categories, namely, dug wells, private tube wells, and public tube wells. The alluvial plains, historically, supported a large number of dug wells, which constituted an important source of water supply for domestic consumption and farmlands. The dug wells continue to be the principal source of supply for domestic consumption, but their use for irrigation declined in the areas served by surface irrigation, as discussed in Chapter 4. The Grow More Food Campaign launched in 1943 and finally integrated with the First Plan laid emphasis on dug wells. However, these were not developed much, as even with loans provided by lending agencies, farmers with small or segmented holdings could not afford them. Those who could afford them went for private tube wells, which became very popular as high-yielding varieties were introduced.

The public tube wells, already developed in Uttar Pradesh, were extended, but the availability of electricity became a big constraint. Until 1966, groundwater development was principally in the public sector, but later, private tube wells became more popular. The public tube wells tap semiconfined aquifers occurring at 50–70 m depth. They may be electric or diesel driven.

The private tube wells were developed primarily in western Uttar Pradesh, but they are expanding rapidly in the eastern parts of the basin as well. They are generally of half a cusec capacity (0.014 m^3/s) driven by electric motors or diesel engines.

The Government of India in 1982 constituted a Groundwater Estimates Committee. It devised certain norms/techniques for the assessment of groundwater potential, which were revised later. According to the latest estimates, the replenishable groundwater in the GBM basin is estimated to be 270.71 km^3. The utilizable surface water is estimated to be 280.81 km^3. It will be seen that the estimated groundwater potential is almost equal to the surface water potential. We will discuss the estimate of potential and other policy issues later, demonstrating that there are serious basic technological and policy errors and that the availability is much higher. The fact remains that the groundwater has a central role and tremendous potential.

6.2.2.2 Development Overview

The development of irrigation in India in which GBM has a major role was impressive. Starting with an irrigation potential of 22.60 mha in 1945, a potential of 93.98 mha and 37.08 mha from major and medium, and 56.90 mha from minor irrigation schemes, with a potential of 44.95 mha from groundwater sources, has been created by the end of the Ninth Plan (1997–2002).

The future plans are even more impressive. Out of the entire geographic area of India, which is about 329 mha, the current cultivable net sown, and gross cropped areas comprise 197, 141, and 176 mha, respectively, representing 60%, 71%, and 125% of the geographic area, cultivable area, and net sown area, respectively. The ultimate irrigation potential in India is estimated to be 113.5 mha, which will be 64% of the gross cropped area. It was expected that this potential will be achieved by 2010.

6.2.2.3 Evaluation

Although important achievements have been made, continuous introspection is necessary on account of the importance of the sector and heavy investments. Unfortunately, this is rarely done. Some analysis was carried out by the Irrigation Commission (1972) and the Agricultural Commission (1976) appointed by the Government of India (GOI), but it is very general and dated. The Planning Commission carries out five yearly reviews in successive 5-year plans, but generally, it is in terms of targets achieved.

Some observations have been made in the National Water Policy (GOI, 1987). More detailed policy analysis has been carried out by Chaturvedi (1985, 1988, 1990, 1992); a review of the projects has also been done (Chaturvedi, 1981, 1986, 1988). The World Bank (1991) has carried out a detailed review of the sector. A National Commission of Integrated Water Resources Development Planning (NCIWRDP, 1999) was constituted by the Government of India to undertake a scientific review, but unfortunately, it could not make much scientific contribution. An overview has been given by Shekhar (2007), who was Advisor in the Planning Commission, GOI. Some observations may be made as follows, on the basis of the studies found in the author's work in the area (Chaturvedi, 2011a, 2011b, 2011c).

As noted in the World Bank study (1991), agricultural performance is fundamental to India's future economic and social development. Agriculture contributes to 30% of GNP and 60% of employment and is the primary source of livelihood in rural areas, which account for 75% of India's population and

80% of its poor. With all arable lands under cultivation, increased agricultural output will depend on raising crop yields, increasing cropping intensity, and diversification to higher-value crops. The performance of irrigated agriculture, which contributes 55% of the agricultural output even with about 30% of irrigated agriculture, will be the most important contributor to these objectives.

With the largest irrigated area in the world, Indian irrigation has much to be proud of. Its development, the product of major efforts and achievements by India's irrigation engineers and of greatly expanded government investment since Independence, has been the principal force behind agricultural growth. Its role will be indispensable to future performance

According to the World Bank study (1991), irrigation improves (i) cropping intensity and expansion of cropped area, (ii) crop yields, (iii) output stabilization, and (iv) crop diversification, leading to impacts on (a) agricultural productivity, (b) incomes, (c) employment, and (d) regional development. The impacts vary depending on the type of irrigation, agroclimate, and, above all, the technical quality and management of irrigation.

While constraints on arable land prohibited the expansion of the net area under cultivation, in the past two decades, the gross cropped area expanded from 166 mha in 1970 to 177 mha in 1985, with an estimated 60% of this expansion due to irrigation. It has contributed significantly to agricultural extension. Still, average irrigation intensity remains low at about 1.29 irrigated crops per annum in 1984. Similarly, although average yields under irrigation are more than twice as high as yields under rainfed conditions (for cereals, 2 t/ha, compared to 0.8 t/ha rainfed), they are considerably low relative to yields under irrigation in other countries or even in better irrigated areas within India, such as Punjab. Irrigation has reduced interannual fluctuations in agricultural output and India's susceptibility to droughts. Irrigation has also been selectively responsive to the needs of diversified agriculture.

On average, farm incomes have increased 80%–100% as a result of shifting from rainfed to irrigated farming. Incremental labor days used per hectare average 60%–80%. Through its influence on agricultural incomes, irrigation development has a multiplier effect on nonfarm incomes. It has been estimated that a Rs.100 increase in irrigated agricultural output stimulates Rs.105 of additional output in manufacturing and Rs.114 of additional tertiary output, translating to a total number of farm output multiplier of 2.19. Expansion in agricultural output as a result of irrigation has also helped keep food prices down. Between 1970 and 1986, food grain prices in India fell by about 20% relative to the price index for all commodities. This had a significant impact on the real incomes of the urban poor and landless rural households for whom food is a large component in consumption. Irrigation is strongly correlated to poverty alleviation. For instance, for districts where less than 10% of the gross cropped area was irrigated, 69% of the population had incomes below the poverty line, while in districts where irrigation covered more than 50% of the crop area, poverty incidence was only 26%.

Thus, although irrigation is crucial to the economic development and it has contributed significantly, the development impact of irrigation is much less than its potential, and deficiencies in implementation have accumulated over time. As the World Bank study (1991) has noted, the sector is now under crisis. Four issues are of particular concern: productivity, sustainability, investment focus and financial discipline, and sector management.

The productivity of irrigation is extremely low and well below potential. In India, which has the second largest area under cultivation, almost equal to that of China, are one of the lowest, almost about half of those in China.

The potential of increase is brought out vividly comparing the current yield of rice with yields in Japan, historically. The yields in India are still what the yields in Japan were in AD 1200 and are about one-fifth the current yields in Japan. The trends in yields in 93 developing countries are given in Table 6.1. The table clearly brings out the poor yields in India and the tremendous scope for improving them.

The sustainability of vast irrigation investment is a serious question, the two important facets being adverse environmental impacts and decline in the maintenance of the infrastructure. Finally, sharp financial deterioration of the sector poses a serious threat. The sector is posting severe financial losses, which continue to increase exponentially.

Irrigation is a vague descriptive expression. What is important are the quantity, timing, and reliability. The latter depends on who exercises the control and management. Irrigation in India was developed by the Government to stabilize sustenance agriculture, providing an apology of irrigation through a constellation of political, technological, administrative, economic,

TABLE 6.1

Trends in Yields in 93 Developing Countries

Crop Type	Yield (kg/ha)			
	1961–1963	1969–1971	1979–1981	1990–1992
All cereals	1171	1461	1894	2466
Excluding China	1116	1271	1557	1951
China	1336	2070	3017	4329
Wheat	868	1153	1637	2364
Excluding China	964	1146	1460	1997
China	673	1169	2046	3208
Rice	1818	2218	2653	3459
Excluding China	1650	1855	2145	2790
China	2355	3281	4236	5722
Maize	1157	1456	1958	2531
Excluding China	1122	1291	1572	1837
China	1265	2005	3038	4545

Source: World Resources 1996–1997, Oxford University Press, New York. With permission.

and financial considerations. All these have changed, but it still remains an apology of irrigation (Chaturvedi, 1992).

Serious environmental problems have already been experienced on account of irrigation, which accounts for about 90% of water use. With the large-scale diversion of surface waters for irrigation, the rivers have been converted into open sewers. With the increasing use of chemical fertilizers and pesticides, the groundwater is increasingly being polluted. The World Bank study (1991) noted that the sector is in a crisis.

6.2.3 Hydroelectric Development

The GBM basin has one of the world's richest potentials of hydroelectric energy, estimated to be about 128,700 MW at 60% load factor, as already brought out in Chapter 5. It is a renewable energy resource, with one of the highest returns on energy. As we will demonstrate later, the estimates are very low, but and if properly developed, the potential will be several times higher. However, there are serious challenges to hydroelectric development in the Himalayan region. Controversies regarding the environmental impacts of dams have surfaced. The scientific position about the matter is brought out by Chaturvedi (1981, 1986, 2011c). Rangachari (2006) and others demonstrate that the controversies are ill founded but have serious implications for implementing sustainable environmental management in developing countries, particularly in the region.

According to current estimates of the total 128,700-MW hydroelectric potential 83,000 MW is in Nepal, 45,700 MW in India, and almost none is in Bangladesh. Of the potential is in India, 10,800 MW is in the Ganga basin, mostly in the Himalayan region, and 34,906 MW is in the Brahmaputra basin. The Brahmaputra basin constitutes 41.5%, and the Ganges basin 12.8% of the total all-India hydroelectric potential of 84,044 MW at 60% load factor, with an annual energy of 422 billion units of electricity (with seasonal energy, the total energy potential is assessed to be 600 billion units per year) in India. It may also be noted that this region is least developed energy-wise even in India, with the eastern region having only 12.7% of the all-India energy consumption. The Brahmaputra region has a mere 1.3% of the all-India energy consumption and a mere 0.55% of development (193 MW out of 34,920 MW).

Only about 2% of the GBM potential has been developed (Narasimhan and Singh, 1994). With the poor availability of other energy sources such as coal and oil in the region and with the extremely low per capita energy use (kilogram of oil equivalent) even in the region (as in Bangladesh) at about 6.3 compared to the figures for industrial countries and average developing countries at 4840 and 550, respectively, hydroelectric development is of utmost urgency and importance in the region (UNDP, 1994).

Hydroelectric development was totally neglected during the British regime, but even after Independence, it has not received the importance it deserves even though, paradoxically, the earliest water resources development activity focused on hydroelectric development, as discussed earlier. While studies of the development of the storage project, which involved the construction of very high dams, were being carried out, development of the hydroelectric potential on the existing canals was undertaken. Falls in the head reaches of canals had been provided to negotiate the difference in the slope of the natural ground, which was higher, and that of the canal, which was lower. These falls were provided in quick succession. The technology consisted of combining them into one, and developing a small hydroelectric power station of the order of about 30,000–40,000 kW. The earliest developments were on Ganga Canal and Sarda Canal. (The author started his professional career designing the Sarda Power House, immediately after passing from the Thompson Civil Engineering College, now renamed as the Indian Institute of Technology, Roorkee.)

Since Ganga, Yamuna, and Ramganga are entirely within India, investigations for multipurpose developments were started immediately after World War II, even before Independence, as discussed earlier. The focus, however, was on irrigation.

6.2.4 Management of Floods

Floods are one of the most serious natural hazards for humankind and the environment. We will briefly provide an overview of the problem, current policies, and efforts to deal with it. Unfortunately, as in other areas, the activities and reports are classified. Some information is available from the reports of the National Commission on Floods (GOI, 1980). The Centre for Science and Environment has made efforts to collect as much information as possible and has published it as *State of India's Environment, A Citizens Report 3* (CSE, 1991).

6.2.4.1 State

Of all the flood-caused deaths all over the world, over 50% have taken place in Bangladesh and 20% in India, mostly in the GBM basin, making it the world's most severely flood-affected area (Hagman, 1984). On the average, as many as 1439 lives are lost in India every year on account of floods, with as many as 11,316 lives lost in 1977 alone (GOI, 1992). Most of these are in the GBM basin. An area of 40 mha, i.e., nearly one-eighth of India's geographic area, is flood prone. The total area affected annually is about 7.7 mha and was as high as 10 mha in the worst year. Most of it is in the GBM basin. Different opinions have been expressed about the increase in the area affected by floods. While the number of people affected, value of property, and crop damage are bound to increase on account of population increase and economic advancement, it is doubtful if the magnitude of floods has increased. It must be clarified that estimates of flood in terms of discharge, area affected, damage to crops, and loss of life and property are very difficult and uncertain. Measurement of discharges even in normal conditions is difficult and is poorly organized. Measurement in flood conditions is almost impossible and is arrived at by measures that entail much uncertainty. The valuation is even more difficult and arbitrary. The records are kept poorly by the revenue authorities, making them very unreliable.

The catastrophic flood proneness of the basin follows naturally from the physiographic and hydrologic characteristics. The Himalayas are the world's highest and extremely steep mountains, geologically very young and erodible. The scene is compounded by heavy monsoon rainfall. The mountains are followed by very flat and narrow alluvial plains, which have indeed been formed by the erosion of the Himalayas and the sedimentation in the region. The flood intensity increases in the Ganges basin from west to east, both on account of its downstream location and increasing precipitation. The flood proneness is even more severe in the Brahmaputra basin, as it is even narrower and the rainfall intensity is even higher. The rainfall in the Meghna region is one of the highest in the world, and the flood proneness is further enhanced.

Similarly, the erosion and sedimentation characteristics of the rivers of the basin are one of the highest in the world. This has led to the formation of the world's biggest delta, which constitutes about 80% of Bangladesh and a sizeable proportion of West Bengal. The geomorphologic conditions also lead to meandering of the rivers of the region, with the meandering increasing as one travels downstream in the Ganges basin. Kosi is a classic example of a meandering river, as shown in Figure 2.12. The characteristics follow readily from the observation that Kosi has to negotiate a difference of elevation in almost one-fifth the distance that Ganges does, as it meets the Ganges. Loaded with a high sediment load, Kosi has been known to meander hundreds of miles, devastating large areas in Bihar. Further down, Tista used to meet Ganges before it meandered far enough to confluence with Brahmaputra. Similarly, Brahmaputra herself has meandered over the north of the delta. Only 200 years back, Brahmaputra used to flow 100 km east of its present course. While all the rivers have very high erosion–sedimentation and meandering characteristics, the dynamics of each river varies depending on the geomorphologic–hydrologic properties of the region. For instance, Kosi in its upper reaches not only has a steep gradient but also flows through a narrow and deep valley, whereby the sediment load at the point of entry into the flat alluvial plains is not only very high but also consists of high bed load compared to other rivers at the same elevation.

Again, as these rivers braid in the delta, the numerous channels meander, get silted, and develop new courses. The history of Hooghly is a classic example. It was the main channel of Ganga on which Calcutta port was established. It was reduced to a trace, which led to the decision during the British regime to divert the flows of the newly energized tributary, leading to the genesis of the current Indo–Bangladesh dispute.

With the heavy population, poverty, and the very seasonal nature of high flows and meandering of rivers, people try to make as much use of the flood plains as possible. These active riverbeds, known as Khadar in western Uttar Pradesh and diara in eastern Uttar Pradesh and Bihar, are the habitat of millions. The inhabitants have to continuously struggle against the vagaries of nature. The problem becomes increasingly difficult as one moves downstream to the water's edge as the population pressure, destitution, and environmental uncertainty keep on increasing. In Uttar Pradesh and Bihar alone, the diara land has been estimated at 2.40 mha (GOI, 1980).

6.2.4.2 Policy

Flood management was almost neglected during the British regime. Commissions were appointed from time to time, but these were merely ritualistic activities. When the British left, there were only about 5290 km of embankments along different rivers of which 3500 km were in the Sunderbans in West Bengal and 1209 km along the Mahanadi in Orissa, providing protection to about 3 mha of land mostly through local people's initiative.

Due to efforts after Independence, water resources development gained high priority, and multipurpose dams were proposed at several locations, as discussed earlier. Flood management was an integral component of these multipurpose projects, with some, like Kosi, proposing highest priority for flood mitigation. Unfortunately, the dam could not be taken up and for flood mitigation reliance had to be on embankment only. In 1954 came a spate of floods, one of the worst in the country. There was considerable euphoria in the country about development, and a national policy statement was put before the Parliament, which claimed, "floods in the country can be contained and managed. The administration and the people have both to undertake tasks of a huge magnitude in order that the country may be rid of the menace of floods." The policy statement also noted, "Provided the enthusiasm of the people can be aroused, as it is stated to have been in China, and their cooperation secured for this work of national importance, it would be possible to complete work on embankments in about seven years if a start is made immediately. The program of flood protection works will incidentally provide, on a tremendous scale, opportunities for employment of a simple character, scattered over large areas. Although embankments do not provide absolute immunity from the floods, they will ensure a very large measure of protection, which given good maintenance, should prove to be of a lasting character" (GOI, 1980).

Several committees were appointed from time to time to develop an appropriate policy for flood management. In 1976, the Government constituted Rashtriya Barh Ayog (RBA; GOI National Flood Commission, 1980) to study the subject and give suitable recommendations. As usual with the Commission reports, a bureaucratic approach was followed. An ad hoc approach of constructing dams, embankments, channel improvements, soil conservation programs, flood plain zoning, and prevention of encroachment upon drainage channels was proposed. The RBA also noted that "in the absence of viable measure one would have to live with floods." In such circumstances, flood plain zoning, which prevents undue occupation of flood plains, adjustment in cropping patterns, and raising of villages above flood levels, would be the main measures available. The Commission also recommended that natural water detention basins like lakes and swamps should not be reclaimed for agriculture and should be protected. Disaster preparedness should be strengthened through a better flood warning system. It was emphasized that the flood-affected states should prepare master plans for each river basin.

The progress up to March 1988 was as follows: (i) embankments, 14,703 km; (ii) drainage channels, 28,837 km; (iii) town protection schemes, 546 km; and (iv) raising of villages, 4701. Up to 1988, an expenditure of Rs.22,970 million was reported to have been incurred on flood management assistance for flood relief provided by the Central Government to various states. The total expenditure on relief by the state governments and voluntary agencies would be much more than this figure.

6.2.4.3 Practice

As noted by Singh (1990), even these recommendations, which were considerably short of modern advances in this area, have not been put into practice. Mostly, the emphasis was on structural measures consisting of embankments followed by channel improvement, including the construction and improvement of channels for surface drainage. Reservoirs with exclusive storage capacity earmarked for moderating floods are very few, although many reservoirs are used for multiple purposes, including flood moderation, and most reservoirs provide incidental flood moderation, especially for minor and medium floods. Ring bunds have also been used for protecting isolated small towns and villages. River diversions, natural detention basins, and emerging flood ways are very rare. Interbasin transfer and underground storage for the purpose of flood protection have hardly ever been used. Bank stabilization and anti-erosion measures, which have no impact on flood mitigation, are quite common and have been consuming a large proper of the funds available in the flood control sector.

Embankments are known to have serious consequences. Embankments in isolation affect the river regime, contribute to siltation, pose serious problems of afflux, affect natural shifting of river courses, and pose a danger of breach of embankments. With the sense of flood security, encroachment in flood plains takes place, leading to the possibility of enhanced damage. Serious problems, of drainage congestion and flooding due to backwaters ensue. Several examples of the negative impacts have been given by Singh (1990) and CSE (1991).

Although considerable watershed management has been undertaken, it has not been integrated with flood management or evaluated in this context.

In general, watershed management has been neglected in terms of flood management.

In short, flood management has been undertaken unscientifically and inefficiently. It has serious human, economic, and environmental limitations.

6.2.5 Inland Water Transport

As noted earlier, inland water transport (IWT) had been almost completely neglected in the past. Some attention has been shown from the Sixth Plan (1980–1985). A policy to declare important waterways as national waterways was initiated, and the Ganga–Bhagirathi–Hooghly stretch between Allahabad and Halida was declared as such. Brahmaputra between Sadiya and Dhubri has also been declared as a national waterway, and a master plan for the development in Brahmaputra has been prepared; navigational facilities are being strengthened, and floating terminals have been set up. The Inland Waterways Authority of India (IWAI) was set up in 1986. Yet IWT forms a very small part of the total transport network of the country. Out of the total freight traffic of about 550 million tonnes by all modes of surface transport, IWT carries about 16.6 million tonnes (GOI, 1992). In terms of tonne kilometers, the share of IWT is less than 1%. The IWT traffic is mainly on account of movement of iron ore on Goa waterways, which form about 96% of the total IWT traffic.

6.3 Land

Land is the founding unit, closely interacting with water, in the terrestrial ecosystems and river basin. By land, what is meant is physical space but also the characteristics that govern the uses to which land can be put. These characteristics include the topography, the quantity and quality of soil, the availability of water, and the nature of the local climates. All these features are interrelated, and together, they influence and are influenced by the vegetation that grows on the soil. Soil itself is a product of the underlying rock, the climate, the topography, and the creatures living on it and in it; the availability of water depends on how much falls as rain and snow, how much evaporates, and how much is retained, where, and for how long; the latter property depends on local soil and vegetation as much as they influence them.

The problem of land management has been studied by several researchers and official agencies (Bali, 1988; Datyae et al., 1988; Yadav, 1988; Agrawala and Narain, 1989; Singh, 1989; Dhruvanarayana et al., 1990; SPWD, 1988; Vaidyanathan, 1991). Following them, we consider the basic issues, overview the developments and the current scene, and undertake a brief evaluation of the management of land.

6.3.1 Basic Issues

While the subject should be considered in terms of integrated watershed management, it is usual to undertake land management in terms of some specific activities for conservation, restoration, and/or development. These are usually grouped under (i) soil and water conservation and (ii) land reclamation and development.

Soil is a crucial life support system. Even under natural conditions of vegetation cover, nature takes from 100 to 400 years or more to generate 10 mm of topsoil. The basin is very susceptible to soil erosion due to the topography of the land, hydrology, and the nature of soils. Out of India's total land area of 3.3 million km^2, 1.4 million km^2 is subject to increased soil loss, while an additional 270,000 km^2 is being degraded by floods, salinity, and alkalinity. An estimated 6000 million tonnes of soil is lost every year from 800,000 km^2 alone, and with them go more than 6 million tonnes of nutrients—more than the amount that is supplied in the form of fertilizers.

The land use pattern of the basin broadly underscores the problems of soil and water conservation. The forest area is very low, and even within this, there is a substantial extent that has been depleted because of overexploitation and uncontrolled grazing. Hilly and mountainous areas are subject to landslides and slips, besides surface waste and gullying, which is a major cause of watershed degradation in the Himalayan region. In the hilly areas of the Brahmaputra basin, where the tribal population predominates, an age-old method of crop husbandry known as shifting cultivation or jhumming is practiced over an area of 2.7 mha, which causes severe soil loss. The area shown as "barren and uncultivable land" constitutes probably the most severely eroded areas.

The magnitudes of "cultivable lands" and "fallow land other than current fallows" indicate neglected land management, which are liable to erosion. The areas categorized as "permanent pastures" and "other grazing lands" represent, fallaciously, some of the worst eroded areas, as they are characterized by unchecked misuse.

As for land reclamation and development, deterioration is considered under the following categories: (i) waterlogged land, (ii) saline and alkali soils, (iii) land infested with shrubs and bushes, (iv) ravines, (v) riverine land, (vi) stony and gravelly lands, (vii) laterite soils, and (viii) high-altitude, steep slopes, and meadows.

Water logged soils are defined when the yield of crops is affected adversely. A water table may be considered harmful depending upon the type of crop, type of soil, and the quality of water. The actual depth of the water table, when it starts affecting yield of the crop adversely, may vary over a wide range, from zero for rice to about 1.5 m for other crops. Wheat and sugarcane are affected when the water table is within 0.6 m; maize, bajra, and cotton are sensitive to the water table within 1.2 m, and gram and barley within 0.9 m. In general, it is considered to keep the water table below 5 m; otherwise, soil salinity and wasteful evaporation may increase. Water logging is entirely on account of not providing proper drainage and is a simple matter to be attended to as part of water resources development and management.

Saline and alkali soils may be found in natural course but have also been developed by a lack of drainage facilities in irrigated areas. Early studies have shown that salts formed by the natural weathering of igneous rocks become evident on the surface layers under certain predisposing conditions like (i) arid or semi-arid climate, (ii) an impervious subsoil or hard pan, and (iii) temporary abundance of humidity in the soil interspersed with dry periods. The soils have been divided into four categories. The four categories are (1) saline, (2) saline–alkali, (3) nonsaline alkali, and (4) degraded alkali. The saline soils could be treated by leaching and providing adequate drainage. Using gypsum could treat the alkaline saline best. Acid or acid-forming amendments would be used provided that alkali earth carbonates are present. Usually, it is very difficult to reclaim degraded alkali soils. It was considered that even though a lot of work on reclamation of saline and saline alkali soils has been carried out and a large volume of scientific data is available, no large-scale successful reclamation project on saline alkali soils has so far been initiated or is currently in vogue. It was considered that this may be attributed either to the immenseness of the problem or due to imperfect approaches adopted in these ventures.

Large areas of land are invested with shrubs and bushes in the basin, but not much recognition has been given to these lands or their treatment undertaken. The main reason for such lands remaining unutilized is deep-rooted grasses and weeds, unhealthy conditions, lack of drainage, low fertility of shallow soil, lack of water supply, salinity and alkali conditions, damage by wild animals, and severe erosion. These lands can be reclaimed by tractorization and made to yield good crops by means of good farm management practices, but by and large, the problem has been neglected.

Extensive degradation of land has occurred in the basin along the banks of the rivers. Erratic short duration and high intensity rainfall, loose and friable nature of the soil, steep slopes and undulating terrain, faulty agricultural practices, illicit cutting of trees and bushes, grazing of land, and similar biotic interferences have combined to aggravate the situation, ultimately resulting in deep gullies commonly known as ravines. High flood level has contributed to the deepening and widening of the ravines. The rivers in high floods back up in the ravines and hasten soil slipping. Prolonged inundation causes damage to protective vegetative cover. Ravine formation has removed fertile lands along the riverbanks of the region. The process, once started, continues with increasing speed. Some ravine reclamation has been undertaken through forest plantations of economical species and by development of fuel/fodder reserve. However, it remains a token activity.

Riverine land is another problem, particularly in the alluvial plains along the rivers emanating from the Himalayas. These carry heavy sediment loads, and as they negotiate steep slopes, meandering action and serious bank erosion ensues. Fertile areas with flourishing crops, orchards, towns, and cities on the banks of rivers are often eroded by meandering streams. Depending on the course of rivers and the speed of flow, they deposit large quantities of sediments on their way to the sea. Gradually, landmasses are formed, which remain, for some part of the season, at least above flood level. Such riverine lands are called Khadar in north India, Diara in east Uttar Pradesh and Bihar, and Char in West Bengal. Another riverine problem is that of deposition of huge sediment load carried by the hill torrents as they emerge in the plains. Called "chos" in the western Himalayan region, where the problem is most pronounced, the deposition is about 20–25 km from the foothills. The third aspect of the riverine problem is in the deltaic region, where, ultimately, the heavy sediment load is deposited. The three aspects of the riverine problems emphasize the river–land interaction with primacy on river management on one hand and watershed management on the other.

Three other categories of land management problems may also be mentioned. One is that of stony and gravely lands. Such lands lying waste are normally found on rolling topography and plateaus, which have been subjected to heavy grazing or indiscriminate felling of trees. Lying without adequate vegetation, they are severely eroded. Soils in such lands are generally acidic and can be made productive by liming. The second category is that of laterite soil. These are highly leached ferruginous soils, poor in fertility, low in water retentivity, and high in phosphoric fixation. As the soil depth becomes shallow, soil–plant–water relationship becomes precarious for successful establishment of vegetation. There is a possibility of growing certain trees and grass in these areas, but no extensive attempt has been made in this direction. The third is that of extensive areas of land on high altitudes over steep slopes or as overgrazed meadows, especially in the Himalayan region. Serious soil erosion takes place under injudicious land use. They constitute the spots, which contribute most to the heavy silt discharges from the Himalayan watersheds. Their treatment is most important, urgent, and remunerative.

6.3.2 Overview of Activities

As planned development started after Independence, agriculture received increasing importance. Watershed management has also been given some attention. One of the early undertakings for water resources development, flood control, and conservation considered integrally was the Damodar Valley development, a small tributary of Ganges at its tail end, on the lines of the Tennessee Valley development. Land and soil management was initiated on a limited but integrated basis. A directorate fo soil conservation was established, and a multidisciplinary organization to deal with soil conservation/watershed management/land use planning in an integrated manner was established in 1949. In the first 5-year plan, a Central Soil Conservation Board was established to look into the problem in its entirety. But this remained an isolated experiment. The necessity of purposive measures to check soil erosion, improving the moisture retention capacity and the natural fertility of the soil, reverse the progressive decline in the extent of forest cover, and so on, had been long recognized.

Successive 5-year plans have included a variety of programs to this end; substantial resources have been allocated for these programs, and the scale of outlays has increased manifold. However, the approach has been to tackle each of these problems separately and on a piecemeal rather than an integrated manner (Vaidyanathan, 1991). Thus, the soil conservation programs of the agricultural department and, subsequently, those meant to raise the productivity of drylands have been confined to agricultural lands; construction and renovation of minor surface irrigation works are the responsibility of the Public Works Department while afforestation and social forestry are taken care of by the Forest Department.

The idea of integrated treatment of watershed was revived in the early seventies in the context of the program to reduce the sediment inflow into the big reservoirs and subsequently to the catchment of selected flood-prone rivers. From the early limited activities of contour bunding and gully plugging, an area saturation approach to treat all types of lands on a complete watershed basis was proposed later. It was estimated that while the areas requiring soil conservation measures in different catchments vary from 11% to 39%, it would be adequate to treat 10% to 15% of the high silt producing areas in the catchment, which are considered critical. The emphasis has been on small watersheds of 1000 to 2000 ha.

A series of integrated watershed development projects (other than those for upper catchments of major river valley projects and flood-prone rivers) was started during the seventies and early eighties: one of these is the Indian Council for Agricultural Research (ICAR) operational research projects for the integrated development of 47 selected watersheds. Scientists from the Centre for Research in Dryland Agriculture (CRIDA), the All-India Coordinated Program for Research in Dryland Agriculture (AICPRDA), and their research centers were to prepare integrated master plans for the selected watersheds, supervise and monitor their implementation by the concerned departments of the state governments, and assess their impact. A second set consisted of projects for watershed development in rainfed areas funded by the World Bank; four of these are currently under implementation. The average size of a watershed under this scheme is much larger (around 25,000 ha) than under the ICAR or the National Watershed Projects, but none of them is in the GBM basin. Then, there is the National Watershed Management Project. Details of the distribution of these projects between states or their organization and functioning are not readily available.

The initiative to promote watershed development has mainly come from the Center. While the response at the state level has been mixed and some states have formulated ambition plans, the GBM basin states have not shown much interest.

Nongovernmental and voluntary organizations have also shown growing interest in integrated watershed development. Some (like Ralegon Siddhi and Jawaja) started out tackling a particular problem (like water management and wasteland development) and came to the realization that a broader, more inclusive program for ensuring optimum use of available land and water resources was necessary. Others (like Sukhomajri, Tejpura, Naigoan, and Daltonganj) were conceived as integrated watershed projects. There have been doubts as to whether the watershed or the village should be the unit of planning, but in either case, the idea of a coordinated plan covering all land seemed to command widespread acceptance. The Society for Promotion of Wasteland Development (SPWD), which started out being concerned primarily with wastelands, has come around to the view that development of degraded and other wastelands cannot be viewed in isolation from that of cultivated lands and that integrated programs covering all categories of land in an area is essential. Nongovernmental activity is, however, quite limited in scale, and their projects are largely dependent on state funding. In any case, the bulk of the integrated watershed projects are conceived, funded, and implemented by the Government directly.

To the original list of departments involved in this activity, namely, soil conservation, minor irrigation, and forests, another has been added viz., the Department of Rural Development. The number of programs under which each one of these activities is taken up has multiplied. The reporting mechanisms are so poor that it is difficult to say with confidence how much of a particular activity has been achieved under each program. In every district, the line departments pursue their own programs largely unmindful of similar activities carried out by other departments. With rare exceptions, there is not even an attempt to avoid duplication of the efforts under different programs. On the other hand, even programs like the ICAR model watershed schemes and wastelands development under NWDB, which are supposed to be "integrated," remain fragmented in terms of funding sources and implementation responsibility. Thus, the wasteland development program draws funds from at least five different sources—the Central Government, the state plans, the Rural Employment Programs, DPAP, and DDP. And different components of the model watershed plans are implemented by the respective line departments. Attempts have been made to streamline the various apex agencies that developed over the course of time, with little advances.

There is another area of activity called Command Area Development which involves land management. The focus of the activity is the management of the irrigated command area to improve irrigation, which intimately involves land and watershed management, but it has been usual to link it with irrigation and it shall, therefore, be discussed under "water."

6.3.3 Highlights of Experience

As noted by Vaidyanathan (1991) and other studies (GOI, 1988, 1989; Puri, 1992), the watershed projects implemented under various auspices have not been an unqualified success: indeed, they have serious shortcomings. The organization and expertise needed for preparing comprehensive plans for the selected watersheds are not available except in the ICAR. Even these, however, suffered from a lack of skilled, experienced, and motivated staff to implement the projects, and this was aggravated by frequent transfers. They also suffered from lack of coordination between concerned line departments, absence of active interest on the part of the district level officers in monitoring the program, complicated financial procedures leading to delays in release of funds, unrealistic norms and lack of flexibility in use of funds, weak arrangements for monitoring and evaluation, and the low level of beneficiary involvement and participation in the whole activity. A few impact studies have been made, but these were not systematic enough. Also, they have been done too soon after the launching of the project to permit confident inferences. The absence of any provision for follow-up action and continuing management of the watersheds after they are developed is clearly a major weakness, which is bound to affect the prospect of sustaining the impact.

Compared to the governmental programs, those involving nongovernmental and voluntary organizations reveal an approach that is very different and much more exploratory and open minded. Marked by diverse social philosophies and approaches to local mobilization, they show much greater awareness of the need for flexibility in both technical and organizational solution in the light of specific local conditions, of the tension between equitable distribution of the benefits from the development and the logic of local social and political configurations, of the problems in evolving community interest and participation and sustaining it, and of the need for innovative adaptations and the importance of learning from experience.

Many of the projects have also shown keen concern for evolving better, more cost-effective treatments and for improving the effectiveness of investments (e.g., through proper choice of species, natural regeneration instead of plantations, reducing costs of raising nurseries, and improving the survival rates of planted samplings). There are also instances of experimentation with new designs and materials to reduce costs of structures, evolving multitier crop-cum-tree systems to make more effective use of soil moisture, simultaneously augmenting the moisture storage capacity of the soil, and developing low-cost water distribution systems using local materials and skills. There is a great deal more appreciation—sometimes to the point of fetish—of the need to utilize local knowledge and experience and the necessity to adapt programs to circumstances, which are highly variable.

Although NGOs are acutely conscious of the active involvement of beneficiary participation, considerable difficulties arise on account of differing perceptions of individuals and conflicting interests of various groups. It is also apparent that the social orientation of the NGO groups is often much stronger than their technical skills/experience. Moreover, a great deal of their effort is spent in mobilizing funds and liaisoning with concerned Government departments, who are not always favorably disposed to such initiatives. These are therefore quite limited in scales, and their projects are largely dependent on state funding. The bulk of the activity for land management program is undertaken by the Government directly.

6.4 Fauna and Flora

The GBM basin was densely forested until recent historical times, but the exploding population and poverty have created serious resource and environmental problems.

The forest area lost for various purposes in India during the period of 1952–1980 before the forest conservation act was enforced is inferred to be near 4.328 mha, with 155,000 ha having been lost annually on an average. An area of 0.19 mha was deforested during 1981–1983 to 1985–1987, implying an annual loss of 47,500 ha. Most of the loss is in the GBM basin. The impact of agriculture has been maximal as exemplified by the diversion of 2.623 mha of forest area to agriculture between 1952 and 1980.

Shifting cultivation (or jhumming) is practiced by tribals in 16 states of India and is most prevalent in the seven northeastern basin states. Essentially, this practice is a low-energy-budget and low-investment system of subsistence husbandry. The intensity of jhumming is on the rise. About 50 to 60 years ago, the jhum cycle was in the range of 20–40 years. It is now 2–6 years. Now that there is not enough area for the system to function, forest degradation is rampant. Forest fires are a most potent source of damage to forests. The forest area subject to annual fires varies from 33% in West Bengal to 93% in Arunachal Pradesh.

The development of roads particularly in the Himalayan belt has caused severe erosion, affecting hill slope stability and causing heavy landslides, injury to forest and other vegetative cover, blocking of natural drainage pollution of streams, and damage to human life and property. It is estimated that the total damage, direct and indirect, from the unplanned system of road transportation in the Himalayas is likely to be more than Rs.1 billion annually.

Forest area is also lost due to hydroelectric and multipurpose developments. The forest area lost due to submergence during the period of 1952–1980 was of the order of 0.502 mha. Besides this, a considerable percentage of the area was lost for infrastructure development, and the rehabilitation of the project affected the people.

Excessive grazing, frequent fires, the ease of cutting small trees for fuel wood, and other biotic factors have adversely affected the natural regeneration of basin forests to a considerable extent. Inventories carried out in few representative localities indicate that natural regeneration is either absent or inadequate over large tracts in India. Only 1% of the forested land carries predominantly young regeneration or pole-sized trees. Available figures, though only indicative, nevertheless point to an alarming situation.

There are also vast areas of degraded forests and wastelands in the basin, which have to be rehabilitated and brought into production to make them ecologically sustainable and economically viable. Spectacular gains have been achieved in the production of wood when trees of species of natural forests are grown in plantations.

Artificial regeneration is also practiced in India while tree cultivation is being extended through the wasteland development program, social forestry, farm forestry, and agora forestry, which envisages growing of trees with the involvement of the people on agricultural and village common lands, government wastelands, rail and road sides, canal banks, etc. Some 3900 mha of plantations has been raised so far in India. The picture that emerges from the overview is that there is a very serious problem of deforestation and that the concerted action required to meet the serious challenge is lacking.

6.5 Environmental Management

The concept of environment is historical in Indian tradition but is very recent in modern science, technology, and management. Environmental management was restricted to the management of forests and conducting botanical and zoological surveys. A need was felt for preventing and controlling pollution. The Water (Prevention and Control of Pollution) Cess Act of 1977 and other acts were gradually enacted in the early period of the Independence phase. Gradually, environmental concerns and activities are being activated. According to the Ministry of Environment and Forests annual reports, the following activities are being undertaken for environmental conservation.

A survey of the natural resources, which currently embraces fauna, flora, and forests, is being undertaken. The conservation of natural resources embraces the conservation of forests, wildlife, biosphere reserves, wetlands, mangroves, and coral reefs. Under the latter, management and regenerating of the Sunderbans have been undertaken. A core committee (of which the author is a member) was set up by the Ministry of Environment and Forests to formulate a National Conservation Strategy (GOI, 1990). Major river valley multipurpose irrigation and hydroelectric projects in the water sector, power projects, industrial projects, and mining projects require clearance from the Ministry of Environment and Forests in terms of environmental impacts.

Environmental appraisal committees have been constituted (of which the author was also a member) to review the projects.

For control of pollution, draft policy statements have been prepared, and efforts are being made in controlling industrial pollution. In an action plan for controlling pollution from the heavily-pollution industries, some critical areas have been identified, of which include Singrauli in Uttar Pradesh, Durgapur and Howrah in the Ganges basin and Digboi and Assam in the Brahmaputra basin. The Central Pollution Control Board conducted studies on Ganga and Yamuna to access the impact of pollution-related activities in these basins. Under the United Nations Global Environmental Monitoring Systems (GEMS), the Monitoring of Indian National Aquatic Resources (MINARS), and the Ganga Action Plan (GAP), 450 water quality–monitoring stations have been established.

6.6 Conclusion

The Ganga–Brahmaputra–Meghna basin, as in India, is home to the world's largest component of humanity. It is a very sensitive environmental area because of its physiographic and hydrologic characteristics. It is one of the poorest regions in the world. A scientific environment and socioeconomic management is urgently required. Unfortunately, it has been badly neglected. It should be undertaken urgently.

Its management depends critically on the management of the environment of the upstream areas, in Nepal and Bhutan, which is largely out of its political domain. It will be in India's interest to actively support their development in all domains, in the area of the management of the environment, viewed in comprehensive terms. This will require very careful management as it is a very delicate subject. It can be easily misconstrued as interference in view of the comparative, vastly varying capabilities and potential. This should be well appreciated by India. India has also to appreciate that environment is not conscious of human political divisions. It is in the interest of India, much less the entire human community of the region, that she pursues the development of the entire GBM basin as one integrated entity, with utmost humility and commitment, putting the interests of the smaller regions first and foremost. The proof of the success of her policy is the reaction of these smaller and weaker components—their perception of the performance of India. In the final analysis, this will be in India's own interest. Thus, we consider the development of the GBM basin as one socioeconomic–environmental activity, irrespective of the recent political divisions, as it has historically been, as Mahabharata brings out.

7

Environmental Management in Bangladesh

7.1 Introduction

Bangladesh is at the tail end of the world's second largest river system. It is the world's largest and wettest and, over most of the year, it is also an arid delta. It is a unique ecosystem. It is a land of very fertile soil with numerous rivers and abundant groundwater. It once had very rich fauna and flora and, even now, it is quite rich in this respect. Until the industrial revolution, it was one of the world's most prosperous regions, which attracted the British trading East India Company to India (Spear, 1978). Currently, it is one of the poorest and is in a desperate state for survival, for which the management of the environment is crucial and extremely challenging, even among all the countries of the GBM basin.

Except for the tertiary hills of the Chittagaon Hill Tracts and Sylhet, the elevation of the country is very low, averaging only between 5 and 6 m above sea level. Bangladesh receives annually, on average, about 1073 km³ of surface water inflow and 203 km³ of rainfall, enough to submerge each unit of land under more than 9 m of water. During each monsoon season of, June through September, one-third of Bangladesh is gripped with flood calamities, which could extend to over half of the country during a severe event. During the dry months, November through May, major river flows decrease drastically. Fresh water becomes scarce for use in agriculture, fisheries, navigation, industries, and even for drinking or domestic purposes.

The rhythm of the annual water cycle, severe flooding during the monsoons, and drought during the dry season has fashioned the environment and the life of the people. Over a long period of time, people evolved a way of life with the vagaries of environment, enjoying the gift and calamities of abundance. Its exploitation started by the mid-eighteenth century, and it became the springboard of the establishment of the British Empire in India. With the establishment of the British rule, peace, law, and order were instituted, but development did not take place. One impact of development was population explosion, which started taking place by the early twentieth century. It could be described as a supersaturated solution of land, people, and water.

The story, unfortunately, does not end here. The groundwater has been found to be severely polluted by arsenic, making it hazardous to drink water or eat the rice grown, as it would lead to cancer. With the forecast climate change, most of the country is estimated to be submerged.

An attempt is made to face the formidable environmental challenges. Starting with a review of the environmental characteristics, with particular attention to floods and a brief overview of the man-made environment, policy perspectives of environment and management are discussed. This is followed by a study of water resources development. Flood management is a crucial issue and is considered separately, although it is an integral part of water resources development. Starting with a brief background on flood management, the recent proposals are considered in detail, as they constitute a novel, though controversial, approach, bringing out the difficult challenges faced for sustainable development in the GBM basin. The attempts to face the arsenic challenge are discussed. Finally, the issue of climate change is considered.[1] These are only some problems relating particularly to management of water. Some wider aspects of environmental management are also considered.

7.2 Country Context

The total area of the country is 147,570 km², 6.7% of which consists of rivers and inland water bodies. Some 88% of the country's total area belongs to the GBM region.

A critical problem faced by Bangladesh, even more than the other riparian countries, is the large population. In 1973, the total population was 74 million, increasing at a rate of 3%, although it has since been reduced to 2%. The estimated goal is to reach a zero population growth by 2045. The estimated population of the country, as of 1999, was about 128 million, with a population density of about 860 persons/km². It is estimated to exceed 176 million by 2025, when the population density will rise to about 1200 persons/km², and to over 2020 million by 2050. The effects of population growth will continue to be most severely felt in the urban sector, where the growth rate during the last two decades has been between 5% and 6% per annum and is likely to be similar for many years in the future. Currently, the urban population is about 30 million, accounting for about 20% of the total national population, and is expected to rise to 53% by 2025, with a population of 10 million in 2050 (Ahmad et al., 2001; WB, 2005).

7.3 Environmental Scene

The environmental characteristics of the GBM basin were presented in Chapter 2, with brief reference to the delta. Some characteristics may be noted in slightly greater detail.

7.3.1 Physiography

Bangladesh has three broad types of land: floodplains (80%), terraces (8%), and hills (12%). Bangladesh's landscape appears as a vast plain sloping gently south and southeastward. It was formed of sediment deposits from the three great river systems, which join in south central Bangladesh and flow into the sea through a broad and complex estuary. This deltaic plain is crossed by a large number of small rivers and channels that form an intricate web, carrying monsoon flows from the upper basins and the surrounding Indian hills and also distributing flows spilled from the main rivers.

The most significant feature of the landscape is provided by the rivers, which have molded not only its physiography but also the way of life of the people. While this is true of the entire GBM basin, it finds its most impressive realization in Bangladesh. The rivers may be divided into five systems: (1) the Ganges or Padma, as the united streams of the Ganges and Brahmaputra are known, and their deltaic streams; (2) the Meghna and the Surma river system; (3) the Brahmaputra and its adjoining channels; (4) the North Bengal rivers; and (5) the rivers of the Chittagong Hill Tracts and the adjoining plains. A general map of Bangladesh showing its relationship to the GBM basin is given in Figure 7.1. The river system in Bangladesh is highlighted in Figure 7.2.

The Ganges is the pivotal point of the deltaic river system. The river and its tributaries enclose a large area covering the districts of Kushtia, Jessore, Khulna, Faridpur, Patuakhali, and Bakerganj in Bangladesh. The Padma enters Bangladesh at the western extremity of the Rajashahi district and forms, for about 145 km, the international boundary between Bangladesh and West Bengal (India).

The Meghna, another mighty river is formed by the union of the Sylhet–Surma and Kusiyara rivers. These two rivers are branches of the Barak River, which rises in the Nagar–Manipur watershed of India. The main branch of Barak, the Surma, is joined near Ajmiriganj in the Sylhet district by the Kusiyara branch. The Dhaleswari, a distributary of the Jamuna River, joins the Meghna for a few kilometers above the junction of the Ganges and the Meghna. As it meanders south, the Meghna grows larger after receiving the waters of a number of rivers, including the Burhi Ganga and the Sitallakhya.

The Brahmaputra and its adjoining channels cover a large area from the eastern parts of the districts of North Bengal to the Meghna River in

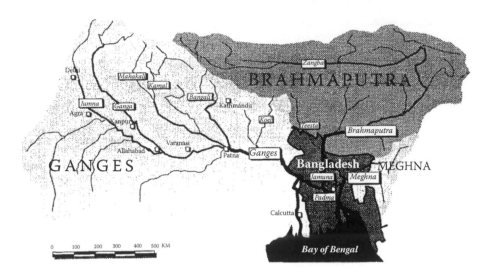

FIGURE 7.1
Map of Bangladesh as part of GBM basin. (From World Bank, *Water Resources Sector Strategy*, Dhaka, 2003. With permission.)

the southeast. The Brahmaputra (Jamuna) receives waters from a number of rivers, especially on its right bank. The river, with its notorious shifting channels, not only prevents permanent settlement along its banks but also inhibits communication between the northern and the eastern part, where Dacca, the capital, is situated.

The Tista is the most important water carrier in the northern districts of Bangladesh. Rising in the Himalayas near Darjeeling (India), it flows southward. After the floods of 1787, the Tista changed its course, moving southeastward to join the Brahmaputra. That shift caused the rivers of North Bengal to be cut off from the upland waters and led to the deterioration of the natural drainage. Similarly, a number of small- and medium-sized rivers in the southwest are silting up and adversely affecting the economic life of that region.

Four main rivers constitute the river system of the Chittagong Hills and the adjoining plains—the Femi, the Karnaphuli, the Sangu, and the Matamuhari. Flowing generally west and southwest across the coastal plain, they empty into the Bay of Bengal.

Rivers in Bangladesh are characterized by severe bank erosion, riverbed silting, and scouring, and the courses of the rivers are extremely unstable, both laterally and cross-sectionally. The main causes of this are the extremely flat and low-lying topography of the plains, the fact that riverbanks consist mainly of sand, the large volume of flood flows, the geological tilting movement of the ground, and the sparseness of vegetation along the riverbanks.

Approximately 200 years ago, the main channel of the Brahmaputra River followed the present course of the old Brahmaputra River to a confluence

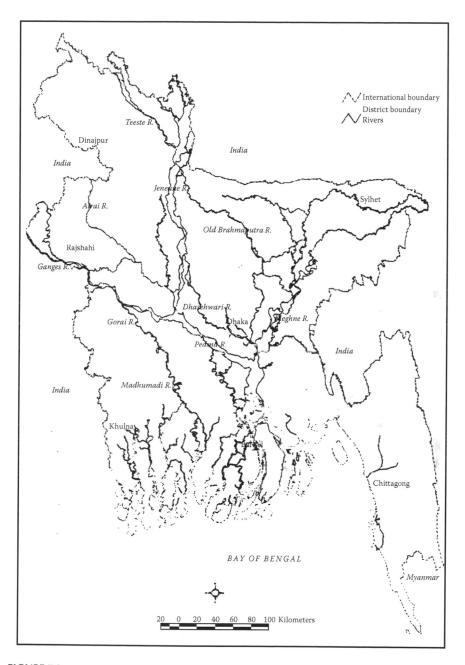

FIGURE 7.2
Map of Bangladesh highlighting the river systems. (From Ahmad, Q.K. et al., *Ganges–Brahmputra–Meghna Region—A Framework for Sustainable Development*, The University Press, Dacca, 2001. With permission.)

with the Meghna River. Apparently, the river started to shift to the west from around 1780 onward, and by the year 1820, its course had changed completely.

Until the late eighteenth century, the main channel of the Ganges River followed the present course of the Arial Khan River. Thereafter, it shifted to the east and converged with the Meghna River. The sedimentation of silt carried by the Ganges and Brahmaputra caused the riverbed to rise downstream from the confluence, and it is estimated that overflow into the Gorai River began sometime between 1820 and 1830.

The slopes of the major rivers are extremely slight. Those of the Ganges and Brahmaputra are of the order of 1/13,000–1/16,000. The Meghna has a slope of about 1/30,000.

The topography of Bangladesh is extremely low and flat. Within the area drained by the three major rivers, the highest point, which is located in the northwest, is only 90 m above sea level. The elevation of 50% of the country is 8 m or lower. This plain is surrounded by the Shillong Hills in the north, the Chittagong Hills in the east, the Rajmahal Hills in the west, and the Bay of Bengal in the south.

Although the topography is very flat, it is quite varied. Physical geographers, utilizing the topographic criteria, have divided Bangladesh into as many as 20 regions. Of these, the following 11 are regarded as the most important. Appreciation of this variation is necessary for environmental and flood management.

(1) The Barmid tract comprises the districts of the Rajashahi division between the Ganges and Brahmaputra. The region, a diluvial plateau, is comparatively elevated.

(2) The Bhar Basin is the depression southeast of the Barhind tract. It includes parts of the Rajshahi and Pabna districts, with its center being in the vast marshy area called the Chalan Bil (bil, lake).

(3) The Brahmaputra floodplains.

(4) The Madhupur Tract in the east between Brahmaputra and Meghna is the second diluvial plateau, with hillocks varying in height from 10 to 20 m. The valleys, mostly flat, are cultivated. The Madhupur jungle contains sal trees. Both diluvial plateaus arose since the Pleistocene Epoch due to an upsurge of tectonic activity in the orogenic zone of the Himalayas.

(5) The Northeastern Lowland is composed of the southern and southwestern parts of the Sylhet district and the northern part of the Mymensin district. It is characterized by a large number of lakes.

(6) The Sylhet hills in the east consist of a number of small hills and hillocks ranging from 30 m to more than 300 m in height.

(7) The Meghna Flood Basin includes the low and fertile Meghna–Lakkhya Doab, enriched by Tistas distributary as well as the diaras

and chars (land areas formed and changed by the deposition of silt and sand in riverbeds) of the Meghna, especially between Bhairab Bazar and Daudkhandi. It was built up by Brahmaputra in its old course.

(8) The Central Delta Basin includes the extensive lakes in the central part of the Bengal Delta in the Faridpur district. The basin's total area is about 8300 km².

(9) The immature Delta is the belt of land in southern Bangladesh bordering the Bay of Bengal. It is lowland of about 7800 km². It contains, in addition to the vast forest known as Sunderbans, the reclaimed and cultivated lands to its north.

(10) The active Delta includes the Dhaleswari–Padma Doab (i.e., the land between those rivers) and the estuarine islands of varying sizes that are found from the Pusur River in the Khulna district to the island of Sandwip in the east.

(11) The Chittagong Region is the area south of Femi River. It is full of hills, hillocks, and valleys and is quite different from other parts of Bangladesh.

7.3.2 Climate and Rainfall

The climate and rainfall in Bangladesh are characteristic of the subtropical belt, as in the rest of the GBM basin, but there are slight variations because of the time taken by the monsoon moving up and down the long basin and the proximity of Bangladesh to the ocean. This has important bearing on the environmental system, agriculture, and natural resources management.

The monsoon season in Bangladesh is spread over May to September. Eighty to eighty-five percent of the annual rainfall occurs in the monsoon season. The intermonsoon period of October to April may be divided into three seasons:

(1) The postmonsoon transition season, October–November, which is warm and humid with rapidly diminishing rainfall. Average rainfall during this period ranges from 7% to 12% of the annual total.

(2) The cool December–March dry season, which is sunny with scattered rainfall that averages 2% to 4% of the annual total.

(3) The very hot transition season, April–May, which is characterized by intermittent rainfall, including severe thunderstorms accompanied by high winds and hail. Rainfall during this period ranges from about 11% to 22% of the annual total.

Besides the tremendous temporal variation, there is also considerable spatial variation in rainfall. Average annual rainfall varies from a maximum of 5690 mm in the northeast corner of the country to a minimum of about 1110 mm over quite a short distance in the extreme west. Mean annual rainfall is 1820 mm in the northwest, 2830 mm in the northeast, 2690 mm in the southeast, 2410 mm in the south central, and 1765 mm in the southwest regions. The 30-year average rainfall over the whole country is 2320 m.

The highest and lowest temperatures on record are 43°C and 3°C, respectively. The records for Dhaka are 39.5°C and 8.3°C. The temperatures permit rice cultivation round the year, with three crops.

The national mean annual lake evaporation is approximately 1.010 mm, just under half of the mean annual rainfall. The minimum lake evaporation rate (1.5 to 2.5 mm/day) occurs during the December–February period, increasing to a maximum rate (3 to 6 mm/day) in the April–May period.

7.3.3 Floods

The physiography and hydrology of the GBM basin lead to severe flooding as a characteristic feature of the environment. This reaches its peak in Bangladesh, the world's most severely flooded region. Each year, about 26,000 km², approximately 18% of the country, is flooded. During severe floods, the affected area may exceed 52,000 km², which is about 36% of the country and 60% of the net cultivable area. In an average year, 775,000 million m³ of water flows into the country from June to September through the three main rivers. The average yearly flooding, with spatial distribution of different types of floods in Bangladesh, is shown in Figure 7.3. It is important to note, as brought out by Rogers et al. (1989), that areas vulnerable to floods from Brahmaputra and Ganges are almost equal—a fact often overlooked.

The annual average flood damage to Bangladesh is estimated at 18 billion Taka (US$945 million) (Siddique, 1981). The flood damage statistics reported in monetary terms cannot indicate the magnitude of disruption of communication, loss of life, and other social consequences to which no cost can easily be assigned.

Floods and chronic inundations in Bangladesh appear to result from a combination of the following factors:

1. Increased flows in the major rivers due to monsoon rainfall in the river basins
2. Discharges due to localized heavy rainfall in hilly areas in the eastern and northern parts of the country
3. Inundation within embankments/dike-protected areas due to lack of sufficient drainage capacity to cope with local rainfall or to higher levels of the outfall rivers
4. Backwater effects caused by the rising water levels along the Bay of Bengal coast due to southwesterly monsoon winds and high tides
5. Bangladesh's extremely low flat topography

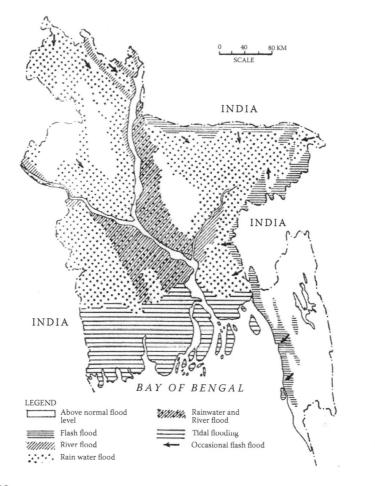

FIGURE 7.3
Spatial distribution of types of floods in Bangladesh. (From Ahmad, Q.K. et al., *Ganges–Brahmputra–Meghna Region—A Framework for Sustainable Development*, The University Press, Dacca, 2001. With permission.)

6. Other causes suggested to have led to flooding:

   a. Coincidence of peak flows on major rivers

   b. Reduction of river flow capacities and the closure of distributary channels due to the sedimentation of silt

   c. Impediments to drainage by roads, railways, and other infrastructure activities

   d. Increased peak flood flows in downstream areas due to the construction of embankments in upstream areas

   e. Increased flood flows due to deforestation in the upstream areas.

There is a particular sequence and type of flooding. The normal sequence of floods starts with flash floods in the streams of the eastern hills during the premonsoon period in the months of April and May. The onset of the monsoon generally occurs in June, and the Meghna and Brahmaputra reach flood peaks during July and August, whereas the Ganges normally peaks during August and September. The Brahmaputra generally peaks about one month earlier than the Ganges, but the gap between the peaks sometimes narrows down to less than 10 days, and at the time of the 1988 flood, the gap was just 3 days. Floods take approximately 2 days to travel along the Brahmaputra River from Gauhati in India to Bahadurabad (approximately 320 km) and a further 2 days to reach the Ganges confluence (approximately 200 km). Floods on the Ganges River take approximately 3 days to cover the distance between Patna in India and the Hardinge Bridge in Bangladesh (approximately 500 km).

Besides the flash floods in the eastern and northern rivers, rain floods due to high intensity rainfalls, and monsoon floods from the major rivers, there are also floods due to storm surges in the coastal areas. Cyclones are most common during the pre- and postmonsoon periods (April–May and October–November, respectively). These have not been known to coincide with monsoon flood peaks.

7.4 Environmental and Water Management

7.4.1 Historical Perspective

Anthropological and historical documents reveal that environmental management, particularly management of water resources, was an intrinsic part of the precolonial socioeconomic system of Bengal as in the rest of India, but the efforts were very modest (Rashid, 1977, pp. 37–72). Environmental management was introduced by the British principally in terms of irrigation in India, but in view of the hydrologic conditions of Bangladesh, there was almost no activity. The colonial administration undertook some projects in Bengal to increase revenue procurement through human occupancy of new land. Such colonization of Southern Bengal, particularly in the Sunderbans (i.e., the mangrove forests), was vigorously opposed throughout the British Raj during the nineteenth century.

The rapid eastward migration of the Ganges and decay of its eastern distributaries resulted in the absence of regular inundation of the zones west of Gorai–Madhumati, leading to widespread incidence of communicable diseases such as malaria and plague and a short supply of silt to replenish agricultural land. In view of these circumstances, plans were made in the earlier decades of the twentieth century to irrigate the moribund delta area of Kushtia, Jessore, and Khulna and to resist salinity intrusion. This multipurpose scheme, known as the Ganges–Kobadak Project (G-K Project), was one of the earliest undertakings, having started in the 1950s. The water resources management was, however, organized gradually.

Concern with riverine hazards control, i.e., floods and riverbank erosion, water resources, and agricultural development, in Bangladesh by the public sector started in the 1950s after its formation as then East Pakistan. During the last four decades, several studies by international consultants and agencies have been undertaken. The focus of the pertinent strategies has oscillated in several directions.

7.4.2 Water Resources Sector

National water planning may be said to date from 1964, when a 20-year master plan was prepared with emphasis on large-scale flood control, drainage, and/or irrigation projects, but continues with lots of twists and turns (WB, 2005). This plan emphasized large-scale publicly financed surface water development, overestimated public water capability, and overlooked the country's groundwater resources. The World Bank conducted a land and water sector study in 1972, which promoted a radical shift in strategy to minor development using low-lift pumps to irrigate with surface water and tube wells to irrigate with groundwater and advocated small- and medium-scale projects through minor irrigation. In 1983, the government initiated a National Water Plan (NWP) preparation exercise, which was completed in 1986 and later updated in 1991.

Bangladesh experienced a disastrous flood in 1987 and even a more disastrous flood in 1988, which were the worst in living memory. The two floods had different natures and origins, but by any standards, they were catastrophic events. Studies by Bangladesh authorities, supported by UNDP and several countries, were undertaken for flood management. Based on these studies, an internationally sponsored Flood Action Plan (FAP) (1989–1994) was prepared. The 1995 Water and Flood Management Strategy synergized the earlier NWP with the outputs of the FAP, promoted a broader technical approach to water planning, promoted a reduced public sector role in water management, identified the need for a national water policy and recommended preparation of a broad-based national water management plan, which would be guided by the policy. In 1999, the government declared the first national water policy. The policy has spawned a number of related initiatives.

Although flood control and drainage rather than integrated water management was the dominant theme, its scope was more comprehensive than the national water planning exercise preceding it. However, the FAP has faced a range of criticisms. Steps are being taken to develop a new NWP (NNWP) (WB, 1998). Following an FAP recommendation, the preparation of a comprehensive National Water Management Plan (NWMP) was initiated in March 1998.

7.5 Water Resources Development

Water resources development for agriculture was an important element of development program of the then East Pakistan from the early 1960s and has continued to receive high priority since the liberation of Bangladesh in 1971.

Of the total irrigable land in the country (7.6 mha), the area irrigated by 1985 reached over 2 mha. About 78% of that area is irrigated by modern mechanized methods, including low lift pumps, shallow tube wells, deep tube wells, manually operated shallow tube wells for irrigation, and large complex canal irrigation schemes. The figure for irrigated area in 1996/1997 was 4.0 mha. Based on NWP estimates, the irrigated area would reach its maximum potential by 2025 (Ahmad et al., 2001). Water resources development for agriculture is critical in Bangladesh on account of land scarcity even for food grain self-sufficiency and has historically averaged about 15% of the annual development program.

Domestic and most industrial water uses, except thermal power plants, rely on groundwater. About 31% of the estimated urban population of 15.1 million in 1985 was served by a piped water systems or standpipes supplied primarily by high-lift deep tube wells. About 43% of the 84.3 million people living in rural areas in 1985 were served by rural water supply development based on suction lift hand tube wells.

Water is important for fisheries, as fish production in inland waters represents 80% of the annual fish harvest of Bangladesh, with fish providing 70% to 80% of the total animal protein in Bangladesh. About 70% of the rural households are engaged in subsistence fishing.

The NWP (1985–2005) was prepared in 1986. The objectives were as follows:

1. Maximize agriculture growth and production and contribute to achieving food grain self-sufficiency
2. Ensure adequate water supplies for fisheries, domestic and industrial use, navigation, salinity control, and environmental management.

7.5.1 Water Availability

For the purpose of these studies, Bangladesh was divided into five planning regions (denoted as northwest—NW, northeast—NE, southeast—SE, south central—SC, and southwest—SW), comprising 60 planning areas (Figure 7.4).[2] There is hardly any storage potential in Bangladesh, except through local ponds. Stream flow is the largest component of Bangladesh's water resources. Available groundwater recharge is less than one-third of the volume of surface water available during the critical January–April period.

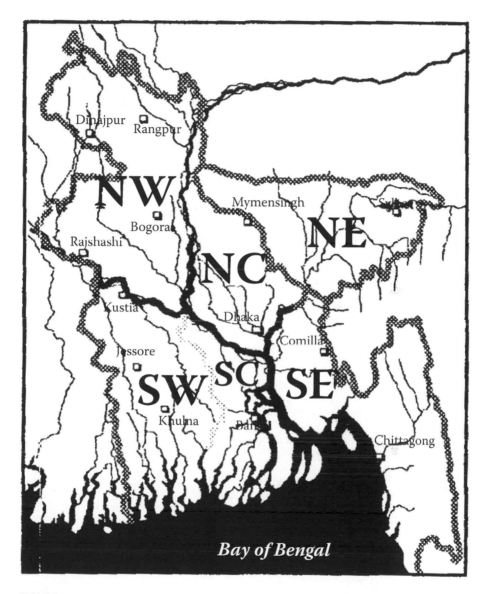

FIGURE 7.4
Regional units. (From World Bank, *Water Resources Sector Strategy*, World Bank, Dhaka, 2003. With permission.)

Each day, on average, approximately 3400 million m³ of water, which is about 3.9 times the average daily rainfall over Bangladesh, is discharged into the Bay of Bengal. About 90% of Bangladesh's total stream flow originates in the upper portions of the Ganges, Brahmaputra, and Meghna rivers outside its borders.

The largest component of the stream flow is the flow in the main rivers. Stream flows in the smaller rivers consist of stream flows from India, spills from the main rivers, and outflows from the groundwater reservoir. They vary considerably spatially. In March, inflow to the smaller rivers from the main rivers is nil in the northwest region, about 30% of the average stream flow in the northeast region, 4% in the southeast region, 80% in south central, and 90% in the southwest region. The regional rivers also carry substantial stream flow from the smaller watersheds across the border in India, which is as much as 90% in March. It is quite variable regionally. The smallest component of the stream flow is generated within the country. Regional surface waters available for development are given in Table 7.1. The Ganges River represents only about 15% of the total dry season flow of all rivers in Bangladesh while the largely uncontrolled Brahmaputra represents more than 65% (WB, 2005). Regional variation is significant.

Salinity of surface water has a major effect upon the availability of usable water resources within Bangladesh's coastal zones. Tolerance to salinity varies widely among different water uses. The threshold value of salinity at which rice, which is the predominant crop in the affected coastal zone, begins to suffer yield loss is 2000 μmhos/cm. Drinking water tolerance is 600–1000 μmhos/cm; for industrial cooling, water uses about 1000 μmhos/cm. The brackish water aquaculture industry, which produces large amounts of shrimp for export, however, requires water in the 6000 to 30,000 μmhos/cm range.

Bangladesh is very rich in groundwater. Bangladesh is underlaid by a deep basin that has been filled with various sediments during previous geologic periods. Groundwater development and related exploration has been confined largely to the shallow sediments and aquifers within 150 m of the surface (except in the coastal areas where information concerning aquifer conditions and characteristics is available for depths in excess of 400 m below the surface). The groundwater storage reservoir has three divisions:

(1) An upper silty clay layer, varying from one to 50 m in thickness

(2) A middle composite aquifer of fine to very fine sands averaging about 20 m thickness

(3) At the bottom, the main aquifer of medium, medium to fine, or medium to coarse sand with layers of clay and silt extending to a depth in excess of 150 m

The capacity of the aquifers soil to store water (specific yield) increases with depth because there are less clay and silt. Near the surface, the specific yield varies from 0.05% to 2.00%, whereas at greater depths, the specific yield increases to 5% to 25%.

TABLE 7.1

Regional Surface Water Available for Development, 1989–1990 (Million Cubic Meters)

Region	Dry Months							Wet Months				
	Nov	Dec	Jan	Feb	Mar	Apr	May	Jun	Jul	Aug	Sep	Oct
NW	2970	1840	1160	780	780	920	2210	5360	13,950	13,520	13,540	6470
NE	5090	1710	710	130	230	2480	9680	22,290	40,840	39,800	28,860	16,350
SE	820	950	750	640	650	630	1090	3170	6450	5480	3190	19,800
SC	8250	4400	3050	2020	2320	3760	7680	14,300	53,370	44,230	40,050	20,960
SW	1180	580	270	140	120	100	230	1080	4400	8220	7280	3050
Total	18,310	9520	5940	3710	4100	7890	20,890	46,200	101,050	111,250	92,920	48,810

Source: World Bank, *India—Irrigation Sector Review*, Vol. 1 and 2, Agriculture Operations Department, India Department, Asia Region, 1991. With permission.

7.5.2 Water Demand and Balance

Water use falls within two categories:

(1) Withdrawal in uses (composed of agricultural use, potable and domestic use, and industrial use)
(2) In-stream use (composed of fisheries, both flowing and static water based), navigation and salinity management, and sediment control

For the former, both surface and groundwater can be used. The projected water supply and demand, based on a March 2018 estimate worked out by the 1991 NWP, are given in Table 7.2.

Like the other riparian countries, Bangladesh is also predominantly an agrarian society. The country has about 8.74 mha of cultivated land, which is about two-thirds of the area. Of the net cultivated area, 33% is single cropped, 45.0% is double cropped, 1.5% is triple cropped, and 10.2% is cultivable waste or fallow. The overall cropping intensity is 176%. Agricultural demand accounts for the major share being as high as about 59% in the critical month of March. The demands have been worked out on the basis of the current cropping patterns and water demands.

There are three crops over the year. The three cropping seasons approximately coincide with the three meteorological seasons: *Kharif* I (premonsoon), *Kharif* 11 (monsoon), and *Rabi* (winter or dry). *Aus, aman,* and *boro* are the three rice varieties grown respectively in these three cropping seasons. The crop calendar (grain-based) is presented in relation to temporal distribution of rainfall and temperature in Figure 7.5. The annual average area (thousand hectares) and yield (9 tons/ha) of Aus, Aman, and Boro for 2001–2005 are as follows: Aus (1204/1.48), Aman (5593/1.95), and Boro (3877/3.22) (WB, 2007). The monsoon season crops such as broadcast and transplanted local and HYV Aman are normally grown without irrigation. The winter food grain crops such as local and HYV Boro rice cannot be grown without

TABLE 7.2

Projected Water Supply and Demand, March 2018

	Water Requirements			Water Supplies	
	Amount (10⁶ m³)	Percent		Amount (10⁶ m³)	Percent
Agriculture	14,290	58.6	Main rivers	11,740	50
Navigation, environment and fisheries	9910	40.7	Regional rivers	6390	27
Domestic and industry	170	0.7	Ground water	5360	23
Total	24,370	100	Total	23,490	100

Source: World Bank, *India—Irrigation Sector Review*, Vol. 1 and 2, Agriculture Operations Department, India Department, Asia Region, 1991. With permission.

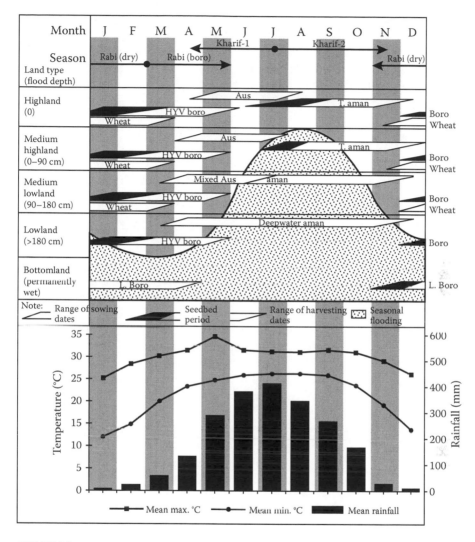

FIGURE 7.5
Crop calendar. (From Ahmad, Q.K. et al., *Ganges–Brahmputra–Meghna Region—A Framework for Sustainable Development*, The University Press, Dacca, 34, 2001. With permission.)

irrigation. The yields are poor. The figures impressively bring out the benefits of irrigation. Only about one-third of the cultivated area in Bangladesh is under irrigation.

The principal in-stream uses in Bangladesh consist of fisheries, navigation and salinity management, and sediment control. There are presently little agreement and few data on the stream flows and volumes of water actually required. Under the NWP, 40% of the stream flows in each planning area (excluding stream flow in the main rivers) during the critical months of February and March are allocated to in-stream uses.

Fisheries in Bangladesh flourish in beels, floodplain depressions, and ponds as well as rivers. Certain minimal conditions are required in each location for fish production to remain high. These requirements often conflict, however, with flood control development and irrigation strategies adopted to increase agriculture productivity.

Bangladesh's dense waterway network has traditionally provided a vital means of transportation. This is especially true in remote areas of the country where no roads exist and in many other areas during the monsoon season when alternative transportation networks become unusable due to floods.

7.5.3 NWP Development Strategy for Flood Control and Drainage

Flood control and drainage received attention primarily from consideration of increased agricultural production. Standards of increasing agricultural production were established for this purpose. The primary standard for project planning and formulation was considered to be the establishment of conditions for the adoption of HYV fertilizer technology. Where drainage requirements to satisfy such a standard cannot be met, schemes would be designed to establish conditions in which the annual depth, timing, and duration of flooding are within a definite range with a high degree of probability, enabling an increase in cropping intensity and a shift to higher production technology such as local improved varieties (LIV) fertilizer technology. The main emphasis under this strategy would be on gravity drainage schemes in shallow to medium-flooded areas and submersible embankments in deeply flooded areas. Greater emphasis shall also be given to other water uses and water management needs in floodplain areas, including fishery development.

7.6 Flood Management

7.6.1 Background

Water resources development in Bangladesh, formerly East Pakistan, got the initiative after the floods of 1954. In earlier reports of foreign consultants, flood management received considerable emphasis even if a cautious approach was stressed in view of the magnitude of floods and complexity of the problem. No heavy floods were experienced for some time after the 1954 experience. Emphasis, therefore, soon shifted to irrigation for food, self-sufficiency, and employment generation. In the NWP developed with assistance from the World Bank, finalized in 1986, flood prevention occupied a low priority. Three distinct goals were identified: (i) to minimize damage and destruction caused by catastrophic floods and storms, (ii) to provide safety to life and prosperity and to minimize damage and disruption of essential economic activities, and (iii) to increase agricultural protection through changes in crop type and cropping pattern.

The gross area protected by flood control and drainage projects has grown steadily since the mid-1960s to 1985 to 2.59 mha, through the construction of an estimated 6.130 km of embankment, 4.521 hydraulic structures, and 985 river closures. This area is equivalent to nearly 20% of the country.

However, as 1987 and 1988 floods showed, this protection was inadequate. The scene changed dramatically. A disastrous flood was experienced in 1987, followed by an even worse flood in 1988, the worst ever in living memory. The extent and complexity of these floods underlined the fact that total elimination of flooding is neither feasible nor a desirable agroecological perspective. However, the protection of people's lives and their places of habitation are absolutely essential as are the commercial and industrial centers. The Government set out "Eleven Guiding Principles" in 1988 as policy directive for flood management. These are (1) phased implementation of a comprehensive flood plan aimed at protecting rural structure and controlling flooding to meet the needs of agriculture, fisheries, navigation, urban flushing, and annual recharge of surface and groundwater resources; (2) effective land and water management in protected and unprotected areas; (3) measure to strengthen flood preparedness and disaster management; (4) improvement of flood forecasting and early warning; (5) safe conveyance of the large cross-border flows to the Bay of Bengal by channeling it through the major rivers with the help of embankments on both sides; (6) river training to protect embankments and urban centers; (7) reduction of flood flows in the major rivers by diversion into major distributors and flood relief channels; (8) channel improvements and structures to ensure efficient drainage and to promote conservation and regulation; (9) floodplain zoning where feasible and appropriate; (10) coordinated planning; and (11) construction of all rural roads, highways, and railway embankments with provision for unimpeded drainage.

The Government prepared a National Flood Protection Program and, with the assistance of UNDP, carried out a flood policy study. In addition, teams of specialists from France, Japan, USA, and China undertook independent studies of flood impact and mitigation.

Emphasis of each study was different and is worth noting. The French study represented one extreme of very high degree of flood control even from abnormal floods by heavy embankment construction. About 3350 km of high embankment at an estimated cost of US$10 billion with an annual maintenance cost ranging between US$200 and $600 million was recommended. The Japanese study also recommended heavy embankment of all the major rivers and some distributaries totaling to 1688 km and costing about $6.7 billion.

The U.S. study perception, philosophy, and approach were entirely different. It emphasized the complexity of the water resource–flood–environment scene, pointed out the difficulty of flood management through a "confrontational approach" to the floods by massive investments in embankments and river training, and advocated living with the floods as proposed in the seven guiding principles, as well as the NWP flood strategy. It was noted that "in few other situations does mankind have to deal directly with the raw power of nature as it does in Bangladesh" (Rogers et al., 1989). To give an idea of the size and power of the floodwaters and limitations of technology, attention was drawn to the vast changes in channel depth and configuration that occur rapidly on these rivers. For example, during the flood season of 1966, just downstream of Faridpur on the combined Ganges and Brahmaputra, the river moved 1500 m laterally in a northward direction, eroding the banks and digging a new channel 30 m deep. This amounts to about 100 million tons of sediment moved per river mile. During the 1988 flood, the outlet of the river system swung about 550 m eastward toward the protective embankment of the Chandpur flood control and irrigation project, where it dug a new channel 45 m deep. This embankment will now have to be abandoned and a new one built behind it. The problem was studied in a wider perspective than mere flood protection, and it was pointed out that "there are narrow limits to what technology can do to control the risks of living in the major floodplains" (Roger et al., 1989). Regarding flood management, the approach and detailed plans of the NWP, which had worked out a comprehensive flood plan in conjunction with the water resources development plan, was recommended. The water resources problem of Bangladesh were studied comprehensively, and several policy directions and programs of applied research were recommended, emphasizing that single-solution approaches should be avoided.

The Government–UNDP study adopted a more aggressive approach, opting for greater flood control to contribute to economic growth by ensuring a flood-safe environment.

The major aid donors to Bangladesh discussed the flood mitigation problem at the G-7 summit meeting in Paris in July 1989. It was decided that the World Bank should coordinate the efforts of the international community so that a sound basis for achieving real improvement in alleviating the effects of flood can be established. Accordingly, in 1990, the World Bank prepared an FAP (WB, 1990).

7.6.2 Flood Control and Drainage: Staged Approach

The essential features of the FAP are as follows. The objectives of the plan are to (1) safeguard lives and livelihoods; (2) improve agroecological conditions to increase crop production; (3) enhance development of public facilities, commerce, and industry; (4) minimize potential flood damage; (5) create flood-free land to accommodate the increasing population; and

(6) meet the needs of fisheries, navigation, communications, and public health.

In this context, the Government established the "Eleven Guiding Principles" summarized as follows: (1) phased implementation of a comprehensive flood plan aimed at protecting urban infrastructure and controlling flooding to meet the needs of agriculture, fisheries, navigation, urban flushing, and annual recharge of surface and groundwater resources; (2) effective land and water management in protected and unprotected areas; (3) measures to strengthen flood protection and disaster management; (4) improvement of flood forecasting and early warning; (5) safe conveyance of the large cross-border flows to the Bay of Bengal by channeling it through major rivers with the help of embankments on both sides; (6) river training to protect embankments and urban centers; (7) reduction of flood flows in the major rivers by diversion into major tributaries and flood relief channels; (8) channel improvements and structures to ensure efficient drainage and to promote conservation and regulation; (9) floodplain zoning where feasible and appropriate; (10) coordinated planning and construction of all rural roads, highways, and railway embankments with provision of unimpeded drainage; and (11) encouragement of popular support by involving beneficiaries in the planning, design, and operation of flood control and drainage works.

The framework for the Government's long-term plan of physical works together with measures to improve preparedness and management of floods is shown in Figure 7.6. In physical terms, a long-run objective is to protect large areas from uncontrolled overbank flooding from the major rivers along with measures to evacuate excess rainfall from the protect areas. Physical works would include structures to control inflow to some areas and flood relief channels to intercept and divert flood flows. The northwest and north central regions (Figure 7.4) are the parts of the country most susceptible to overbank flooding, mainly from the Brahmaputra, and therefore, most of the flood control works would tend to be concentrated in these areas. A large area in the northeast is flooded every year to a considerable depth. There is no practical solution to this monsoon flooding, so the farmers have adopted a system that concentrates on dry season, crop production for which limited protection against early flooding can be provided by submersible embankments. Parts of the southwest and southeast are affected by river flooding from Ganges and the Lower Meghna, but the main problems in these areas are drainage congestion, salinity inflows in the dry season, and exposure to cyclones. The south central part is affected by flooding, from Padma, which might be alleviated once the flow of the Ganges, the Brahmaputra, and the upper Meghna is brought under control. Drainage congestion at the confluences adds to the problem of flooding which will also have to be addressed.

The action plan covering the 5-year period, 1990–1995, was the first of several stages in the development in the comprehensive system of flood control and drainage.

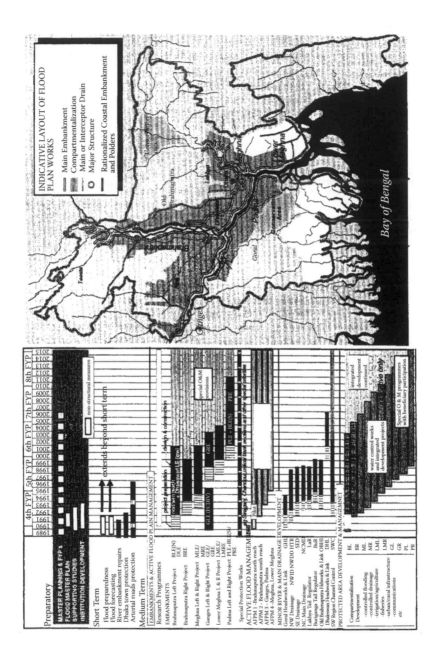

FIGURE 7.6
Indicative layout of flood plan works (WB, 1990). (From Bangladesh Flood Policy Study. UNDP - Govt. of Bangladesh, May 1989.)

7.7 Water Resources Management in Bangladesh: Steps toward a New National Water Plan (NNWP)

Water resources development has been an issue of major concern since the formation of then East Pakistan, and since then, numerous plans have been made. Steps are being taken to develop an NNWP (WB, 1998). It is briefly as follows.

7.7.1 General Issues

Water resources planning in Bangladesh has to take into account a variety of geographic, economic, and environmental factors. Bangladesh is a lower riparian country. Most of it is locked within the floodplains of three great rivers—the Ganges, the Brahmaputra, and the Meghna—and their tributaries and distributaries. The river systems drain a total area of about 1.72 million km^2 in India, China, Nepal, Bhutan, and Bangladesh. Only 8% of the catchment lies within Bangladesh. As a result, huge inflows on water over which Bangladesh has no control enter the country. The lack of control is a critical problem because Bangladesh has an agrarian economy dependent on water. At different and unpredictable times, it has too much and too little water. The intricate network of alluvial rivers carries a huge annual discharge and sediment load, causing channel shifting and bank erosion. Withdrawals in upstream areas seriously affect socioeconomic growth, the environment, and ecology. The habitat of fish, a major source of protein for the poor, is under threat from the increasing conversion of land to agricultural use. Inland navigation is hindered by blockages in the river delta. Meanwhile, the need for pure water is increasing along with salinization of the coastal belt and the degradation of ecosystems. The study, focusing on steps to be taken within Bangladesh in preparation for the water plan, leaves out one critical issue—the regional context.

The critical problems that any water planning system will have to address include the following:

- *Flooding and drainage congestion.* The annual minimum low of rivers near the Bay of Bengal is only one-twentieth of rivers' peak monsoon discharge. During the monsoon season, water is slow to drain, and banks overflow. Flood damage from the Ganges, Meghna, and Brahmaputra is catastrophic, and after the peak season, water continues to impede drainage.
- *Drought.* Parts of Bangladesh suffer droughts during the year that adversely affect agriculture and other economic sectors.
- *Siltation.* Bangladesh is the outlet for all the major rivers upstream of it. The average sediment load that passes through the country to

the Bay of Bengal is 0.5 to 1.8 billion tons. Part of the load is deposited on the floodplain, gradually changing its topography and seriously reducing the carrying capacity and navigability of drainage channels.

- *Riverbank erosion.* Large seasonal variations in river flow and gradual shallowing of riverbeds cause bank erosion and channel migration. Between 1982 and 1992, roughly 87,000 ha was lost to erosion, and thousands of people lost their homes and agricultural land.

- *Salinization.* Environmental degradation caused by the intrusion of salts is a major problem in the southwestern region of Bangladesh. The process has been exacerbated by the Ganges's reduced flow during the dry season because of diversion of water by the Farakka Barrage by India. Since 1975, this reduction has caused the salinity front and tidal limits to move northward, threatening urban and industrial areas like Khulna and the economic viability of over 25,000 km^2 of area dependent on the Ganges.

- *Pollution of surface and groundwater.* Unchecked dumping of industrial and municipal waste and agricultural chemicals has contaminated both surface water and groundwater. Pollution is fast becoming a serious problem for the human population as well as aquatic life.

7.7.2 Water Management in Wet Season

Floods are a recurring phenomenon in Bangladesh. A detailed coordinated action plans to deal with floods was developed with the support of international agencies and governments, as discussed in Section 7.6. Some controversies have emerged about the proposed flood management plans.

7.7.2.1 Shortcomings of Flood Action Plan Studies

Embankments for containing floodwater within river channels have been criticized for the following reasons: loss of land due to acquisition, breaches in embankments due to river erosion and ultimate failure of the investment, adverse effects on fish in the floodplain, and the possibility of worsened flooding downstream.

In most FAP studies and projects, the role of sediment and subsiding land was either completely ignored or marginally considered, although they play an important role in preserving the delicate balance of the floodplain and the delta.

Most of Bangladesh is still going through active land building. Sedimentation on the floodplain, tidal flats, and delta front and accompanying channel shifting are all parts of this process. A considerable part of the country is subsiding because of tectonic adjustments in the surrounding areas, including the Eastern Folded Belt, the Himalayas, and the Indian plate. Normally, the sediments deposited by the major rivers partly compensate for the subsistence, but river embankments and polders may prevent this, leading to water logging in the polders and between river embankments. In FAP studies, subsidence was not carefully considered in engineering solutions to the flood problem.

Although the studies focused on flood mitigation, they later shifted to integrated water management. The resulting management plan did not properly account for the role of groundwater, and thus, full water potential was not determined.

Water management and environmental management were not integrated in FAP studies. No acceptable method emerged to determine the water requirements for maintaining the ecosystems during the wet and dry seasons. The FAP addressed environmental concerns mainly by preparing a manual and guidelines and recommending environmental impact assessments for most environmental concerns.

Most FAP studies and projects excluded stakeholders from discussions of planning options. Local constituents were not fully consulted, and the process tended to be confrontational.

The components of the FAP were not sufficiently synchronized. All supporting studies should have been considered together with the pilot projects before regional studies were undertaken.

7.7.2.2 Flood Damage Mitigation: Options for Bangladesh

Bangladesh needs to adopt a balanced rational approach that combines both structural and nonstructural measures. Moreover, flood control and mitigation cannot be considered apart from the other issues of water resources development. They should be part of an integrated water management plan embracing environmental concerns as well as various sectors of the economy such as navigation and fisheries.

Three lessons can be drawn from humankind's protracted struggle against floods. First, there are some palliatives from the problem of flooding, but no full cure is available. Flooding cannot be tamed by ad hoc or short-term measures. Therefore, it is necessary to draw up long-term plans to mitigate the problem. Second, the adverse effects of floods are compounded by myopic human actions. Unless undesirable human interference with nature is halted, floods are likely to increase in intensity and frequency. Third, complete flood control is neither feasible nor desirable. Flooding is nature's way of maintaining and rejuvenating floodplains and deltas.

Without detailed studies, it is difficult to say what method of flood mitigation would be appropriate for which part of the country. However, it is safe to suggest embankments, designed for controlled flooding, for major rivers such as Ganges and Meghna as well as for many small rivers. Over 40,000 km of roads, including rural roads and embankments on which they are built, would have to be integrated into the flood protection structure.

Riverbank protection and river training will be required for rapidly eroding areas like the banks of Brahmaputra and Meghna. There, the "hard point" solution of the FAP could be evaluated to manage segments of rivers where major erosion is taking place. For floodplains, compartmentalization combined with controlled flooding and flood proofing, may be appropriate. In coastal areas, polder projects should be modified to work with controlled flooding measures. In some coastal areas, flood proofing may be a viable option.

7.7.3 Water Management in Dry Season

7.7.3.1 Balancing Demand and Supply

The projected water supply and demand, based on a March 2018 estimate worked out by the 1991 NWP, are given in Table 7.2. It has been considered that there are shortcomings in the developments of water resources development and accordingly in the estimates proposed in Table 7.2, which need to be rectified (WB, 1998). Some suggestions are as follows:

1. Re-estimating and updating surface flows on the basis of the Indo–Bangladesh joint water accord.
2. Estimating alternative flows from India on the basis of water released from India.
3. Estimating alternative groundwater technology on the basis of different extraction technologies.
4. Updating data on cropping and irrigation coverage as well as existing water use for irrigation, fisheries and navigation, salinity control, industrial water use, and urban and rural drinking.
5. Projecting long-term requirements for both groundwater and surface water in each sector, on the basis not of availability but of desired levels of development. To account for overlapping uses, care must be taken in the demand model to integrate absolute consumption (as in irrigation) and use (as in navigation and fisheries).
6. Generating data on the quality of groundwater and surface water, and on the sources of pollution, in individual planning units.
7. Preparing alternative pictures of water balance and water availability for development, based on the preceding steps, for the next 25–50 years.

8. Developing downstream institutions (in villages and thanas) for managing use by different constituencies such as farmers and fishers. These would be democratic institutions involving users in joint decisions.

In addition to the above steps, public water institutions will require fundamental change, as will be discussed later.

The dry season in Bangladesh, which normally lasts from November to April, is characterized by a drastic reduction in the discharge of major rivers, drying of water channels, falling water tables, and salinity encroachment, particularly in the southwest. This is the season of greatest demand for groundwater, which has powered the production of cereal production over the last decade.

7.7.3.2 Water Availability in Dry Season

From December through March, when rainfall is low, Bangladesh has to depend on other sources of water. Considerable scarcity has been felt regarding surface water and groundwater.

SURFACE WATER

The 1996 water accord with India is likely to give Bangladesh a minimum of 35,000 cusecs or 50% of the water available at Farakka if the Ganga's flow is less than 70,000 cusecs during the dry season (Table 7.3). This should increase the availability of surface water in the dry season and create a new water balance.

The Government of Bangladesh is considering building a barrage on the Ganges at Pangsha to manage water in the southwest region. The project has a wide range of objectives aimed at sustainable development of the southwest region and long-term environmental improvement.

GROUNDWATER

There are two interrelated issues concerning groundwater. The first is the basic question of availability. The second issue is arsenic. Has arsenic become a problem owing to overuse of groundwater, and does arsenic force

TABLE 7.3

Main Features of Indo–Bangladesh Treaty on Sharing Ganges' Water

- The treaty is for 30 years and covers the period from January 1 to May 31 each year.
- India is to make every effort to maintain inflows to Farakka at or above the 40-year average.
- Bangladesh is to receive a guaranteed minimum of 35,000 cusecs of water as long as availability at Farakka does not fall below 70,000 cusecs.
- Not more than 200 cusecs is to be drawn off by India downstream of Farakka.
- India is to release not less than 90% of Bangladesh's entitlements at all times.

a limitation on the use of groundwater for irrigation so that a supply can be maintained for drinking?

There are differences in estimates. According to current evidence, in the 1994–1995 irrigation year, a total of 2.2 mha was irrigated by groundwater, and there was enough to irrigate an incremental 5 mha. Thus, it is considered that Bangladesh agriculture can and will largely depend on groundwater irrigation. The economics of groundwater irrigation further support the view.

7.7.3.3 Water Demand in Dry Season

The principal use of water in Bangladesh is for irrigation, which has increased significantly since the mid-1970s. With only about one-third of the cultivated area in Bangladesh presently under irrigation, this trend is expected to increase. At the same time, the demand for other uses will increase with population, urbanization, and economic development.

Salt water is a problem that tends to increase as flow diminishes, especially in the southwest region of Bangladesh. With the drastic fall in surface flow during the dry season, the coastal areas of Bangladesh face three types of saline water intrusion: in streams, groundwater, and soil.

Another critical issue for dry season water management is silting in river channels. Estimates have varied, but about 1.67 million tons of suspended sediment is discharged annually through the Ganga–Brahmaputra Rivers in Bangladesh annually. Bed load sediments are extra, which may account for as much as 50% of the total sediment load.

Then, there are the dangers posed to fisheries and navigation. Fish are a vital source of protein for Bangladeshis, especially those in rural areas, and plans to increase fish production are dependent on an assured supply of water. A sure supply is also critical for navigational waterways, which must have a minimum depth of flow to accommodate boat and ship traffic.

To manage water resources more effectively in the dry season, a balanced set of policies and institutional reforms should be sought that will harness market forces and at the same time strengthen the capacity of the government to carry out its essential role. The emphasis should move away from developing new water supplies toward comprehensive management, economic discipline, policies to overcome market and government failures, and incentives to promote more efficient water use.

7.7.4 Improving Water Management through Institutional Development

Institutions and the way they are set up determine the long-term ability of a nation to manage its water resources. Water policies and strategies express the nation's priorities and how those priorities should be achieved: institutions are the tools for realizing them. The institutions that deal with water in Bangladesh with water resources have serious flaws. On the basis of extensive dialogues with experts both inside Bangladesh and out, as well as with open forum discussions with the public, suggestions for improvement have been made (WB, 1998).

7.8 Structure for Future National Water Management System

To address deficiencies in current water management, detailed suggestions have been made (WB, 1998). Meeting the future challenges calls for new policies, institutions, and programs. A bird's eye view follows (WB, 1998).

7.8.1 First Step: Developing Policy and Institutional Framework

An NWP is a public sector investment program based on policies whose aim is to develop and manage water supply and demand. Bangladesh will have to shift its focus toward integrated management. The first step would be developing a policy for an NWP and designing institutions to support it. This would be followed by an investment program.

The first step is to develop an appropriate national institutional setup. Detailed suggestions have been made (WB, 1998).

Formulating a national water policy, specifying institutions, and encouraging the participation of stakeholders are prerequisites of an NWP. Policies must take into account user's rights, allocation principles, quality and pollution control, pricing principles, and cost sharing for private users.

A water code and rules and regulations for implementation need to be prepared right after policy framing. Conflicts among users, though not common now, are bound to surface sooner or later. The code should establish the principles and framework for appropriating, controlling, and conserving water resources; define the rights and obligations of users; and identify administrative mechanisms for enforcing the code. The policy and the institutions supporting it would serve as the framework for the water code and its implementation. The National Water Council will have to approve the code and procedures for its implementation.

The water policy will have to be updated periodically. For instance, the water policy of the Netherlands is updated every 5 years and periodically comes before national parliament for approval. Periodic updating and review will require an institutionally stronger WARPO. Indeed, one important direction for the future is capacity building, something that has not received due attention.

Tailoring institutions to the needs of water management is a critical part of the government strategy. All major public institutions have to refocus their objectives and develop the capability for joint management of the water projects with private users. The new organizational culture should view water as a common property to be developed, used, and reintroduced into the ecosystem through joint endeavors. This concept should be an integral part of the institutional design, supported by appropriate legal and regulatory provisions and clearly defined responsibilities.

There will have to be a way of ensuring the involvement of local people in all stages of the project—planning, design construction, operation, and maintenance. Public participation requires institutional encouragement, genuinely from the bottom up and as transparent as possible.

7.8.2 Second Step: Preparing a New National Water Plan

An NNWP has to be formulated in light of the suggestions received and further spelled out (WB, 1998). The NWMP, which was slated for completion by 2001, would give Bangladesh a long-term investment program with a specific portfolio of projects. To ensure integrated water resources management, major projects should be taken up only after the plan is completed. However, in the meantime, Bangladesh should complete the high-priority projects that were identified for early implementation in the Bangladesh Flood and Water Management Strategy of 1995.

7.9 Service Sectors

7.9.1 Irrigation and Drainage

Currently, about 4.5 mha of the total 8 mha of cultivable land is irrigated, and the area under irrigation contributes an estimated 13 million metric tons of cereals (mainly rice) annually to Bangladesh. About 90% of this irrigation is provided by the private sector, mainly from groundwater. Diesel is used to power about 90% of the pump sets. It is estimated that a further 1 mha could be brought under irrigation through groundwater development. The NWMP suggests that government has a role to play in the development of surface water irrigation systems, where feasible, and to focus on the conjunctive use of groundwater and surface water (WB, 2005). Further gains are also possible by diversification to crops requiring less water. However, the large-scale surface water irrigation projects have not performed well.

7.9.2 Water and Sanitation

Nearly 97% of the rural population is served by over 10 million hand tube wells. On average, more than half the urban population has access to water supply. The four largest cities, Dhaka, Chittagong, Khulana, and Rajashahi, have piped water systems that serve 70%, 33%, 51%, and 40% of the population, respectively. Almost 40% of the population has low income, and this group is largely unserved. In addition, only 100 of over 250 municipal towns have piped water systems, and these primarily serve the core urban population. The urban population in the slums and fringes of medium and small towns rely on hand tube wells. Nevertheless, squatters and those living in urban slums are without easy access to water and sanitation.

Sanitation coverage is even worse. Recent surveys place sanitation coverage at 33% in rural areas and slightly over 50% in urban areas. The popular technology in rural areas is the pit latrine, with or without a water seal. The sanitation coverage in large towns and cities is between 65% and 76%. Pit latrines and septic tanks are the technology of choice in urban areas. Poor drainage decreases soil permeability, making dense urban areas unfit for on-site sanitation options. Pits fall too fast, and effluent from septic tanks flows into open drains, aggravating environmental pollution. Only Dhaka has a waterborne sewerage system, and it serves only about 27% of the population.

7.9.3 Inland Water Transport

Transport accounts for about 8% of the GDP, and water transport accounts for about 15% of the total transport GDP. Inland water transport moves about 30% of the total freight and 14% of the passengers, compared with 7% and 13% for rail and 63% and 73%, respectively, by road. Water transport depends entirely on adequate water depths and is thus particularly vulnerable to the effects of siltation, declining river flows, and flood management infrastructure constructed across watercourses. Inland water transport is the least expensive form of transport in Bangladesh and, in the southwest part of the country, most cost effective. There are presently about 6000 km of navigable waterways in the summer and 3800 km in winter. This is significantly lower than the figures in 1970, when the navigable waterways during the summer were about 13,500 km and the winter waterways were 8500 km in length. An estimated 3 million people rely on inland water transport for their livelihoods.

By 2015, the actual volume of inland transport freight and number of passengers is forecast to increase by 70% and 50%, respectively, even as its share of total freight and passengers is forecast to decrease to about 17% and 8%, respectively (notwithstanding its cost advantages and livelihood benefits).

Waterway maintenance will be essential, given the expected rapid increase in actual freight and passengers. However, it has declined.

In promoting development and more efficient resource utilization in the sector, a number of challenges need to be addressed. The navigability of many of the rivers is declining due to a combination of reduced dry season discharges and siltation in the channels. The response has been less than effective because of poor management.

7.9.4 Fisheries and Environment

The fisheries sector accounts for 6% of the GDP, providing 9% of the country's employment, and shrimp export is the second most important source of foreign exchange. Fish account for 65% of the country's animal protein consumption.

The four major fish subsectors with their yield (million tons/year) are inland aquaculture (850,000), brackish water aquaculture (94,850), inland capture fisheries (750,419), and coastal/marine capture fisheries (589,500), for a total of 2,284,499 in the year 2002.

The fish habitat has to be carefully managed. Unfortunately, the rivers around all major cities are heavily polluted. Wetlands provide essential environmental services to all Bangladeshis in the way of flood protection, water purification, and groundwater recharge, among other services. In addition, wetlands serve as nursery and breeding grounds supporting fish production. Wetlands seasonally cover 60% or more of the country. Permanent wetland areas in the country have declined by 50% and continue to decline. Current development practices do not aggressively support the maintenance of the wetlands and revitalization of rivers around cities (WB, 2005).

7.10 Arsenic Challenge

Besides the challenge of managing the surface waters, the groundwaters of Bangladesh are posing extremely serious problems. It has been found that the groundwaters of almost the entire country suffer from a high degree of arsenic contents. High levels of arsenic (over 0.05 mg/l) in groundwater have been detected in 59 of the 64 districts of Bangladesh, especially in the southern regions. Arsenic in groundwater has affected an estimated 35 million people both in urban and rural water areas. Between 10,000 and 12,000 cases of arsnicosis have been identified so far, and undoubtedly, many unidentified cases exist. Figure 7.7 provides a profile of the arsenic occurrence.

This has serious implications for the domestic water supply as well as for the agriculture sector because of transfer of arsenic into the food chain through irrigated crops. This has been confirmed by recent studies (HT, 2007).

The Bangladesh government has launched a 4-year Bangladesh Arsenic Mitigation Water Supply Project (BAMWSP) with a view to providing arsenic-free water supply to rural and urban communities. The US$44.4 million project is funded by the World Bank and the Swiss Development Corporation in (i) identifying the causes of arsenic contamination, (ii) determining alternate sources of water supply, (iii) awareness building about the

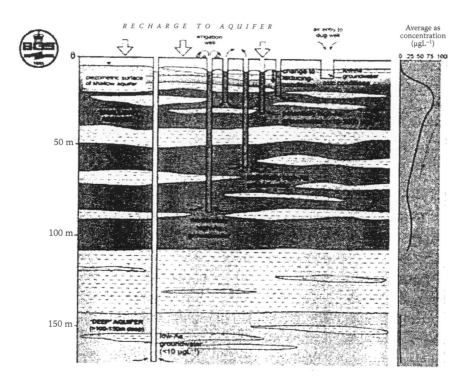

FIGURE 7.7
Profile of arsenic occurrence in groundwater. (From World Bank, *Bangladesh Country Water Resources Assistance Strategy*, Bangladesh Development Strategies, Dhaka, 2005. With permission.)

arsenic hazard, and (iv) preparing detailed proposals for a national program for arsenic mitigation (Ahmad et al., 2001).

7.11 Environmental Management

Environmental management, so far, as per usual practice, has been considered only in terms of water resources development and flood control. It must be undertaken in totality, as use of environmental vectors as resources is important for economic growth, particularly in poverty eradication; their conservation is also vital for economic growth and even more so in terms of adverse impacts to the poor, who are the most vitally affected by the adverse effects. Some issues that have been particularly concerned of late are briefly discussed. More comprehensive consideration is, however, required as emphasized (WB, 2006).

Natural resources productivity is under threat. With population density approaching 1000 people/km², among the highest in the world, and growing at 1.7% annually, pressure on natural resources in Bangladesh is necessarily very high. Two-thirds of the land area is under crops, the highest proportion being in South Asia, and the share of forest cover is the second lowest in the region. Protected areas cover the smallest share of any country in South Asia, at less than 0.5% of the total area compared with an average of nearly 8% for Asia as a whole, although pressure on biodiversity is close to the regional average, with approximately 17% of the mammal species and 4% of the bird species under threat. Pressure on wetlands and aquatic life is also a particular concern in Bangladesh, as booming urban and industrial growth leads to land reclamation and pollution, and fishing efforts increase to meet the growing demand, with fish continuing to provide more than half the animal protein in the national diet.

Agriculture accounts for some 23% of the GDP and more than half of the total employment, so the ability of soil to sustain agricultural production is an issue of national concern. Cropping intensity in Bangladesh is the highest in South Asia, having increased from below 140% to 175% over the last 30 years, as the introduction of the short-duration cereal varieties and winter irrigation facilitated double and triple cropping. Pressure on agricultural land is further intensified by urbanization and infrastructure development, as a result of which the cropped area is declining at about 1% annually. To support this intensity of cultivation, rates of fertilizer application became the highest in the region; nevertheless, surveys and soil testing provide evidence that agricultural inputs are imbalanced and that nutrient mining is occurring.

Bangladesh has limited natural forest cover, at about 10% of the land area, down from 20% in the 1960s. Large areas of the Chittagong Hill Tracts have been degraded, the Sundarbans mangrove is in decline, and most parts of the plain land Sal forest are now shrublands. The deforestation rate in the 1980s was 3.3% per year, but in 1989, the Government issued a moratorium on felling in natural forests, which continues to date. As a result of this ban and social forestry initiatives, official figures indicate a net rate of reforestation of about 1% annually during the 1990s. The moratorium also stimulated private sector timber production, which now accounts for over 80% of local timber marketed in Bangladesh, and satellite imagery suggests that as much as 35% of the land area has 10% or more tree cover of all types, including private plantation.

Water quality is a key determinant of outcomes in many sectors of the Bangladeshi economy, from industry and transport, through agriculture and fisheries, to human health. Much emphasis has therefore been given to the management of water quantity, but the importance of water quality has been lacking. It has particularly significant impact on fisheries and human health. The urbanization poses a very serious challenge for provision of water supply for drinking and sanitation.

Soil conservation or management has not been considered, as soil was abundantly renewed every year by the annual floods and often by the high floods. Salinity intrusion at the water's edge is, however, recognized as a serious problem. The problem of salinity intrusion has particularly been faced on the western side by natural and man-made activities. River Ganges used to flow down to the sea on the western side through River Bhagirathi, now in India, until recently when it shifted to its present course. The Ganga–Kobadad Project in Bangladesh and the Farakka Project in India were envisaged long back in undivided India to counter this problem. The former was developed both to irrigate the moribund delta area of Kushtia, Jessore, and Khulna and to resist salt water penetration into their cultivable land (Huque and Rahman, 1994). The Farakka Barrage Project was developed to divert water in Hooghly to maintain the navigability of the Calcutta Port. Although the feasibility studies of both were undertaken in the beginning of this century, construction was started and projects were completed only after partition, leading to international disputes regarding the latter activity. This has affected the management of the salinity intrusion problem.

This variation of hydrologic conditions in the coastal areas causes continuous salinity fluctuations over time and space, affecting about 26 million km^2 (Nishat and Chowdhury, 1987). Flows of the Ganges, Brahmaputra, and Meghna moderate salinity shifts. The location of outflows has varied on account of natural and man-made events. For instance, the main flow of Ganges was on the western side until about two centuries back. On account of its shift to the east, the salinity intrusion is much higher on the west, which is further exacerbated by diversion of flows at Farakka. There have been little efforts to manage the salinity problem other than the construction of polders.

Air quality, particularly in the urban areas, is a matter of increasing concern. Up to 10% of respiratory infections and diseases in Bangladesh may be attributable to urban air pollution. While the problem is most severe in Dhaka, both because air quality is worse and because more people are exposed, air pollution is a growing concern in other major cities. A much stronger program is required than that of the present. Such actions should focus on gross diesel polluters, fuel quality, and the expansion of air quality monitoring.

7.12 Vulnerability to Climate Change

Climate change is an extremely serious threat to humanity. The GBM basin is particularly threatened, as it will increase the already high risk of disasters and exacerbate existing vulnerabilities. Even in the GBM basin, Bangladesh is most severely vulnerable.

Global warming will cause changes such as higher temperatures, sea level rise, and changing rainfall patterns, as well as abrupt effects such as increase in the intensity and frequency of extreme events such as floods, storm surges, and cyclones. An increase of 1°C in ocean temperature could increase tropical cyclone intensity by 10% (WB, 2000). Temperature alterations associated with climate change are already affecting the rate of snowmelt in Himalayas, which is expected to lead to increased flooding. It is predicted that by the year 2030, an additional 14% of the country will become extremely vulnerable to floods, and currently vulnerable areas will experience a higher level of flooding. Indeed, significant areas may be permanently inundated (BCAS and DOE, 2001). At the same time, some areas of the country may be at greater risk of drought and food insecurity during the dry season, and agricultural productivity in coastal areas may be compromised by increasing salinity. There is the serious danger of large parts being submerged by increasing sea levels.

7.13 Management in Long-Term Perspective

A long-run perspective of the society has been undertaken, and management of water in this context has been discussed (WB, 2007). It is considered that commitment to rapid economic growth, as achieved by many high-performing economies, is imperative if the trap of poverty, exploding population, urbanization, and economic degradation is to be avoided. Nearly half the population lives in abysmal poverty, and only 20% is functionally literate. Creating a viable vision of the future is the first task in this context.

Population control is the first challenge. The stark demographic reality is the driving force of the future policy. Having to accommodate 1100 persons/km^2 and providing them with a decent living, jobs, and sustainable environment are a daunting challenge. Its population of 813 persons/km^2 is already one of the highest. Dhaka, with over 125,000 people/km^2, is ranked as the third densest urban center in the world—after Hong Kong and Nigeria. Unless the population challenge is met, all dreams of a sonar Bangladesh will be shattered.

Improved governance will be critical to the future. From the current scene of a government that is apathetic, nonfunctioning, and unaccountable, one that is effective, responsive, and accountable has to emerge urgently. This is particularly important in the public sector. It will not be able to meet the challenges unless it is managed by trained and committed personnel. Efficient and responsive local governments are urgently required.

Development of the human sector is another priority area. Health, education, and basic amenities are the most urgent areas of concern. Growth is the surest remedy for poverty. Rapid economic growth is a basic imperative, and development of people's potential and capability is of utmost importance in this context.

Where will that élan vital come from? Not from agriculture. It contributed to 28% of the country's growth. The share has already dropped to 11% and is still dropping. Seventy percent of the country's land is already under cultivation. Yields could be increased through greater use of irrigation, but the key to stronger performance in this sector lies with rapid and more extensive introduction of new technologies and exploration of opportunities for growing livestock, fish, fruits, and other products for both domestic consumption and export.

An estimated 10% of this land suffers from significant environmental damage. An equal amount on the hillsides is highly eroded. Roughly 3 out of 5 acres of the rainfed cropland is considered vulnerable to droughts.

A slower but no less serious environmental damage is endangering the productive wetlands and the fish grown on them.

7.14 Conclusion

Bangladesh faces one of the most severe challenges of environmental management. A revolution is needed to meet the formidable challenges.

Notes

1. The study is based on the World Bank studies with additional inputs from Ahmad et al. (2001) and other references cited, to whom acknowledgment is made.
2. The regional units shown in Figure 7.4 correspond to flood studies discussed later and are slightly different from those of the NWP studies for monthly river flows discussed in Section 7.5.1.

8

Indo–Nepal and Indo–Bhutan Interaction

8.1 Introduction

Nepal and Bhutan are the uppermost riparians in the GBM basin. Nepal straddles the Himalayas, with water flowing down the four major tributaries, Kosi, Gandak, Ghagra, and Mahakali, which form the boundary to meet the Ganges closely downstream. Indian state regions of Uttaranchal on the west and Sikkim on the east in the Himalayas flank it. Bhutan is comparatively very small, and the tributaries discharge into Brahmaputra, flowing through the states of West Bengal and Assam in India. Again, it is flanked by the Indian states of Sikkim and Arunachal Pradesh on the west and the east, respectively. The environmental and the environmental perspectives in Himalayas are almost the same irrespective of different political sovereignty.

Water is the dynamic environmental interlinkage between the upstream and downstream countries. Environmental processes do not recognize man-made boundaries and perceptions. The resource availability and the environmental impacts, positive and negative, are not equitably distributed spatially. It therefore becomes necessary for the riparian countries to arrive at agreements to develop and conserve the environment. We consider Indo–Nepal and Indo–Bhutan interactions from both environmental and socioeconomic considerations, with emphasis on the former.

The coverage of the Indo–Nepal interactions is in the following sequence. The socioeconomic scene is first reviewed from the context of interrelations. This is followed by the agreements already arrived at and developments undertaken. Next, the development under consideration or as perceived under various activities is studied. This is followed by perceptions of developments in the two riparian countries, considered separately. The same scheme is followed for the Indo–Bhutan interaction, but the coverage is brief.

The environmental dynamics and developments in Nepal and Bhutan also affect Bangladesh. The subject of trilateral interaction is covered in the Indo–Bangladesh interaction study in Chapter 9. Nepal, Bhutan, and India also have riparian interactions with China, as some tributaries arise in Tibet. This issue is almost inconsequential, has not arisen so far, and is therefore not considered.

8.2 Indo–Nepal Interrelation

8.2.1 Indo–Nepal Relations

India and Nepal were one entity, with varying political control of the two regions. Mahatma Buddha was born in the present Nepal. Hinduism was the dominant religion of the region. By some measures, Hinduism is practiced by a larger majority of people in Nepal than in any other nation. Buddhism, though a minority faith in the country, is linked historically with Nepal. Indo–Nepal relations have been very variable. Nepal was a British suzerainty until India's independence, and the early work on Sarda (Mahakali) was undertaken without any difficulty.

Indo–Nepal relations were extremely intimate and cordial after Independence. India was instrumental in the restoration of the monarchy and in the starting of the democratic process. A friendship treaty was signed between the two countries in 1950. India has been giving considerable financial aid to Nepal.

The warm relations continued until the death of King Tribhuvan in 1955. Conditions between the two countries, however, remained friendly up to 1960, when there was a royal coup in Nepal. The first popular Government was dismissed, the Prime Minister and his colleagues were imprisoned, the constitutional system of parliamentary democracy was suspended, press was controlled, and direct rule by then King Mahendra was imposed. To sustain aggregate and legitimate power, King Mahendra painstakingly and vigorously pursued four sets of strategies, one of which was the mobilization of international diplomatic and economic support for his domestic policies and programs. Relations with India, which had expressed its sympathies and support for the democratic forces in Nepal, became the worst casualty of this strategy for several domestic and diplomatic reasons (Muni, 1986).

Some disconcerting issues from India's point of view were severely strained trade relations between India and Nepal, introduction of work permits for all foreigners in Nepal, Nepalese importation of some arms from China, and and Nepal's awarding of a contract to China to complete the last segment of its east–west highway. India's refusal to recognize Nepal as a zone of peace was disconcerting from Nepal's point of view. At the bottom of it was Nepal's psychological need to assert its independence and equality in contacts with India. Absence of a free press in Nepal until 1990, when a popular ministry was reinstalled, tended to vitiate the atmosphere. From the Indian point of view, it seemed as if that "at times almost an orchestrated attempt was made to denounce India" (Kaushik, 1992).

With the installation of a popular ministry in 1990, the situation appears to have improved considerably. The Indo–Nepal Subcommission for Water Resources Development met in December 1988 and April 1991. After the Indian and Nepali Prime Minister's, meeting in June and November 1990 and again in February and December 1991, considerable progress appears to have been made concerning water resources development projects. In December 1991, it was agreed to (1) finalize the project parameters for the Karnali Multipurpose Project, (2) prepare a detailed project report for the Pancheshwar Project (3) carry out expeditiously the investigations to finalize the parameters of the Kosi High Dam Project, and (4) extend the embankments at the border along Kamla, Bagmati, and other rivers (Chitale, 1992).

As pointed out by Muni (1986), the future of Indo–Nepalese economic relations would not depend on economic logic or rationality. Vested interests with powerful political links, even those with foreign connections, vitiate an objective evaluation of mutual benefits. Thus, the nature of political dynamics becomes a major factor in India–Nepal bilateral relations. Those dynamics are not conditioned by the geographic, sociocultural, and economic imperatives alone. In a more significant way, these dynamics are affected by mutual fears and suspicions, which are rooted in the nature of domestic political concerns in Nepal and strategic perspectives in India.

Introduction of Article 126 in the constitution is a case in point. In the wake of the visit of Nepal's Prime Minister, Mr. K.P. Bhattarai, to India on June 10, 1990, in the joint communiqué, it was stated that "the two leaders declared their common intention to usher in a new era of cooperation, particularly in the harnessing of waters of the common rivers for the benefit of the two peoples and for the protection and management of the environment." The use of the expression "common rivers" raised a storm of protest in certain quarters in Nepal, and Mr. Bhattarai's government was accused of selling out Nepal's rivers to India. The upshot was that the constitution committee then working on the draft of a new basic law for Nepal added an Article 126, which lays down that any treaty or agreement on several matters, which includes "natural resources and distribution in the utilization thereof," shall be done by a two-thirds majority if it affects "the country in a pervasive grave manner or on a long-term basis." The article is so widely cast that any activity in the environmental area can be argued to affect the country in a "pervasive manner or long-term basis" requiring two-thirds majority, which will be very difficult to obtain.

The Tanakpur controversy, described briefly, following Verghese and Iyer (1992), is a case in point. The Tanakpur Project, located a little above Sarda Canal Headwork, is constructed to generate hydel power and divert water for irrigation in Nepal. Of the hydel power generated, 10 million units shall be given to Nepal free of charge. Provision has been made to release up to 28.3 m³/s (1000 cusecs) for irrigation in Nepal. Release of 4.2 m³/s (150 cusecs) will be made in the first instance until the Pancheshwar Dam is put up. A barrage has been made, which is entirely on Indian soil. A short 577-m afflux bund, occupying some 2.9 ha of land, has been constructed, with the consent of the Nepalese Government, tying the barrage to high ground in Nepal as a flood protection measure. The afflux bund is in Nepal, remains under Nepalese sovereignty, and does not involve any cession of Nepalese territory to India. The waters diverted in the channel are for nonconsumptive use of power generation and are returned to Sarda River upstream of the Banbassa barrage.

Certain Nepalese critics have interpreted the Indo–Nepal Joint communi-qué of December 1991 as a treaty and have taken the matter, including that of the Tanakpur Project, to court. The leftist opposition also raised the matter in the national assembly as a violation of Article 126. However, a joint select committee of the assembly has ruled against the point raised.

The issue is that agreements, institutions, and the environment have to be developed for expeditious collaborative development.

8.2.2 Agreements and Development

A number of agreements have been reached between India and Nepal cover-ing joint use of water for certain rivers and basins. Some of these agreements cover specific projects, and some more broadly cover use from an entire river. For waterways and usages not covered by these agreements, either the issue has not been raised, or agreement has not been reached, or reliable data is lacking on the impact of possible use. Agreements have been reached to regulate the use of Mahakali (August 23, 1920, and October 21, 1921), Kosi River and its tributaries (April 25, 1954, as revised on December 19, 1966), and Gandak River and its tributaries (December 4, 1959, as revised on April 30, 1964). Agreements have been reached for the future implementation of the Karnali Multipurpose Project to be located on Karnali River (April 4, 1978) as well as for the Pancheswar Multipurpose Project to be located on Mahakali River (December 1977). These are briefly discussed as follows:

8.2.2.1 Mahakali River

The first agreement between India and Nepal was made in 1920, when the British Government decided to build the Sarda Canal taking off from Sarda or Mahakali, as called in India and Nepal, respectively. Nepal was under British suzerainty at that time. The headworks were located at Banbassa, a few kilo-meters below the point where it debouches from the hills. At this point, the river forms the boundary between India and Nepal. Under the agreement, the Nepal Government exchanged a small piece of territory with the Government of India to enable the left abutment of the barrage and the left bank works to be located within Indian territory (GOI, 1972). The barrage and the main canal, with a capacity of 396.5 m^3/s (14,000 cusecs), were to convey water to the Indian area are located on the right bank of the river. The Sarda Irrigation Project was built in 1928. The required head regulator and structures and an additional main canal on the left bank of the river with a capacity of 28.3 m^3/s (1000 cusecs) to convey water to the Nepal Mahakali Project were also built. The agreement stipulated for Nepal from May 15 to October 15 a water supply of 13.0 m^3/s (460 cusecs) and provided a surplus supply of up to 28.3 m^3/s (1000 cusecs). The agreement further stipulated that from October 15 to May 15, the left bank canal would be alternatively closed and opened for 10 days at a time, running at 8.5 m^3/s (350 cusecs) (Agarwal, 1991).

8.2.2.2 Kosi River

The second agreement relates to Kosi River. Kosi is one of the most highly silt-laden rivers in the world. The resultant lateral migration of over more than 100 km has caused considerable misery in Bihar, and because of this, it has come to be known as Bihar's "river of sorrow." The project, as originally designed in 1950, proposed to have a 239-m-high dam at Barakshetra in Nepal and a downstream barrage at Chatra, also in Nepal. The project with storage of 50,000 ha m had considerable flood attenuation capacity, reducing a peak flow of 24,050 m^3/s to 5660 m^3/s, thereby providing substantial flood control benefit. The project was planned to generate 1800 MW of power and irrigate large tracts of land in Nepal and Bihar.

The Kosi Project was dropped by India in 1958 mainly because the economy was not envisaged to absorb the large amount of power generated and also because it could not possibly utilize the irrigation potential created (Sain, 1978). There was, however, considerable political pressure in India to control Kosi River, especially after a major flood in 1954. An agreement was reached between India and Nepal dated April 25, 1954, regarding use of waters of Kosi River and its tributaries. Indo–Nepal relations were extremely cordial at this time. A Kosi River barrage was built below Chatra in Nepal, 3 mi. upstream of the Hanuman Nagar town along with a pair of 120-km-long embankments to confine the Kosi to its existing course. Eastern and Western Kosi canals have been constructed to provide irrigation to 1.04 mha in Bihar and 12,150 ha in Nepal. The project also generated 20 MW of power of which 10 MW is supplied to Nepal. All the costs were borne by India. Besides irrigation, flood protection, and power benefits, Nepal also got the benefits of the bridge over the barrage, which facilitated east–west communication in that part of Nepal.

The 1954 agreement was amended on December 19, 1966. Section 4 (i) of the Agreement, as revised, provides that "HMGN shall have every right to withdraw for irrigation and for any other purpose in Nepal water from Kosi River and from Sunkosi River or within the Kosi basin from any of the tributaries of Kosi River as may be required from time to time." The agreement also entitles HMGN to use up to 50% of the total hydroelectric power generation by any powerhouse situated within a 10-mi. radius from the barrage site and constructed by India. The tariff rates for the electricity to be supplied to Nepal are to be fixed by mutual agreement between the parties. The agreement also provides for the establishment of the Indo–Nepal Kosi Project Commission to discuss problems of common interest in connection with the Kosi Project. The revised version of the agreement is valid for 199 years.

8.2.2.3 Gandak River

The third agreement relates to the development of Gandak/Narayani and its tributaries. An agreement was reached on December 4, 1959, but was amended on April 30, 1964 (The Gandak Irrigation and Power Project

Agreement). The Indo–Nepal relations were still very cordial and straightforward. Under the aforementioned Agreement, India undertook to construct, at its own expense, a barrage on Gandak River at the border of India and Nepal, a canal head regulator, a powerhouse with an installed capacity of 15,000 kW, and various irrigation schemes commanding an area of about 1.2 mha in India in the state of Bihar and 60,000 ha in Nepal. In addition, India agreed to construct, at its own cost, (i) the Main Western Canal, including the distributary system thereof, down to a minimum discharge at 0.57 m^3/s (2 cusecs) to serve, in the states of Bihar, a gross command area of about 16,190 ha (40,000 acres); (ii) the Eastern Nepal canal from the tail end of the Don Branch Canal up to Bagmati River, including the distributary system thereof, down to a minimum discharge of 0.57 m^3/s (20 cusecs) to serve a gross command of about 41,600 ha (103,500 acres) in Nepal, and (iii) transmission lines to supply power to the Bihar Grid from the power house located on the Main Western Canal in Nepal territory up to the Indo–Nepal Border.

The agreement provides that HMGN will be responsible for the construction of a channel below 0.57 m^3/s (20 cusecs) capacity for irrigation in Nepal, but GOI will contribute a reasonable amount (not specified) to meet the cost of such construction. Regarding the supply of power, the agreement stipulates that GOI shall supply to HMGN an aggregate maximum of 10,000 kW up to 60% load factor. The charges for supply at the power house would be the actual cost of production or the cost of production plus the cost of transmission on terms and conditions to be mutually agreed upon for power supplied at any point of the grid up to and including the power of Raxaul.

The agreement also states that HMGN will continue to have the right to withdraw for irrigation or any other purpose from the river or its tributaries in Nepal such supplies of water as may be required by them from time to time in the valley.

8.2.2.4 Other Developments

In addition to the three major developments on Mahakali, Kosi, and Gandak, in which benefits were provided by the GOI to Nepal, Trishuli and Devighat hydroelectric projects on River Trishuli, Chatra Canal, and renovation and extension of Chatra Canal were completed by the GOI who bore the complete costs of these projects.

8.2.3 Development of Multipurpose and Hydroelectric Project

Nepal has a large potential of hydroelectric energy. Nepal cannot use it, except for a small fraction. India, on the other hand, needs this energy and is a convenient and sole prospective customer. From Nepal's point of view, hydroelectric energy could be Nepal's most profitable export. It could be what oil is to Middle East countries, with the further advantage that it is nondepletable (Chaturvedi, 1976, 1985). The revenues could be a critical source of

Nepal's growth, besides providing employment in this sector. There will be certain adverse environmental impacts like displacement of people and problems of their resettlement and submergence of land and forests, which will primarily impact Nepal, but these can be adequately managed as discussed in Chapter 12. Nepal has proposed developing navigation although this has yet to be technologically, economically, and environmentally explored. For India, besides energy, there are immense usual advantages of storage projects for irrigation and flood mitigation.

Thus, both countries have much to gain directly by collaborative development of multipurpose projects. Besides, the direct economic benefits there will be a very positive fall out on trade and closer socioeconomic interaction. Yet no substantial and concrete advances have been made so far. A brief analysis follows (Malla, 1991; Verghese and Iyer, 1992; Rao and Prasad, 1993).

Developing the energy potential according to some novel technologies proposed by Chaturvedi (1998, 2007c) will enhance the benefits manifold for both countries. It, however, requires close collaborative development. Perhaps this may contribute to collaborative development. It is discussed in Chapters 12 and 13.

8.2.3.1 Karnali Project

Nepal and India have viewed the storage project on Karnali at Chisapani with interest for a long time. A Japanese firm, Nippon Koel, investigated the Karnali basin in 1962–1966 on behalf of the Nepalese Government and investigated 10 sites for hydropower development.

India has also shown considerable interest in the project. India's Prime Minister, Mrs. Indira Gandhi, expressed India's interest in Karnali when she inaugurated the Indian-assisted Sundarijal water supply project near Kathmandu in 1966. A year later, then Deputy Prime Minister Morarji Desai, offered a technical appraisal of the Karnali project to determine India's interest in it. In 1968, President Zakir Hussain stated Indian willingness to assist Nepal with this project in a joint communiqué issued at the conclusion of a visit to Nepal.

Nepal continued examination of the hydroelectric projects by foreign experts on its own, with Karnali project as its centerpiece, in a fitful manner. The Snowy Mountain Authority reviewed the two best Karnali sites. In 1977, Norconsult and Electrowatt of Norway reviewed these sites again.

Development of Nepal's hydropotential with particular interest in Karnali continued in a fitful manner. In January 1976, the Indian Foreign Minister, Y.B. Chavan, met his counterpart, Mr. Anjal, in Kathmandu. In a statement in the Indian Parliament, he announced that "both sides agreed to continue the work on Karnali project as under the existing arrangement and to undertake, at the earliest possible, the joint investigation of the Pancheshwar Dam Project and the Rapti Flood Control Project." Little action followed until on March 22, 1984, when the World Bank announced an $11 million IDA credit to enable Nepal to examine the technical viability and economic justification of constructing a major dam and powerhouse at Chisapani on the Karnali River. The World Bank short-listed a number of international consultants, and in consultation with India, Nepal finally chose four Canadian–US firms who formed a consortium called Himalayan Hydrodevelopment Company (HHDC). A joint committee consisting of representatives from both countries directed the feasibility study of the project. The HHDC produced the feasibility report in December 1989.

The original Karnali dam proposal suggested a height of 240 m for the dam and a power potential of 4500 MW. Its cost was estimated to be about US$3.7 billion in 1983. HHDC proposed a 270-m-high gravel fill embankment dam with a total storage of 28 km^3 and a live storage of 16.2 km^3, total installed capacity of 10,800 MW with 18×600 MW units, an irrigation potential of 191,000 ha in Nepal, and much more in India through supplementation of the existing Sarju and Sarda Sahayak schemes. Flood attenuation benefits would be obtainable through the large degree of flood storage available, which reduces peak flood discharges to about one-third of the peak flood. However, these have to be considered as incidental, as firm operation is based on power generation. The capital cost was estimated at $4.89 billion at 1988 prices. HHDC also suggested phasing of the Karnali Project with an upstream run-of-river hydroelectric project and storage development on the Bheri and Seti rivers, which are tributaries of the Karnali. These supplements to the Karnali Project would generate an additional 5400 MW of electricity.

It is understood that the Karnali Project will have positive and negative environmental impacts. The most important positive environmental impacts are contribution to the mitigation of poverty, which is the worst environmental hazard and outweighs all negative environmental impacts. The most adverse environmental impact is that it involves the dislocation and rehabilitation of 60,000 persons. The submergence of 33,900 ha of land, including 5000 ha of agricultural land, and adverse effects on the wildlife and water quality of the reservoir and releases downstream are other serious environmental impacts.

The Karnali Project has been very conservatively designed for a maximum credible earthquake (MCE) with the magnitude of 8 on the Richter scale involving a peak horizontal acceleration of 0.6 g in contrast to MCE of 7.0 on the Richter scale and a peak horizontal acceleration of 0.25 g.

It has to be noted that the installed capacity of Karnali or the power benefits arising from Karnali could be increased from the earlier estimate because Karnali could be planned as a peaking power station as part of the Indian power system, thereby taking advantage of the thermal hydro integrated functioning. About 95% of the power generated is committed to India, with Karnali designed as a peaking station with a load factor of 20%. The project economics is very attractive by all measures. Unit cost of generation is estimated to be 3.3 U.S. cents per kWh as compared to 4.5 U.S. cents per kWh from a comparative coal-based thermal generation in India. Comparing the benefits with and without the Karnali Project, the total net discounted benefits for the project amount to US$9947 million and the discounted benefit/cost ratio is 2.46. The economic internal rate of return on additional investment requirement compared with the base case alternative is 27%. Carrying out the sensitivity analysis, the benefit/cost ratio under worst-case alternative is found to be 2.21. According to the Karnali feasibility study reports as noted by Matta (1991), "the attractiveness of the Karnali Project hinges mainly on the power and irrigation benefits that India can avail from the project. Thus, India's cooperation in the execution of the project is not only desirable but indispensable."

Despite the attractive benefits and urgency of undertaking the project in the context of Nepal's economic development and India's starvation for electrical energy in the region, no action has followed. Some political factors were important hindrances, as discussed later. Of late, the high level Karnali Committee and Karnali Coordination Committee have taken some follow-up action. Besides certain technoeconomic differences, India has since expressed a preference for a lower dam at Chisapani that might yield around 7000 MW sooner and at a reduced cost, which would be easier to finance and would produce fewer environmental impacts and hazards. It has also indicated two or three other upstream projects on Karnali River that could be taken up later as the load develops.

8.2.3.2 Development of Other Major Multipurpose Projects

The formation of a popular government in Nepal and with the Indian Prime Minister from Bihar, with whom the Nepalese politicians had excellent rapport, Indo–Nepal cooperation in water resources developed momentum. The Indian and Nepalese Prime Ministers, Mr. Chandra Shekar, and Mr. K.P. Bhattarai, met in Kathmandu on February 15, 1991, and set up a program for economic cooperation. It was agreed to activate the Karnali and Pancheshwar Committees and to consider the possibility of undertaking several attractive hydroprojects on the Bhuri Gandak, Kali Gandak, and Seti rivers and a joint survey for flood control, especially on a number of smaller rivers responsible for flash floods. Talks at other levels brought up the idea of a high dam for flood storage on the Kamla while the UNDP offered to assist Nepal in commissioning a feasibility study of a 90 MW run-of-river hydel project on the Kali Gandak.

The visit from the Prime Minister of Nepal, Mr. G.P. Koirala, to Delhi in December 1991 carried matters further. The Pancheshwar Project, which involved a 262-m straight gravity concrete dam with a gross storage of 6800 million m³ (5.50 m a f) on the border of India and Nepal on Mahakali (Sarda), designed to generate about 2000 MW and provide irrigation benefits, has been under investigation by both governments independently, particularly by the Indian engineers for a long time. It was agreed to propose a joint project report by June 1993. Other agreements included joint studies and investigation of the Kosi High Dam (likely to be completed by 1994–1995), a joint field survey of the Bhuri Gandak Project (600 MW), and Nepalese investigations of the Kamla and Bagmati reservoir schemes by 1993, all three projects being candidates for bilateral cooperation (Verghese and Iyer, 1992).

8.2.3.3 Sharing of Costs and Benefits

The perceptions of cost–benefit and priorities by Nepal and India are bound to be different. They will require careful resolution.

Current Nepalese perceptions, as stated by Malla (1991), appear to be as follows:

> "India undoubtedly is the greater beneficiary of the potential benefits accruing from three multipurpose water resources projects, irrespective of whether these benefits come from power, irrigation, or flood control. India needs to derive these benefits as early as possible because these are likely to be available at a price that is more attractive than the best alternative options India can undertake on its own. Since these benefits fit within the demand parameters of India, India can afford to put in a significant investment in the development of these projects.
>
> Looking from the Nepalese perspective, these projects have a potential to render benefits far in excess of what it can use for itself. However, Nepal wishes to avail of the benefits from these major projects because these are very attractive. However, the investment required for the implementation of these projects is far beyond what Nepal can afford. Nepal also wishes to earn through sale of power to India and thus improve its trade balance of payments position with India; it also desires to gain some proportion of the benefits that India as a lower riparian country exclusively gets as a result of the implementation of these projects.
>
> Judging against the above background, multipurpose water resource development projects can be viewed as investment projects with tremendous export potential. His Majesty's Government of Nepal has a declared policy of promoting joint investment projects, which can earn foreign exchange. Thus, it is recognized that these projects have a great potential for promotion as joint investment program since these serve the interest of both the countries. Details about sharing of costs and benefits will constitute an integral part of such joint investment programs. Consideration may also be given toward Nepal mobilizing resources prior to project implementation in the form of advance payment against future sale of electricity to India as well as Nepal's share of incremental benefits to India as a direct consequence of the project. These were precisely the steps Canada took prior to the construction of various dams in the upstream reaches of the Columbia River in accordance with the provisions of the Agreement on the Columbia River development between Canada and USA.
>
> In negotiating the price, Nepal should attempt to obtain for power exports to India a price just under India's opportunity costs. The question is not what India's opportunity cost is today but what it will be 12 to 20 years from now. Circumstances can change radically in such a time

period. The price contract should negotiate the principles on which price should be based when the plant comes on stream. The price should be subject to an escalation clause whose principles can be determined in advance.

The negotiation of fixed price contracts and inflexible long-term contracts has posed problems in both the developing and the developed countries. The lessons learnt elsewhere suggest the adoption of the basic pricing agreement, which should be a formula and not a fixed price containing clauses which reduce the risk of (i) fluctuating exchange rates, which affect the principal and interest payment, (ii) escalation clauses, and (iii) changed demand or changes in operating mode."

Nepalese scientists also consider that determination of costs and benefits are difficult and must be undertaken carefully. It is considered that the actual social costs in terms of resettlement are likely to far exceed the estimated costs. This is more likely in major projects, which involve the relocation of a great number of people. Similarly, economic estimates of the environmental impacts are difficult to make. Estimation of allocation of joint costs is also difficult. Equally difficult is the problem of estimation of benefits (Malla, 1990).

The issue of estimation of costs and benefits has not been explicitly stated on the Indian part. Study of the large number of projects prepared in India brings out that often this has not been done properly. As regards price of electricity, it is usually taken as actual cost plus adequate financial returns. Indian perception of pricing of electricity from these mega projects will be according to this practice. As we will discuss later, this is the central complex issue to be resolved. With the proposed water revolutions, the scene will change completely.

8.2.4 Navigation

One of the Nepalese goals in water resources development has been stated to be "inland navigation connecting Nepal to nearest seaports for relief of the landlocked economy" (Pradhan and Shrestha, 1992). There have been statements that navigation could be developed in Nepal also when India undertakes development of waterways, and hopefully this will be kept in mind (Sharma, 1983). In addition, it is considered that Nepal should develop transportation capability in cooperation with other countries of the Himalayan block viz. Bangladesh China and Pakistan. A 16-km-long tunnel link between the eastern part of Nepal passing through India at a depth of 300 m and coming out at the western border of Bangladesh, providing land transportation, has been proposed (Sharma, 1983). These proposals can at best be considered as fancy ideas.

In its 1978 proposal to India for augmenting the lean season flows of the Ganga, Bangladesh proposed a series of high dams in Nepal. It was suggested that a navigation canal could be undertaken along the Tarai from the waters of Gandak and Kosi through the Mahananda and Karatoya–Atrai–Baral basins in northern Bangladesh, draining into the Ganga. Such a waterway and its connecting route through India and Bangladesh, created for increased dry flows, could also serve as an international water route (Indo–Bangladesh Joint River Commission, 1985). Reference has been made in the 1985 updated proposals of Bangladesh of an international water route as one of the benefits, but details are missing. Some populist Nepalese writers have also written about it (Kurve, 1981). This is just a fanciful idea, totally ignoring technological, environmental, and economic realities.

8.2.5 Watershed and Flood Management

India and Nepal face a very serious problem of floods, soil erosion, and sedimentation. The severe flooding in the alluvial plains and delta results from the unique physiographic and hydrologic interaction of the mountains and plains. Nepal faces a very serious problem of soil erosion, which will affect Nepal's interests in several ways, besides the loss of one of the most precious and almost nonrenewable environmental vectors. For instance, the valuable hydropower projects proposed to be developed through storage reservoir and other benefits will be seriously curtailed with the consequent sedimentation of the reservoirs. While the physiographic setting of Nepal is a great asset in terms of the hydropower potential, it is also a great liability from consideration of soil erosion, sedimentation, and floods. The impacts are borne principally by India and Bangladesh.

So far, these adverse impacts may predominantly be on account of natural factors but human intervention and developmental efforts in Nepal are likely to increasingly accelerate and significantly contribute to these environmentally disruptive activities, which have serious consequences for the downstream riparian countries. This could be a serious cause of conflict in the future. Therefore, from all considerations, it is important that the problem of environmental degradation and natural disasters be addressed urgently and collaboratively. Unfortunately, hardly any action has been taken so far. Only recently (1991), Nepal and India have agreed to set up a flood forecasting and warning system in Nepal. Forty-five such systems are to be installed, with 15 of them due to be operational by the 1992 session.

For undertaking environmental management efficiently, it is important that the ideas regarding all vectors and particularly the hydrological–meteorological data be collected systematically and on a long-term basis. Unfortunately, little attention has been paid to this serious matter (Dixit, 1991). Measurement of rainfall was started in Nepal in 1947 by the British. The Indian Government continued the practice. Rainfall data were collected by several agencies until 1966 when the centralized and planned meteorological data collection program was initiated under the auspices of the United Nations. Currently, there are 39 agrometeorological, 44 climatological, and 170 precipitation recording stations (Dixit, 1991).

Stream flow and sediment data, however, were not measured until very late. Before 1960, very little was done in this field. A project to establish a nation-wide network of river flow measurement stations in Nepal was started in 1961. After several administrative transfers and amalgamations, the program in the current form dates only from 1988. It is very inadequate (Dixit, 1991).

8.2.6 Perceptions of Collaborative Development

Perception of developmental activities differs among engineers of the two countries. How far these views represent the opinions of decision makers is an open question.

Regarding the agreements carried so far, some Nepali engineers feel that they have not been treated equitably. Agrawal (1991), writing about the Mahakali agreement, states that it is "not based on the sharing of benefits through a comprehensive agreement. In this agreement, Nepalese interests are not well reflected." Sharma (1983), writing about the Gandak Project, states that "had Nepal been economically strong, it would have constructed the entire system and sold the benefits at a fair price to India…" The feelings run quite deep. It is felt that at present, the entire watershed of Nepal is working to tap water for the benefit of the lower riparians (Sharma, 1983). Finally, it is Sharma's (1983) opinion that provision of extra benefit to Nepal at the time of the Kosi and Gandak agreement "would have helped to solve all the prevailing misunderstanding which unfortunately have permeated down to people."

Sharma (1983) further states that Nepal entered into the Sharda Project (1920), Kosi Project (April, 1954), and Gandak Project (1959) as "a gesture of friendship for the well being of the people of India even though Nepal received less benefits from the projects constructed under these agreements." Malla (1985–1986) writes that "Nepali experience with India in this field has not been very satisfactory... The Kosi agreement in particular has been considered almost an 'unequal' and 'unfavorable' agreement signed by Nepal with India. Although a list of improvements has been made in Nepal's favor in the Gandak agreement, the Nepal Congress government, which had concluded the agreement in 1960, and the Indian Government have been accused of ignoring Nepal's interest. There is a deeply rooted feeling in Nepal that it has been 'cheated' by India in agreements like the Kosi and Gandak. There is enough apprehension that the same might be the case of Karnali too."

The views of the Indian engineers are entirely different. According to Chitale (1992), who was Secretary of the Ministry of Water Resources, GOI, writing about the Sarda, Kosi, and Karnali projects already executed, "all these projects fully funded by India have provided large benefits to the two countries."

Nepal has long felt that "there will not be a level playing field and that giant India will take advantage in a way that cannot be acceptable to Nepalese as nationalists" (Rogers et al., 1989). These perceived suspicions and apprehensions were one of the additional reasons that led to Nepal's insistence on an outside consultation during 1970s and 1980s, in the development of a water resources project almost to the total exclusion of Indian engineers and technicians. There are two additional reasons, which are often not fully appreciated. First, the foreign donor governments and organizations like to use the aid or loans to be channeled back to their own countries. Second, there are kickbacks involved, which the politicians and officials of the developing and developed countries alike find very convenient and rewarding. Indian engineers develop simplistic reactions such as the "Nepalese who, rather than do their own work, have expensive Western consultants to do even simple things…" (Kumar and Sinha, 1992). It is not realized that unless the Nepalese and Indians develop the projects themselves, the full benefits will not go to them but will be skimmed off by the foreign construction firms as the Karnali feasibility study brings out.

With the installation of a popular ministry in Nepal, changes have taken place, as we have discussed earlier. Although some engineers seem to be satisfied with the present rate of those projects, the Nepalese perception appears to be quite different. Regmi (1992) writes that "For both Nepalese politicians and the public who thought that the country's new political environment would in itself be sufficient to receive magnanimous support from India in Water Resources Development, the outcome came as a surprise. Though, at a political level, understanding between the two countries was improved, the much needed goodwill for water resources sharing was not apparent. The business-as-usual attitude in dealings with Nepal, ingrained in Indian water bureaucracy, has led to uncertainty not only over the question of water resources but also even over Nepal's emerging politics. A simplistic and ideal scenario, the deal showed, did not exist in the real politic of resource negotiation."

As Rao and Prasad (1994) point out, "this is a strong statement and reflects Nepal's frustrations. The factors, which govern the response of Indian water bureaucracy, are not well defined by Regmi... It may be a combination of arrogance of power, inability to accommodate a neighbor's viewpoint, a well-known secretive attitude on their part, or more sloppiness in their dealings with others. The callousness with which the Indian bureaucracy treats Indians is well known (Tewari, 1992). It would not be surprising if this attitude manifested itself in dealings with a neighboring country also."

Thapa and Pradhan (1995), while reiterating the usual Nepal view, analyze the scene and give suggestions for efficient and mutually satisfactory development. According to them, the factors which contributed to poor development in the past are (i) inadequate consultation during planning stage, (ii) lack of conscientious effort to generate consensus on both sides, (ii) inappropriate mechanism of project management, and (iv) sharing of benefits and costs. The proposed future possible approaches are (i) development of strong motivation to cooperate, (ii) formulation of a mutually acceptable framework of cooperation, (iii) meaningful consultation during the planning stage, and (iv) generation of consensus through a deliberate action program.

8.2.7 Conclusion

The sum and substance of the analysis are that although there are tremendous benefits from Indo–Nepal collaborative development of the natural resources and utmost urgency to mitigate the apocalyptic environmental and socioeconomic futures perspective, the scene is depressing. There are little perception of the urgency, little achievement, and effort, and considerable misunderstanding and suspicion. Serious and sustained effort will be required to develop an appropriate climate of collaborative sustainable development (Bajpai, 1986). Some approaches to contribute to implement collaborative development will be proposed in Chapter 12, particularly in terms of some novel technologies.

8.3 Indo–Bhutan Interaction

8.3.1 Sociopolitical Scene

Bhutan was a small isolated country under British suzerainty until India gained Independence. Bhutan is an independent country that has attained UN membership in 1971 with India's support and help.

Bhutan started its development in 1961. The first four 5-year plans were almost entirely funded by India. Bhutan is gradually moving to achieve self-reliance in development, with funding from other countries and agencies as well. India remains a close ally.

Following Das (1986), the emerging sociopolitical scene may be described as follows. Bhutan's present policy can be divided into three basic areas: (i) internal consolidation, (2) gradual extension of diplomatic and economic relations, and (3) active involvement with a noncontroversial regional consciousness imperative for stability and identity. The ethnic problem and economic backwardness constitute the major constraints. The danger could be of finding a scapegoat for diverting attention from these issues. Raising the national fervor to a pitch where it must find a proper outlet and satisfaction is a difficult task. India becoming a whipping element for failure may cause a serious setback to the present evolutionary process of national personality.

The Bhutan of today has moved away from its historical past and has entered an era of realization, assertion, and the development of its sovereign personality. The King's leadership is forward looking, and the new generation reflects a sense of pride and achievement. It will be in Bhutan and India's interest that bilateral and regional linkages are developed, avoiding the pitfalls of a small and a big country syndrome. Both have a stakes in developing the Himalayan resources and conserving the fragile environment.

8.3.2 Development

Concentration has been almost entirely on hydroelectric development. Bhutan launched its first 5-year plan in 1961. The first major hydel project on the Wangchu or Raidek River was undertaken and was fully commissioned in 1989. It has been financed, designed, and executed by India at an estimated cost of Rs.2.45 billion on the basis of a 60% grant and a 40% loan. Bhutan's demand is only of the order of 25 MW, and the rest of the energy is being sold to India at a flat rate of 30 paise. An autonomous Indo–Bhutan Chuka Project Authority Chaired constructed the project by Bhutan with an Indian engineer as general manager.

A 45-MW run-of-river hydel station is to be built on the Kun Chu in eastern Bhutan within the next few years. A small storage dam with an installed capacity of 120 MW is proposed to be constructed at Prunatha, upstream of Chuka I, on the Wangchu. It will augment the capacity of Chuka I by 33 MW. Besides this, two larger power export projects are planned on the lower Wangchu, namely Chuka II at Tala and Chuka III, the Wangchu Reservoir Project, near the Bhutan–India border. Chuka II will be a 1000-MW run-of-river project, but Chuka III will be a storage project with 900-MW capacities. India has proposed to build these projects and to absorb whatever is generated as surplus to its domestic requirements. Full development of these projects within a decade could raise Bhutan's installed capacity on the Wangchu cascade to some 2390 MW. This is only a fraction of Bhutan's potential of about 20,000 MW.

9

Indo–Bangladesh Interaction

9.1 Introduction

India and Bangladesh, the latter formerly being East Pakistan, have a historical, cultural, and political identity that is based on environmental integrity.
The partition of British India in 1947, on the basis of Hindu or Muslim majority, into India and Pakistan shattered this comprehensive integrity. Conflicts
were the natural aftermath. With the partition of land and water, environmental conflicts continue. The central point is the issue of waters of River
Ganges, with focus on the Farakka Barrage in its many facets.

We consider the subject from the viewpoint of how to promote collaborative
environmental management. In this context, we consider (i) Indo–Bangladesh
relations, (ii) an overview of the disputes, (iii) the technoeconomic environmental aspects of Indo–Bangladesh interactions and efforts to resolve them,
and (iv) the perceptions about problems and also the efforts to resolve them.
We believe that integrated sustainable management of the environment is
a natural demand and is in the best interest of both countries. The conflicts
are totally meaningless and arise out of ignorance and ego. The issue of policies and approaches to promote collaborative environmental management
for rapid socioeconomic development and sustainability will be presented in
Chapter 12.

The subject of the Indo–Bangladesh river water dispute has been studied
by several authors and has been brought out in some government reports.
We owe to these studies even though our orientation is different.

9.2 Indo–Bangladesh Relations

The formation of East Pakistan was based on Hindu–Muslim animosity,
developed toward the end of the British reign, even though it was preceded
by centuries of Hindu–Muslim fraternity, as the two communities were part

of the same cultural history, with some getting converted to a new religion. The animosity was exacerbated by the partition.

During the period of Bangladesh as East Pakistan, the bitterness was there but not as much as with West Pakistan or the Center. There are good historical reasons for this feature. For one, East Bengal was never the epicenter of Pakistan's two-nation theory; rather, it was the reluctant periphery. Secondly, India has never treated East Bengal as a point of confrontation in its overall estrangement with Pakistan. As a matter of fact, East Pakistan had its own serious internal problems of East Pakistan versus the Central Pakistan Government, which had imposed martial law in the region.

Following the liberation of Bangladesh in December 1971, in which India had an important role, its relations with India remained very close for a few years. In fact, in the very first year, Sheikh Mujib himself enthusiastically went in to sign a Treaty of Friendship with India. However, the psychological need of the Bangladesh government to distance itself from India and the requirement of external aid and foreign investments from Western and Islamic countries led to a perceivable loosening of Indo–Bangladesh bilateral ties. This process received a fillip in the aftermath of the overthrow of Sheikh Mujib. Internal changes in the Bangladesh polity were reflected dramatically in its external relations as well. There was a sudden warming up of Bangladesh's relations with the United States, China, and the Islamic world, particularly Saudi Arabia and, to a lesser extent, Pakistan. Relations with India became correspondingly sour.

The points in dispute between India and Bangladesh are of two categories: (1) those rousing passions temporarily, such as the issue of maritime boundary and storm over New Moore island, barbed wire fencing on the Assam border corridor through Tin Bigha, and the Chakma insurgency, and (2) those with long-range implications, such as the Farakka water dispute and sanctuary for insurgent groups in northeast India.

The items in the first category are generally orchestrated whenever a political group interested in discrediting the ruling establishment uses it for domestic purposes or as a sort of diversionary move. For instance, New Moore (or South Telpatty) is a deltaic island, which has emerged almost on the international border in the Sunderbans. India has occupied the island, and its reluctance to refer the matter for arbitration compounds the fear that it is inclined to coerce and dominate Bangladesh. Again, the agreement to exchange southern Berubari for the permanent lease of a tiny corridor through Tin Bigha was reached between India and Bangladesh in 1974. India appropriated Berubari, but the corresponding obligation to provide Bangladesh with a passage to its enclaves of Dehagram and Agorapota through Tin Bigha was fulfilled only in 1992.

The Chakmas of the Chittagong Hill Tract suffered displacement as a result of Bangladesh's land settlement policies. Many, who have sought refuge in India, agitated against the intrusion of outsiders in their tribal homeland, a phenomenon well known in India. Bangladesh blames India for conniving in the Chakma insurgency. On the other hand, India faces the problem of illegal infiltration by Bangladeshis in India and their gradual permanent settlement. There is considerable resentment in the neighboring state of Assam, and it is proposed that barbed wire fences be erected to control infiltration as well as drug smuggling. This naturally causes resentment in Bangladesh.

The long-range issues are provision of sanctuary to insurgents from or to northeast India. This is a problem that keeps on arising. The second more complex problem is the sharing and management of the water resources. Farakka occupies a center stage in this context. It is a vital and complex problem, which is examined in detail.

It is unfortunate that the positive issue of the vital necessity of regional collaboration, in which environmental management is central, is rarely discussed. It will be shown in Chapter 12 that if the central issue of the urgency and vital necessity of collaborative sustainable management of the region is considered, all these disputes will emerge as a criminal breach of common sense, in the context of the future apocalyptic scenario.

9.3 Indo–Bangladesh Waters—Official Interactions

The Farakka Barrage is the central issue in the Indo–Bangladesh Waters dispute, but it has recently been expanded to the sharing of the waters. An overview is first given. It may be considered in five phases: (1) Phase I—perpetration and Indo-Pak phase, (2) Phase II—1972–1977, (3) Phase III—1977–1984, (4) Phase IV—1985–1987, and (5) Phase V—1988 onward. The overview will be brief because as Mehta (1992), former Secretary of Foreign Affairs, Government of India (GOI), and leader of the meetings from the Indian side, states, these are largely dialogues of the deaf. Similar sentiments are expressed by Abbas (1982), who led the Bangladesh side for a long time. For details, reference may be made to Rangachari and Iyer (1982), Crow (1985, 1990, 1995), and Chaturvedi (2007c). For specific observations, Crow (1995) and Rangachari and Iyer (1992) have been followed, as even though the details are classified, as official Indian representatives, they are expected to know them and to be reporting them correctly.

9.3.1 Phase I: Perpetration

The Farakka Barrage has a long history. River Ganges bifurcates near Farakka into the Bhagirathi and the Padma. While the Bhagirathi (Hooghly) runs to the sea in India, the Padma crosses into Bangladesh, where it is joined by the Brahmaputra and the Meghna to debouch into the Bay of Bengal. Until some two centuries ago, the Bhagirathi used to carry the bulk of the Ganga flows, which led to the establishment of Calcutta when the British East India

Company started the trading operations in India. Later, the Padma became the main carrier channel. Concern was felt about the deterioration of the Hooghly, and the first governmental inquiry was held in 1853 followed by several others during the British period (Cotton, 1854; Leonard, 1864; Vernon-Harcourt, 1896, 1905; Stevenson–Moore Committee, 1916–1919; Wilcoks, 1930; Oaz, 1939) and Webster (1946) (which also reflects the findings of earlier studies). The position today is that there is practically no dry weather flow from the Ganges into the Bhagirathi or any of the spill rivers in the Nadia group, but in the absence of systematic records, it is not possible to state definitely that progressive deterioration of these feeder channels is going on. No apology is made, however, for re-emphasizing the need to take all possible steps to improve the headwater supply of the Hooghly; otherwise, it will be useless, considering schemes for the development of the Port of Calcutta, which, as has already been pointed out, is dependent for its existence upon the maintenance of an efficient waterway between the docks and the sea.

In the context of impending Indian independence and the consequent partitioning of Bengal, the need for the Ganga Barrage to resuscitate the city and the Port of Calcutta was strongly argued before the Boundary Commission presided over by Sir Cyril Radcliffe. India's contention was that this was justification enough to depart from the principle of "contiguous Muslim majority" areas to constitute the new state. This argument was equally vehemently opposed. In his award, Radcliffe posed two questions concerning Calcutta: (1) To which state was the city of Calcutta to be assigned, or was it possible to adopt any method of dividing the city between the two successor states? (2) If the city must be assigned as a whole to one of the states, what were its indispensable claims to the control of territory, such as all or part of the Nadia river systems of the Kulti rivers, upon which the life of Calcutta as a city and port depends? His decision was that Calcutta could not be divided, and he awarded the Nadia headwaters (the Bhagirathi, Jalangi, and Nadia) and Kulti along with Calcutta to India. This also entailed transferring the Muslim-majority Murshidabad district to India and, to balance things, the Hindu-majority district of Khulna and certain other areas to Pakistan. It can be said, as Crow (1995) has observed, that the partition award does not constitute an endorsement of river diversion but recognition of the important issue of water control.

After Independence, the matter was taken up seriously and with urgency. As Indian capabilities were developing, extensive hydraulic investigations were carried out (Joglekar, 1955). All the available information was considered by an Indian Expert Committee (Man Singh Committee), and the conclusion was that "it is clear that the entire river system shows signs of deterioration, on account of lack of perennial upland discharge as also to the play of natural tidal forces. Natural restoration of dry weather flow of upland water is definitely impossible" (India, 1961). The Committee also examined the other proposals of (i) a ship canal to link Calcutta Port to Diamond Harbor, (ii) dredging, and (iii) an alternative port (Crow, 1995). The decision was positively to divert water from River Ganges, through the construction of the Farakka Barrage, into Bhagirathi–Hooghly. Further investigations and consultations by an expert, Dr. Walter Hensen, confirmed the decision. The new evidence can be summarized as follows: (i) the sustained high discharge of water flowing from the Ganges into Bhagirathi was calculated to have fallen by about 45% between 1936 and 1956; (ii) the Hooghly capacity was shown to have reduced by between 1% and 0.3% per annum; and (iii) despite intensive dredging, the depths over the river's bars have substantially decreased (India, 1961). As Henson recommended, "The best and only technical solution of the problem is the construction of a barrage across the Ganga at Farakka with which the upland discharge into the Bhagirathi–Hooghly can be regulated as planned, and with which the long-term deterioration can be stopped and possibly converted into gradual improvement" (India, 1961).

The Government of Pakistan (GOP) continued to watch the development and voiced apprehensions about the proposed Farakka Project. In as early as 1951, it wanted India to consult it "before putting into operation any scheme, which may tend to prejudice the interest of East Bengal." Thus, even though the main focus of the earlier exchanges centered on the Farakka Project, Pakistan's concern extended to all schemes for using the waters of the Ganga. References to several schemes far upstream were made by the GOP to the GOI. Similarly, GOI voiced its concern about schemes being undertaken or being investigated by GOP such as the Ganges–Koladak Project, Teesta–Barra Project, and Karnaphuli Project.

By 1956, both countries had agreed on a cooperative approach on a reciprocal basis. In May 1957, Pakistan proposed that the technical services of a UN body should be secured, but India did not consider it to be necessary, as the already agreed upon approach on the cooperative basis was the most practical way of dealing with the issue.

Accordingly, five meetings of technical representatives took place between June 1960 and May 1968. Even after these five meetings, India believed that a large number of points remained to be discussed and further meetings were proposed. Pakistan held the view that "no useful purpose would be served by continuing the technical level exchanges" and proposed an early meeting at the Ministerial level with a view to finding a mutually acceptable solution to the problem of apportionment of the waters of the Ganges.

Ministerial-level meetings of the secretaries of the two governments started in December 1968. Five meetings were held until July 1970. By the end of the fifth meeting, India summed up its position by stating that Pakistan's needs for the water from the Ganga were minimal while India was overwhelmingly dependent on the Ganga. The amount of water that India could reasonably be expected to release from the Ganga could be quantified if realistic projects were formulated in Pakistan, taking into account all relevant factors. Pakistan disagreed and argued that it needed to have an idea of the "guaranteed supplies" of water to be made available to it before any formulation of its Ganges Barrage Project was made. Similar differences existed in other projects on the Ganga/Tista and other rivers. Further meetings were proposed, but momentous political developments were taking place in Pakistan, leading to the emergence of Bangladesh in December 1971.

Reference may be made to the report by two American experts who were engaged by the Pakistan government to study the problem (Ippen and Wicker, 1962). Their findings were in sharp contrast to those of Hensen, who was engaged by the Indian government. While Hensen thought that the Farakka Project was the best and only solution to the deterioration of Hooghly, Ippen and Wicker thought that the project would exacerbate the problem. Hensen thought the project would also contribute to the solution of a whole range of secondary problems, while Ippen and Wicker said it would only help to reduce the salinity of Calcutta's drinking water, leaving other conditions unchanged. Upon examination of the two contradictory reports, it was suggested that these might be a "technical ambiguity," and doubts were expressed toward the theory of sedimentation in Hooghly as propounded by the Indian government (Crow, 1995). The fact of the matter, as brought out by Myrdal (1968), is that even in science, so-called foreign experts are sensitive to their masters.

9.3.2 Phase II: 1972–1977

The second phase started with the emergence of Bangladesh, led by Sheikh Mujibur Rahman. Relations between the two countries were at their peak. As noted by Crow (1995), "In fact, the independence of Bangladesh brought, almost unnoticed at that time, a resolution of the first phase of the Ganges waters conflict. The Bangladesh government accepted in general, that India had a right to use the Farakka Barrage. In return, the Indian government recognised Bangladeshi right to negotiate the sharing of the water." After March 1972, a Treaty of Friendship, Cooperation, and Peace was signed between India and Bangladesh. The Treaty, interalia, envisaged studies and joint action in the fields of flood control, river basin development, and the development of hydropower and irrigation. A Joint River Commission (JRC) consisting of engineering officials from both countries was set up to conduct joint studies.

Some studies were conducted. Work on the execution of the Indian and Bangladesh projects of mutual concern, however, went on unhindered. In the joint Indo–Bangladesh Declaration of May 16, 1974, the prime ministers of the two countries took note of the fact that the Farakka Barrage would be commissioned before the end of the year. They recognized that the fair weather flow in the lean months would have to be augmented to meet the requirements of the two countries. They accordingly decided that the JRC should study the best means of such augmentation through the optimum utilization of the water resources of the region available to the two countries. They also expressed their determination to arrive at a mutually acceptable sharing of waters in the lean season.

An agreement was reached on April 18, 1975, on the operation of the Farakka Project, with agreed upon withdrawals during the lean season coupled with joint observations and studies on the effect of such withdrawals. The Commission was able to exchange data and work out the water availability at Farakka in different periods (according to which the 75% dependable flow in the lowest 10-day period was 55,000 cusecs [1556.5 m^3/s]). Diversions from Farakka commenced in April 1975 on the basis of agreed upon withdrawal.

JRC held discussions on the augmentation of the lean season flows of the Ganga, but there was wide divergence. The Indian point of view was that in view of the high demand in Ganga basin, there was no prospect for augmentation even through the construction of storages on River Ganga. On the other hand, Brahmaputra had a considerable amount of water, which could be used to augment flows in lower Ganga. Bangladesh considered Ganga and Brahmaputra as two different rivers, and augmentation from Brahmaputra was considered unacceptable as well as infeasible. It considered that augmentation was possible from the actual and possible storages on River Ganges in India and Nepal.

Soon, certain political developments of a far-reaching nature occurred in Bangladesh. Sheikh Mujibur Rahman was assassinated in August 1975. Considerable political uncertainty prevailed for about a year until military dictatorship was established. Accusations and counteraccusations were exchanged between India and Bangladesh regarding the sharing of water. The latter took the matter to the UN General Assembly in 1976, but saner counsels prevailed, and it was agreed to revert to bilateral discussions, which resumed by the end of 1976.

9.3.3 Phase III: 1977–1984

Discussions were continued, and on November 5, 1977, an agreement on "sharing of the Ganga Waters at Farakka and augmenting its flows" was signed, which was to remain in force for 5 years but was extendable by mutual agreement. The Agreement had two parts, one on sharing at Farakka and the other on long-term augmentation of the flows. The arrangements for sharing at Farakka related to the five lean months, January 1 to May 31, in accordance with agreed upon 10-day schedules as shown in Figure 9.1. Both sides made substantial concessions. In the driest 10-day period, Bangladesh gave away 20,000 cusecs of the historical flow, while India could divert only 20,500 cusecs, compared to the 40,000 cusecs that had been demanded.

A JRC was made responsible for the implementation of these arrangements. The investigation and study of schemes relating to augmentation were to be carried out by the JRC and completed within 3 years. Two reviews of the Agreement were provided at the end of the 3.5 years prior to its expiry.

The Indian and Bangladesh sides exchanged their respective proposals for augmentation in March 1978 as shown in Figure 9.2. (These are later considered in detail.) India considered the Ganga–Brahmaputra–Meghna Basin as

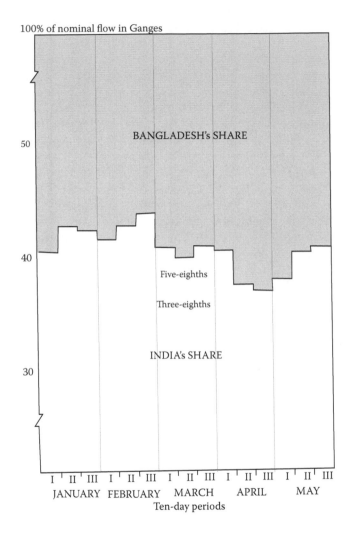

FIGURE 9.1
Arrangements for sharing at Farakka related to the five lean months, January 1 to May 31, in accordance with agreed upon 10-day schedules. (From Crow, B., in *Strategies for River Basin Management*, p. 120, D. Reidel, Dordrecht, Holland, 1985. With permission.)

an integrated system in which the densely populated Ganga subbasin had relatively less water availability, less storage potential, and greater demands for irrigation, whereas the sparsely populated Brahmaputra and Meghna subbasins had relatively plentiful water availability, large storage possibilities, and less demand for irrigation. The Indian proposal claimed that it approached the augmentation issue within a framework of optimum utilization of the water resources of the region available to India and Bangladesh. This envisaged storages on Dihang, Subansiri, and Barak in the Brahmaputra–Meghna

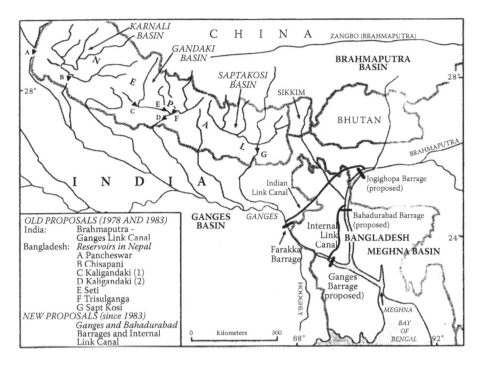

FIGURE 9.2
Indian and Bangladesh proposals for augmentation (March 1978). (From Crow, B., in *Strategies for River Basin Management*, p. 187, D. Reidel, Dordrecht, Holland, 1985. With permission.)

system and transfer of waters from Brahmaputra to Ganga by constructing a link canal taking off Brahmaputra at Jogighopa, passing through both India and Bangladesh, and joining the Ganga at a point just above Farakka.

Bangladesh, on the other hand, considered that Ganga and Brahmaputra are two different rivers. According to them, enough potential existed in Nepal and India, within the Ganga system itself, to augment flows at Farakka. It envisaged a number of storages in the upper catchments of Ganga in Nepal and in India, with the stipulation that releases from them shall be used specifically for augmentation and not meeting the demands in India.

Both sides had serious objections to each other's proposals. The final review of the 1977 Agreement in June 1982 left the matter for decision between the two governments at a high political level. President Ershad visited India in October 1982. In a meeting with India's Prime Minister, Mrs. Indira Gandhi, a Memorandum of Understanding (MOU) was signed by the foreign ministers of the two countries under which the JRC was directed to decide on an optimum solution within 18 months, that is, before April 1984. The JRC decided to update the 1978 proposals, but these remained essentially the same. No agreement was reached, but the releases continued as before even during the dry season of 1985 on an ad hoc basis even though the period of the MOU had expired.

9.3.4 Phase IV: 1985–1987

The MOU came to an end with the dry season of 1984. In November, Rajiv Gandhi became the Prime Minister of India after the assassination of Mrs. Indira Gandhi. He and the President of Bangladesh had the opportunity of meeting at Nassau, Bahamas, in October 1985. As a result, the process was again started to (i) study the sharing of the waters of the common rivers and augmentation of the Ganges waters and (ii) execute the signing of an MOU for the sharing of the Ganga waters at Farakka for a 3-year period commencing with the dry season of 1986 on the same terms as those of the 1982 MOU.

The MOU was accordingly signed in November 1985, more or less on the same lines as the 1982 MOU, with minor variations as follows. Without going into the details of the figures for 10-day periods from January 1 to May 31, at the lowest level of flows, i.e., 55,000 cusecs during the period April 21 to April 30, the withdrawal by India would be 20,500 cusecs and the release to Bangladesh would be 34,500 cusecs. There was also an understanding regarding the contingency of "exceptionally low flows" (Rangachari and Iyer, 1992).

A Joint Committee of Experts (JCE) was appointed, hopefully to get over the past hangovers and to examine the sharing and augmentation as a joint team. The same difference regarding "augmentation" continued. Joint meetings with Nepal were held, but even there, the difficulties cropped up. The Bangladesh team wanted a tripartite approach in studying their proposals. Nepal welcomed the joint approach, including Nepal, on the basis of "mutual benefits," but it insisted that these should be spelled out at the outset.

The new initiative undertaken with the resumption of office by Rajiv Gandhi also ran out of steam without any progress. The 3-year period of the 1985 MOU came to an end in 1988, and it was not renewed. Releases continued to be made on an ad hoc basis. They were slightly reduced, perhaps to give an indication to Bangladesh that the MOU could not be taken for granted.

9.3.5 Phase V: 1988 Onward

Serious climatic impacts and political disturbances were taking place in Bangladesh. In the wake of the severe floods in Bangladesh in 1988, an Indo–Bangladesh Task Force was set up to study the Ganges and Brahmaputra waters jointly for flood management and water flows. It operated for about 2 years, and some limited agreements were reached on monitoring arrangements, information sharing, flood forecasts and warning, and so on.

In April 1990, after a 3-year gap, the JRC met again. "It was the story of the Joint Committee of Experts all over again.... Despite several meetings, the Secretaries Committee was unable to achieve anything useful" (Rangachari and Iyer, 1992).

As reported in a study by the Government of Bangladesh (1993), "Bangladesh has been urging India to settle the issue of sharing the Ganges flows on a long-term basis. But Bangladesh has failed to receive a positive response from her neighbor. In this context, it may be stated that the rightful share of a country to the existing flows of an international river can in no way be denied on the plea of augmenting its flows first. Augmentation can only add to the shares and itself be apportioned on principles of equity that are relevant."

The issue of the Ganges sharing was discussed between the prime ministers of India and Bangladesh during the Bangladesh Prime Minister's visit to India in May 1992. According to the Communiqué (GOB, 1993), "The two prime ministers noted that...the flows available in Ganga/Ganges and Teesta would fall short of the requirements of the two countries. They agreed that an equitable long-term and comprehensive arrangement for sharing the flows of these and other major rivers evolved through mutual discussions could serve the interest of the people of the two countries...."

The concerned ministers accordingly met on August 26–27, 1992. An Indo–Bangladesh JCE headed by concerned secretaries of the two governments was again constituted. The JCE met in November 1992 but could not come to any understanding as usual. The second meeting was held on March 30–31, 1993, and no headway was met in this meeting, as usual.

The prime ministers of Bangladesh and India had a bilateral meeting during the Seventh SAARC Summit on April 11, 1993, and it was agreed that a joint meeting of the prime ministers of the two countries should take place. No such meeting could take place, and the Prime Minister of Bangladesh raised the matter at the 48th session of the UN General Assembly on October 1, 1993. It was stated that "we have not yet succeeded in effectively convincing India about our fair share of the water resources of the common rivers flowing through the two countries. Our two countries share 54 common rivers....

India could have played an important role in strengthening mutual trust between the two countries, if they had lived up to their pledges on the question of water sharing... It is perceived that this is still possible. Bangladesh wishes, therefore, to draw the attention of the world community to this issue in the interest of establishing human right, protecting the right to natural resources and ensuring the process of development. Something must be done urgently to end this inhuman treatment towards the people of Bangladesh. We firmly believe that arrangements must be made to ensure a fair sharing of the water resources of the Ganges by signing a permanent agreement immediately."

9.4 Augmentation Proposals

As discussed earlier, the Ganges Waters Agreement entrusted the Indo–Bangladesh JRC with the responsibility to prepare proposals for augmentation of the dry season flows. India and Bangladesh exchanged their respective proposals in March 1978. They agreed to have these updated and examined in a Commission meeting in July 1983. The two proposals and their review of May 1985 as presented by the two sides are as follows. There has also been some unofficial thinking regarding a Brahmaputra–Ganga link within Bangladesh. It is also briefly discussed. The proposals are shown in Figure 9.2, whereas the Indian proposal is detailed in Figure 9.3.

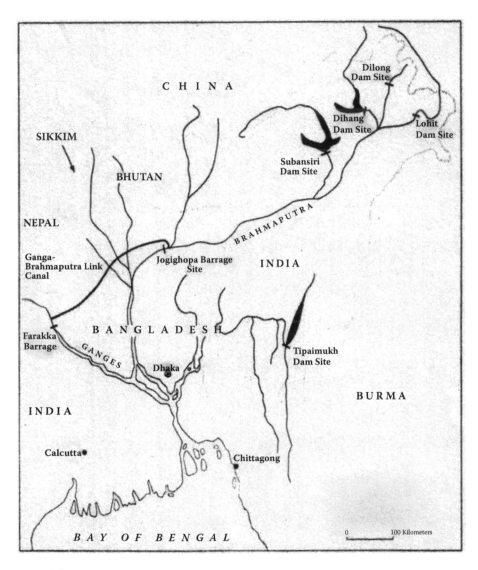

FIGURE 9.3
Indian proposal. (From Crow, B., in *Strategies for River Basin Management*, p. 165, D. Reidel, Dordrecht, Holland, 1985. With permission.)

9.4.1 Updated Bangladesh Proposal

The Bangladesh proposal envisages optimum development of the surface water resources of the Ganges basin, which amounts to 550,000 million m^3 annually, through a coordinated basin plan to be taken up jointly with the cooperation of the co-basin countries: Bangladesh, India, and Nepal. The total demands of the three countries were estimated to be 189,700 million m^3. To meet these demands, Bangladesh proposed the construction of seven dams in Nepal at Chisapani, Kaligandaki 1 and 2, Trisulganga, Seti, Sapt Kosi, and Pancheshwar. These seven dams will increase the present dry season (January to May) availability of the Ganges by 70,000 million m^3. In addition, 51 storage dams in India would increase the availability of the Ganges during the dry season by 63,000 million m^3. Of these 51 dams in India, 29 have been completed from which India is utilizing 30,000 million m^3. Besides, India is also utilizing 36,500 million m^3 of surface water through different completed irrigation projects in the Ganges basin. The existing natural dry season flow of the Ganges at Farakka is about 26,000 million m^3. Thus, against the total requirements of 189,700 million m^3, the updated Bangladesh proposal envisages a total availability of 195,500 million m^3 for utilization during the dry season.

The requirement of 55,000 million m^3 of Bangladesh is based on the irrigation requirements of 3.2 mha out of 5.1 mha of land in the southwestern region of Bangladesh together with the maintenance of navigation, domestic and municipal water supply, industries, etc. Indian demand was based on Indian estimates of irrigating the 55 to 60 mha of land, which was stated to be 91,000 million m^3, plus 14,700 million m^3 for flushing Calcutta Port. The total water requirement for Nepal for all purposes was estimated to be 29,000 million m^3.

The total average annual runoff at Farakka is 550,000 million m^3, while the availability of the Ganges at Farakka is 459,000 million m^3 after upstream uses of which 26,000 million m^3 is the dry season (January to May) flow and 433,000 million m^3 is the wet season (June–December) flow.

Bangladesh made an inventory of the potential storage in the subbasins of the Ganges in India and estimated the potential to be 63,000 million m^3. The potential outflow over a 5-month period was estimated at 4830 m^3/s. India is already using more than 50% of the storage potential. In addition, the groundwater potential in India is estimated to be 108,500 million m^3.

India, in that proposal, estimated that 19.5 mha is presently under irrigation, of which 50% is irrigated by groundwater and the rest by surface water. The utilization of surface water was estimated to be about 66,500 million m^3.

In Nepal, the dams that were already planned could provide an augmentation of 3680 m^3/s. By further raising the seven dams, they could be augmented by 5340 m^3/s, releasing 70,000 million m^3.

The cost of construction of seven dams in Nepal as proposed by Bangladesh was estimated to be US$17.1 billion at 1984 international construction rates, augmenting the flows by 5340 m^3/s. The Indian proposal for augmenting

TABLE 9.1

Salient Particulars of Proposed Dams in Nepal

		Mean Flow (m³/s)		Cost Million
Dam	Height of Dam	Natural	With Dam	Dollars
1. Chisapani	265	1977	1546	3381
2. Kali Gandaki 1	102	482	380	2857
3. Kali Gandaki 2				1488
4. Trisulaganga	284	1305	1055	2600
5. Seti	180	246	192	1003
6. Sapt Kosi	327	2411	1943	4510
7. Panshwar	140	363	223	1313
Total	1445	6784	5339	

Source: Ministry of Flood Control and Water Resources, *BUP (Bangladesh, Updated Proposal) for the Augmentation of the Dry Season Flow of the Ganges*, Ministry of Flood Control and Water Resources, Dhaka, 1985. With permission.

the flows by 2830 m³/s through the construction of three dams at Dihand, Subansiri, and Tripaimukh, a barrage, and a 320-km link canal would cost US$33 million at 1984 international construction rates. The Bangladesh proposal to construct dams in Nepal was claimed to be technically feasible and economically viable and would give multiple benefits to the people of the three countries, Nepal, India, and Bangladesh.

Some salient particulars of the seven proposed dams in Nepal are given in Table 9.1.

9.4.2 Updated Indian Proposal

The genesis of the Indian proposal is that the Brahmaputra–Meghna subsystem has relatively more abundant flows to serve the land and the dependent people than the Ganga has (Table 9.2).

TABLE 9.2

Salient Statistics of Water Scene

S. Water No.	Basin/ Subbasin	Geographical Area (000 km²)	Cultivable Land (mha)	Average Annual Runoff (mha m)	Population (million)	Runoff (m³)	Per Capita Availability
1.	Ganga	1076	65.3	38.0	303	1254	0.58
2.	Brahmaputra	582	9.2	62.6	59	10,610	6.80
3.	Meghna	88	4.5	14.3	32	4469	3.18
4.	Total (GBM)	1746	79.0	114.9	394	2916	1.45

Source: Ministry of Flood Control and Water Resources, *BUP (Bangladesh, Updated Proposal) for the Augmentation of the Dry Season Flow of the Ganges*, Ministry of Flood Control and Water Resources, Dhaka, 1985. With permission.

In the Ganga subbasin of India, out of a geographical area of 86 mha, nearly 61 mha is cultivable. As per the 1978–1979 statistics, the gross irrigated area was 19.5 mha, about half of which was by surface water and the remainder by groundwater. There is already a shortage of water within the Ganga subbasin. The very high demands for Ganga water and the prevailing low intensities of irrigation in the various subsystems bring out the urgent need to use all available waters in the subbasin.

Preliminary studies indicate that after using all the possible groundwater, India's need was around 9061 m³/s in Rabi and 5663 m³/s in the summer. In addition, a minimum need of 1133 m³/s (40,000 cusecs) for the preservation of Calcutta Port was indicated. Further, the need for adjoining drought-affected and other water-short areas in nearby regions also existed.

The total potential storage in India, about half of which has already been developed, would not exceed 50 billion m³. Possibilities of storages in Nepal exist, but the augmentation from them would not exceed 700 m³/s. It is, therefore, not possible to provide the needed augmentation using water from the Ganga subbasin.

On the other hand, Brahmaputra has considerable surface water. The timing of flows is also favorable. The Brahmaputra (at Bahadrabad) carries the lowest flow, in the order of 3964 m³/s (140,000 cusecs), in February or early March, whereas the Ganga (at Hardinge Bridge) carries the lowest flow of 2066 m³/s (73,000 cusecs) at the end of April or early May. The Meghna (at Bhairab Bazar) carries the lowest flow of 481 m³/s (17,000 cusecs) in February. The minimal flow of the entire system at the confluence of Meghna at the Bay of Bengal is around 7180 m³/s (253,000 cusecs) by the end of February. Brahmaputra and Meghna are practically virgin rivers, and they offer the most attractive feasible sites for storage dams.

The updated Indian proposal, in brief, comprises a barrage across River Brahmaputra at Jogighopa in India with a 324-km link canal having a maximum carrying capacity at head of 2832 m³/s (100,000 cusecs). This component of the project, which can be taken up immediately in the first phase, envisages irrigation to 4.5 billion ha of land, augmentation of flows at Farakka, navigation, and hydropower generation among many other incidental benefits. It shall have a power plant with a 300-MW capacity and is estimated to cost Rs.42.74 billion.

According to the design, it is claimed that the link canal avoids thickly inhabited areas, marshy lands, and depressions and generally follows the contour. It could be lined or unlined as would be mutually determined, but tentatively, an unlined canal has been proposed. At the major river crossings, level crossings and drainage siphons have been proposed. The link canal is proposed to be completed in 12 years.

The proposal also envisages the construction of dams across the Dihang River in Arunachal and Subbasiri in Arunachal, both being tributaries of the Brahmaputra, and on the Barak river in Manipur/Mizoram at Tipaimukh. The characteristics of the dams are given in Table 9.3. The integrated Indian proposal will create an aggregate effective storage capacity of 54.5 billion m³

TABLE 9.3

Salient Characteristics of Proposed Dams in Brahmaputra Basin

S. No.	Characteristics	Dihang	Subbasiri	Tipaimukh
1.	Height (m)	262	242	161
2.	Storage capacity gross	47.0	14	15.9
3.	Storage capacity effective	35.5	10	9
4.	Installed capacity (MW)	20,000	4800	1500
5.	Area submerged (km²)	233	n.a.	n.a.
6.	People displaced (No.)	35,000	n.a.	n.a.
7.	Cost (Rs. billion)	82.97	25.75	10.50
8.	Cost of power (considering only power benefits)	16.70	15.9	31

Source: Ministry of Flood Control and Water Resources, *BUP (Bangladesh, Updated Proposal) for the Augmentation of the Dry Season Flow of the Ganges*, Ministry of Flood Control and Water Resources, Dhaka, 1985. With permission.

Note: Capacities are in billion m³.

in the Brahmaputra–Meghna subsystem. This will effectively harness the Brahmaputra–Meghna, allowing significant flood regulation and management and vast multipurpose benefits. The combined installed hydrocapacity will be about 26,600 MW. The augmented waters could extend irrigation benefits to around 8 mha of land in India and Bangladesh.

The Dihang and Subbasiri rivers contribute around 44% of the flows in Brahmaputra. Along with the two other tributaries, Lohit and Dibang, the four contribute about 60% of the yield of Brahmaputra. Dams on these two are also being proposed. It has been estimated that with these four dams, the maximum flood will be reduced by more than 1.5 m in the case of the 1961 flood. The flood peak will not exceed 60,000 m³/s even once in 6 years. The number of flood peaks above 50,000 m³/s would be reduced significantly from the currently estimated figure of 8.

This will show that there will be substantial flood relief. In addition, there will be vastly improved navigation facilities, pisciculture, and prevention of salinity ingress in the coastal areas of Bangladesh and India.

The estimated cost of the Indian proposal was US$16 billion (Rs.161.96 billion) at 1984 international construction rates.

9.4.3 Review of Indian Proposal by Bangladesh

Bangladesh criticized the Indian proposal in principle, technologically, environmentally, and economically. First of all, with the issue of augmentation of Ganga waters, first the possibility of augmentation from the basin should be exhausted before interbasin transfer is considered. In view of the possibility from sources within the Ganges basin itself, as demonstrated by the Bangladesh proposal, there was no rationale for creating storage in the Brahmaputra basin.

Technologically, feasibility is discounted on several counts: (1) the proposed gravity canal lies against the lay of the terrain and will totally disrupt the natural drainage system; (2) it involves overcoming unprecedented engineering problems, including those of intersections and cross-drainage works on major rivers, further complicated by substantial differences in the relative elevation of their water levels; and (3) it predicates the construction of what would be the world's largest canal with a volume of nearly 1000 million m^3, to be constructed substantially below the water table of a potentially good aquifer, crossing 14 major rivers with highly seasonal discharge and high sedimentation consideration in an area characterized by fluvial instability and high seismic risk.

It will have far-reaching negative consequences affecting the entire environmental and ecological balance of the region of Bangladesh served by Brahmaputra. These are (i) immeasurable increase of the hazards attendant on flood drainage and consequent inundation, (ii) obstruction of groundwater movement and enhancement of water logging, and (iii) depletion of waters and consequent adverse effects on the rivers' regime, including sedimentation and salinity intrusion. The environmental consequences are far reaching. The negative impact on the environment and economies are extremely serious, affecting the entire Brahmaputra region, which covers 50% of the country and 60% of the population.

The sheer magnitude of the Indian proposal makes it unwise in terms of contemporary international feasibility. It is uneconomical with a cost of US\$4.5 per m^3/s as compared to the US\$0.97 in the Bangladesh proposal.

9.4.4 Review of Bangladesh Proposal by India

India's rejection of Bangladesh's proposal was equally severe. First, it was emphasized that the issue was that of optimum development of the GBM region as a whole rather than focusing only on Ganges for augmentation of the Ganges flow. With the per capita and per unit of land water availability in the Brahmaputra and Meghna basin 4 to 12 times higher than those in the Ganga, development and transfer for augmentation have to be sought in the Brahmaputra and Meghna basin rather than in the Ganga basin.

India is by any reckoning the major riparian country in the Ganga waters, in terms of the catchment area (99%), population (94%), and the area dependent on its waters (94.5%). For 90% of its length, Ganges flows in India. India is, therefore, preeminently entitled to its waters on an equitable basis and does not have to ask for any permission in undertaking development in the Ganga basin from lower riparian regions. The availability in the Ganga is scarce, and all the development is fully committed to meeting local needs. It is therefore preposterous to suggest that the operation of storages in the Ganga basin should be used for augmentation. On the other hand, the Brahmaputra–Meghna basin has abundant water and very promising storage sites.

The proposed developments in Nepal of raising the height of four high dams and using the three high dams for augmentation are fictitious and infeasible for several reasons. First, the proposals for seven of the world's highest dams have been made without carrying out any detailed studies and, therefore, are technologically inadmissible. Second, there are indications that raising is not technologically feasible. Third, the socioenvironmental costs have not been considered. For instance, it is estimated that the proposed dams will involve the rehabilitation of a very large numbers of people and the submergence of 300,000 ha of land. Fourth, concurrence of the Nepal government has to be taken. Finally, all the waters are already fully committed for use in the Ganga basin itself. The augmentation portion may not be more than 700 m³/s after meeting Nepal's needs and the local needs.

It was pointed out that there are several errors regarding hydrological data. The water availability in Ganga is 390 milliard m³ and not 550 milliard m³ (milliard m³ = billion m³) as stated in the Bangladesh proposal. Similarly, the estimates of water demand are also grossly unequal and erroneous. Moreover, the water requirements for irrigation for India had been wrongly estimated. These were 9061 m³/s during Rabi and 5663 m³/s during the summer months, besides the requirements of 425 m³/s for irrigation of the former Farakka.

The cost estimates have also been arrived at on an ad hoc basis. The claim of generation of 10,000 MW from these dams was not supported by any data. Furthermore, the cost of the Indian proposal was only US$16 billion and not US$33 billion as reported in the Bangladesh proposal.

In short, the Bangladesh proposal is only "a conceived idea."

9.5 Developments from 1987

Both sides completely disagreed with each other's proposals, considering them conceptual and impractical. According to two leaders of the Indian team over the long period, the meetings "were largely dialogues of the deaf" (Mehta, 1992a) and "prolonged and fruitless series of talks over the years" (Rangachari and Iyer, 1992). The assessment made by Bangladesh leaders was similar (Abbas, 1982).

In 1987 and 1988, the annual monsoon floods were unusually severe. Foreign attention was particularly focused on the subject, and several technical reports by foreign consultants were produced. On their basis, a World Bank Action Plan (1989) was also prepared. Concern about floods transcended most other development objectives in Bangladesh in this period.

Regional cooperation was resumed only in 1990 when the JRC met after a 3-year gap. However, some momentous political developments took place that shook the entire activity. In 1990, Rajiv Gandhi was assassinated and President Ershad was overthrown. Considerable political unrest ensued in Bangladesh. The political scene in India, under Rao and subsequent prime ministers, had other preoccupations.

9.5.1 New Visions

However, gradually, the political atmosphere of uncertainty ended. Earlier, neither side could persuade the other, causing an unfortunate impasse. A major impediment to cooperation was overcome with the signing of the Ganga Water Treaty in December 1996.

As Rangachari and Verghese (2001) stated, "The Treaty delinks sharing, a right, from augmentation, a wish. It assumed average flows at 70,000 cusecs at their lowest at Farakka (on the basis of 50% rather than 75% dependability as hitherto). In order to optimize benefits, the flows are to be shared in accordance with a stipulated formula subject to each side being guaranteed a minimum of 35,000 cusecs over six alternating 10-day periods during March–May. This sudden alternation is not without certain technical problems, but the politically sanctified figure of 35,000 cusecs has rendered it acceptable.

The Treaty is subject to periodic review and liable to renewal after 30 years. India has undertaken to make every endeavor to protect existing lean flow arrivals at Farakka in conformity with average availability over the 40-year period between 1949 and 1988. This will entail disciplining the upper riparian states against excessive abstractions or groundwater pumping that could diminish average stream flows down the river during the critical dry season. The objective of the long-term augmentation remains, but this additionality is to be sought as and when possible and to the extent feasible, without impinging on the basic sharing formula."

Since then, political developments have taken place in Bangladesh that have put these matters completely in the background.

9.6 Overview

It is unfortunate that long years have been wasted regarding a very important issue. However, considering the history and past achievements, the development may be interpreted more realistically as trying to find a technical solution to the problem while waiting for an appropriate political environment.

10

Institutional Setup

10.1 Introduction

Human activities get organized through social organization, as cooperation is required and conflicts have to be resolved. Practices, procedures, organization, and institutions evolve over time, and attitudes and culture of activity develop. Natural resources management was the predominant activity in agricultural societies, and social organization developed particularly in that context. Evolution of the village communities was a natural corollary. These could be further organized in terms of increasingly larger organizations of kingdoms and empires, with the latter concerned primarily with issues of protection of life and collection of land revenues. With an almost quasi-stable population and technology, similarly stable institutional setups and practices for environmental management evolved.

Although foreign rulers had come to India, they generally settled down in the country. The advent of the British regime was an altogether different historical event. Some commercialization of agriculture, some advances in technology, and changes in policies also took place. Almost a new institutional setup and management culture were introduced. Further revolutionary changes took place after Independence. The future challenges are even more formidable, and besides new policies and technologies, new attitudes, institutional setup, and management practices will also be required. These have, however, inertia and historicity, which have to be taken into account. Therefore, we review the institutional setup and management in the following phases: (1) the pre-British period, (2) the British period, and (3) the post-British period. The first period was common to the entire basin, and the second period was common to India and Bangladesh. The post-British period is considered separately was common to India, Bangladesh, and Nepal.

Our concern is with environmental management embracing natural resources development and conservation focusing on land, water, and the natural habitat of forests. The associated economic activities are equally important. Agriculture was the predominant activity, and, until recently, management of land was synonymous with agricultural management. We therefore also look into the institutional setup and management of the economic activities from the point of view of environmental management.

Again, environmental management is part of the overall administrative setup and culture. It is therefore also briefly reviewed at the outset, as it impinges dominantly on the attitudes, culture, institutional setup, and management in this area.

10.2 Pre-British Period

10.2.1 General Administration and Resource Management

The countries of Asia and Africa were from the earliest times administrative states, as they were based on land revenue administration (Panikkar, 1959). Suggestions have been made that control of water for irrigation and navigation was an important part of state administration, and they could be called hydraulic civilizations (Wittfogel, 1956, 1957). In Indian literature, particularly the epics Ramayana and Mahabharata, abound in suggestions and advice on government and administration in general and personnel administration in particular, which shaped values and concepts (Paipanandiker, 1966). Kautilya's Arthshastra, an extensive treatise on government and administration, had been written as early as 300 BC. By and large, the tenets and structures of personnel administration detailed in the Arthshastra formed the general basis of the classic system of Indian public and personnel administration (Altekar, 1958).

While these administrative tenets and structures remained basic to the governmental system of India, drastic changes were brought about by the Moguls in some respects in the fifteenth through nineteenth centuries. The Moguls were "kings by profession," coming from a different culture and religion who settled in India. They brought with them the practice of dividing the two political functions, the governorship and treasury. The provincial governor (called Subedar) had control over the military command, police, and penal jurisdiction only. Alongside with him was the head of revenue administration and civil jurisdiction (called Diwan). The Subedar was the first in authority, but he was equal in rank with the Diwan. The two officers provided an effective system of checks and balances.

Introduction of foreign cultural and religious personnel itself introduced a different value system at the top. New styles of consumption were also introduced, which seemed to be the crucial link between the changing political and moral order and the changing economy. Most important was what might be called the Muslim ashraf style. It is a specifically Islamic notion of right conduct and outward observance in contrast to the Hindu conceptions of "inner righteousness"—purity and restraint—embodied in the equivalent term "bhadralog" (cultured people) (Bayly and Bayly, 1986).

Although a schism was created in the society with the introduction of a new culture, religion, and political–administrative elite, the Moguls had come to stay and adopted the country as their own. The bureaucracy also was integrated with the society. Thus, while there were drastic changes in the political setup, administration, and the value system, the land revenue system of Mogul India is noted for its long and close adherence to the old practice, procedure, and even tradition of the country. The old Hindu revenue officials were retained, and there was seldom any interference in the functions of the department.

At the lowest organizational level, the Indian village had evolved as a social unit engaged principally in the use and conservation of the natural resources. The land belonged to the person who first started using it, but there was no private property in land, and as Marx sarcastically commented, this is the real key even to the Oriental heaven (Aveneri, 1960). It has been suggested that the subcaste system evolved from consideration of environmental conservation and efficient resource use (Gadgil, 1985). Anyway, perhaps an Indian village is a perfect adaptation for survival. Over centuries of upheavals of empires and onslaught of foreign invaders, the Indian village survived without even a consciousness of these changes, using the natural resources as best as it could. The establishment of the Muslim and later Mogul regimes made no difference.

10.3 British Period

10.3.1 General Administration

By the end of the eighteenth century, the Mogul power was undermined, and by the early nineteenth century, the East India Company, which had entered India as a trading concern under a charter originally granted by Queen Elizabeth of Great Britain in AD 1600, had firmly laid the foundations of the British Empire in India. It operated with three presidencies, Madras, Bombay, and Bengal, with the headquarters of the latter in Calcutta, which also later became the capital of India. Its administrative activities started when Clive won the battle of Plassey, defeating and killing the Nawab of Bengal in 1757. The Great Mogul in Delhi, who had lost all control of Bengal long ago, thought that he could reassert his influence by coming to terms with Clive. He offered the Diwani of Bengal (revenue administration and civil jurisdiction) to the East India Company, leaving only the military command and penal jurisdiction to the Nawab. The foundation for the plunder of Bengal and subsequent conquest of India was laid.

The East India Company found it expedient to maintain the traditional Indian system of administration and revenue collection. The method of auctioning the right of collection of revenue to the highest bidder was introduced at that time by the Nawab of Bengal, long before the British adopted it too. It slowly instituted a crude form of civil service system, which became the forerunner of the administrative system that was duly started from 1855.

The bureaucratic system developed in India as it was being developed in Great Britain. Bureaucracy is a modern-world phenomenon, but the historical socioeconomic political factors in India were entirely different from those in the West, giving it a unique culture. In the latter, despite national differences and departures from Weber's ideal type, the modern bureaucracy emerged as an antithesis of feudalism, a result of economic progress and intellectual development, as a middle-class bourgeois concept functioning with the elective elements of government for public administration. Both are jointly responsive to public needs and services, being part and parcel of society, although functionally, they differ considerably. In India, there was no parallel economic–political development. Instead, the Government was alien and despotic, designed to rule rather than to govern. The bureaucracy was, thus, the ultimate government. Derived from the military background, most of the civil functions were performed by military personnel. Until 1858, when the British Indian Government became directly responsible to the Crown instead of the East India Company, the ratio between civil and military personnel remained almost half-and-half. In the bureaucracy, the senior officers belonging to the Indian Civil Service (ICS) were all British, with a lower service unit known as "uncovenanted civil service" that employed Indians. As has been observed, the ICS was neither Indian, nor civil, nor a service. In 1892, provincial and subordinate services were established by a subdivision of the existing uncovenanted service. Even with the marginal Indianization of superior services in due course, alienation and aloofness from the people were emphasized. The concept that the services represented what the government was meant to rule was the central philosophy, which was abetted by the schism of the society culturally and economically, with even a different language for the elite and the rest. It was topped by racism, with Indians being considered almost untouchables socially.

With the role of the Government to preserve law and order and collect resources, the corresponding ICS and Indian Police Service had preeminence. Each district had a bureaucratic setup consisting of the District Magistrate and Collector, the Superintendent of Police, and the Civil Surgeon.[1] Representatives of other activities such as judiciary, irrigation, buildings and roads, and agriculture could also be located if they had appreciable activities in the district.

Four classes of services were gradually established. One was the Indian services with the preeminence of the ICS. Services in other areas such as police, education, engineering, forests, agriculture, and so on were created, and some were dropped with the transfer of some of the functions to the states. In 1947, after Independence, a rationalization was tried to be introduced through the nomenclature and pay scale of Class I, Class II, Class III, and Class IV services, each representing the central, provincial, clerical and similar services, and, finally, peons and the like.

10.3.2 Administration and Management of Land and Agriculture

Administration and management of land and agriculture in the British rule was a follow-up of the Mogul system. The concern was only for collection of revenues and maintenance of law and order. However, fundamental changes in land proprietorship were introduced, which brought about revolutionary socioeconomic and environmental changes in the village scene.

Until the British reign, there was no private property in land. The Zamindar was only a tax collector. The agricultural land belonged to the tiller, who could not be evicted. The pasturelands, marginal lands, and village forests were common village resources governed by established practices.

A system of granting the Zamindari on the basis of the highest bidder for land-revenue collection had already been introduced before the British by the Nawab in Bengal. However, there was an intimate relation with the Zamindar as part of the village community. Even revenues could be reduced taking into account bad rains. The British introduced the permanent settlement of Bengal in 1793. It was based on the rule that the Zamindar, who used to be only a tax collector, became a landlord in the British sense of the term. The peasants were now called tenants, and they paid rent rather than revenue to the landlord. The settlement was fixed once and for all and was not at all generous by the standards of that time. The Zamindar would keep 10% of what he collected from the peasants. Land could be sold, and the tenant could be evicted. The common resources became the property of the Zamindar.

By injecting the dynamics of British contractual law into the traditional Indian land revenue system, the new rulers toned up the system in a big way but in their interest and not for the welfare of the people or efficient resource use and environmental conservation.

Administratively, a collector was appointed who was the chief district officer of the Government. Introduction of a land tenure system, opening up of road and rail transportation systems, and promotion of export trade in certain agricultural commodities within the framework of the free trade policy of the colonial power were important steps taken by the British Government, but there was no agricultural policy. A separate Department of Agriculture was created in the Government of India in 1871, and subsequently similar departments were set up at the provincial levels. They were skeletal departments mostly collecting statistics.

Some activities started by the beginning of the twentieth century. Directors of Agriculture were appointed in the provinces, and a few agricultural colleges and experimental farms were established. A Royal Commission was appointed in 1926 under the chairmanship of Marquise of Linlithgow, who was later to become a viceroy, but scientific management or development of agriculture remained a peripheral activity except for the establishment of the Imperial (now Indian) Council of Agricultural Research in 1929. There was no concern for management of land; agricultural departments were looked down upon, as were all developmental departments, the primary emphasis being on administration and police.

A Department of Food was created in 1942 to cope with the difficult food situation. The Department of Education, Health, and Lands was trifurcated to form a separate Department of Agriculture. An all-India policy on agriculture was formulated, and the Government of India issued a statement of agricultural and food policy in India in 1946.

10.3.3 Water

Management of water has been emphasized since prehistoric times, but institutional arrangements were developed only in the British period, starting with the repairs and remodeling of the Western Yamuna Canal in 1835. Gulhati (1965) has given a brief historical background of the administrative system. Water management was conceived in terms of irrigation, and like all other public works and railways, it was the responsibility of military engineers of the three engineering corps of the East India Company, working under the control of the Military Board, and subsequently of the Corps of Royal Engineers. The Military Board was abolished in 1854 and the Public Works Department came directly under the Government of India. In 1866, the Public Works Department was divided into three branches, a Military Works branch, a Civil Works branch (which included irrigation), and a Railway branch. By 1895, the Military Works Department was separated from the Public Works Department. The latter was, henceforth, in charge of all public works, civilian in nature. Even so, most of the senior engineers continued to be from the Royal Engineers. Under the direction and control of the Public Works Department of the Government of India, there was a public works department (PWD) in each of the provinces. The two major activities, "irrigation" and "building and roads" became the two branches of PWDs, or separate departments, varying from province to province over time.

The actual execution and management of irrigation works has thus always been a function of the State Government. Until 1921, however, the Government of India and the Secretary of State in London exercised powers of superintendence, direction, and control on all irrigation activities of the State PWDs, whether technical or administrative, through an Inspector General of Irrigation and a Public Works Secretariat of the Government of India, with an engineer of experience as its Secretary.

In 1921, after the introduction of the first installment of constitutional reform, irrigation became a provincial but "reserved" subject. In other words, the rigid control until then exercised by the Secretary of State and the Government of India was abolished, but with the expenditure on irrigation development not having been made subject of the vote of the provincial legislature, the administration of irrigation works was reserved to the Governor in Council and was therefore under the ultimate control of the Secretary of State.

Since the Government of India ceased to have any direct responsibility for irrigation works, the Public Works Secretariat of the Government of India was abolished in 1923, and the post of Inspector General of Irrigation was converted into that of Consulting Engineer. It was arranged that such advice as the Government of India might need would be obtained from a committee of two or more members of the Central Board of Irrigation (CBI), which had been constituted in 1927. The CBI was a forum in which all the chief engineers (irrigation) of the provinces would meet together with a view toward coordinating work done in the provinces, particularly to coordinate and guide research and bring about improvements in irrigation practice and methods of construction.

From April 1937, with the inauguration of provincial autonomy, irrigation became a transferred subject and the provincial governments got full powers over irrigation works and development. The Central Government was no longer concerned with the development of irrigation except when a province or princely state would make some objection to some development in an adjoining province and the dispute could not be resolved by mutual agreement. In that case, the dispute was required to be referred to the Governor General, who was empowered to give a decision in the matter in his direction, after investigation by a commission to be specially appointed for the purpose. He could, alternatively, refer the matter to His Majesty in Council.

The Government of India experienced difficulty in dealing with water problems and, in April 1945, set up the Central Waterways, Irrigation, and Navigation Commission to initiate, coordinate, and, in consultation with the Provincial and State Governments concerned, further undertake schemes for the control, conservation, and utilization of water resources throughout the country for the purposes of water power generation, irrigation, navigation, and flood control and, if so required, to undertake the construction of any such schemes.

Other areas of water did not attract much attention. Flood control was a negligible activity and was looked after by the Irrigation Department. Hydroelectric development started with the construction of a small run-of-canal scheme through the conversion of the falls on the upper reaches of the main canals from the early 1930s. The activities were also looked after by the Irrigation Department, hiring a few electrical engineers working under the officers of the Irrigation Department, who were all civil engineers. Similarly, the tube well development, which consisted essentially of public tube wells of $0.02832 \text{ m}^3/\text{s}$ (1 cusec) discharge capacity, were also undertaken by the Irrigation Department, hiring a few mechanical engineers to assist the regular officers of the department.

Water supply and sanitation were little developed and were under the control of the municipal agencies of the towns. Gradually, direction was provided by a senior officer under the Local Self Department of the State Government.

The professional responsibility and competence of the engineers were very limited. The task was essentially that of managing the water supply for irrigation. The work was carried out by civil engineers as officers, with some support by the revenue officers for collection of irrigation charges. The officers therefore had magisterial power, but essentially, they were bureaucratic officers rather than engineers or scientists. There was occasionally some engineering construction activity, but it could be carried out reasonably well, with these officers basically, if not very efficiently, providing administrative and technical supervision.

The senior army engineers were mostly British, educated in England. Civil engineering schools were established in India with the undertaking of the construction activities. The first was the Thompson College of Engineering established in 1846 at Roorkee in the then northwestern province near the headwork of the Ganges Canal. The college was essentially an officers' training school, with a Diploma of Civil Engineering being awarded at the end of a 3-year educational cum training program and a certificate that the candidate was fit to be employed as an Assistant Engineer in the Public Works Department. (The author finished the program as part of the last batch in 1946, after which, with India's achievement of Independence, Thompson College was converted into Roorkee University.) The faculty essentially consisted of officers from the state engineering departments on deputation. A school for training overseers was also established at Roorkee. Similar schools were opened in the other states, e.g., Bengal, Madras, and Bombay, in due course.

There was little research. Research stations were established from the early twentieth century, but they were essentially testing stations. The Central Board of Irrigation and Power (CBIP), with headquarters in Shimla, was essentially a club of chief engineers meeting at the hill station during the summers. It must, however, be stated that the science of civil engineering, hydraulics, hydrology, and water resources, even internationally, was still in an elementary stage.

10.3.4 Forests

Greatest importance to fauna and flora has also been accorded in Indian culture. Several communities developed valuable ecological practices (Gadgil, 1985). There was a superintendent of forest in the reign of Chandragupta Maurya (300 BC). However, management of forests was neglected under the Mughal rule. Organized management of forests can be attributed to the British regime. The Forests, Department was created in 1864 under an Inspector General of Forests and the Indian Forest Act 1865 was enacted for the management of forests, bringing them under proper management and empowering the local governments to draft rules for enforcement in the respective regions. The Act was revised in 1878. A National Forest Policy was enunciated in 1894, laying down the general principle for the management of forests.

In 1921, the forest administration was vested in the provincial governments. With the passing of the Government of India Act of 1935, forests became entirely the concern of the provinces, and the role of the Center was limited to the common or general aspects of forestry viz. forest research, forest education, soil conservation, etc.

10.3.5 Environment

The importance of environment has been recognized in India since time immemorial, but organized institutionalized activities were initiated only during the British period. The Botanical Survey of India (BSI) was established in 1890 with the objectives of surveying the plant resources of the country. Similarly, the Zoological Survey of India was established in 1916. However, active environmental management was not undertaken.

10.4 Post-British Period—India

After Independence, India had a revolutionary change in objectives. All that was available were a microscopic committed leadership of the elite and a colonial bureaucracy. Independence was essentially a transfer of power to them. In the planned development that was proposed to be undertaken, the State had a prominent role particularly in the development of natural resources and the infrastructure, which are public sector activities. New attitudes, concepts, capabilities, work culture, and institution were required to implement the challenging tasks. Unfortunately, the revolution that was required has not taken place. Mahatma Gandhi was lost by the nation soon after Independence. After a brief period of euphoria and direction by few committed leaders, an increasingly inefficient and corrupt set of people occupied the seat of government as the people's representative, though through a duly elected process called democracy. Economic and social change, which professes to bring in development with equity, is repeatedly proclaimed, but the rich, powerful, and corrupt increasingly prosper. The pattern of development has been called bureaucratic socialism (Bhagwati, 1994) by a soft state (Myrdal, 1971). Institutional setup and its dynamics that have emerged in various areas are briefly reviewed.

10.4.1 General and Developmental Administration

With the Government by the elected people's representative, the public services had a new role. Furthermore, with the objectives of development and the policy of state-sponsored developmental activities and increasing wide range of activities, public services had another imperative of a transformation. Their reorganization with new objectives, culture, and capabilities was urgently required. Unfortunately, this was not undertaken, and the colonial bureaucratic service continued, with new names, if at all at an ever-growing scale. Instead of a government by the people, of the people, and for the people, it was a government by the government, of the government, and for the government. Government signified the political elite and the bureaucracy.

As the planned development was undertaken, the paraphernalia of Class I, II, III, and IV services was required at an exponential scale. Recruitment to Government of India services had been stopped in 1942 as it was proposed to fill the services with those who had offered to serve in the war. New administrative services like the Indian Administrative Service, Indian Foreign Service, and Indian Police Service were started. In addition to a law and order state, it also became a "permit and quota raj," as all economic and developmental activities came under governmental control. As the committed nationalist leaders gradually departed, there was continuous and rapid erosion in the efficiency and integrity of the political leaders. Increasing criminalization of politics to politicization of criminality took place, leading to rampant corruption and inefficiency. Public sector activities suffered the same fate. With increasing governmental activities in each phase and an exploding bureaucracy, the government has become a "corrupt functioning anarchy," indifferent to people's needs and woes. The scale and scene deteriorate at an exponential scale.

10.4.2 Land Management and Agriculture

Agriculture received the highest emphasis in the planned development. Several schemes were undertaken. The management of the schemes was generally done by the administrative civil servants who were appointed and transferred from time to time in accordance with the colonial practice. There is no system of accountability and rewards and punishment with even the lowest rank of bureaucrats, with the peon, on a permanent appointment, getting the advancement in a grade scale. The higher the service, the higher and faster the promotions and increments in emoluments and perks, with no correlation with performance.

Management of the natural resources and environmental processes is an almost neglected or secondary concern. As noted by the National Commission of Agriculture 1976, "in most of the States with the possible exception of Uttar Pradesh, there is no specialist staff, either at the state or the divisional level, who could make proper assessment of soil and moisture conservation measures and help farmers solve problems of soil deterioration and waste. There is also none to carry out functional field inspection of various programs and practices in respect of their technical adequacy and working efficiency. In most of the states, at the headquarter levels there is no planning cell to collect and compile data on the problems, achievements, and benefits of soil conservation. There is no specialist at the State level to provide information and publicity.... In view of constant shifting of technical staff from one area to another...provisions regarding maintenance have not been enforced fully" (GOI, 1976). The Commission had adverse observations regarding institutional and management aspects of all phases of natural resources development and environmental conservation. The position has not changed much since and if at all has deteriorated.

10.4.3 Water

The concept remained water resources development with focus on irrigation and multipurpose developments. Under the new constitution of independent India, promulgated in 1950, water resources development has continued to be a State subject, but the Central Government has been charged with the "regulation and development of interstate rivers and river valleys to the extent to which such regulation and development under the control of the Union is declared by the Parliament by law to be expedient in the public interest." The only law so far enacted under this constitutional provision is the River Boards Act of 1956, which authorizes the Government of India, in consultation with interested States, to set up advisory River Boards on interstate rivers. No such board has, however, been formed. Since the Ganges, Brahmaputra, and Meghna rivers have international issues, Commissioners for Ganga and Brahmaputra were appointed in the Ministry of Irrigation, Government of India. The designation has now been changed to Commissioner of Bangladesh and Nepal.

As part of the planning process, established in 1951 in connection with the 5-year plans, the State Governments do not undertake any new major or medium project, unless it has first been cleared by the Planning Commission and the Government of India and included in a 5-year plan. All central loan assistance is subject to this overall condition, and this enables the Central Government to indirectly influence the pattern of development and technological efficiency. The Central Water and Power Commission (CW and PC) were constituted as a central coordinating agency in respect of water resources development in India. It cleared the projects prepared by the States technologically for consideration in the plans and undertook preparation of projects on behalf of the States, if so requested. One point needs to be emphasized. Although an overview is understood to be maintained of the technological and environmental activities as well as of progress, this is reduced to the state of recording as to what is going on. Serious technological and environmental shortcomings can be pointed out in almost all projects and there is hardly any economic and financial discipline. However, this goes on, as there is hardly any independent, even academic, review of any project or activity. It is all a closed classified activity.

To start with, the Irrigation Department dealt with hydroelectricity, as it was mostly the States that were involved in hydroelectric energy, with thermal generation being under the private sector. With increasing developments in electric generation, autonomous State Electricity Boards under the Ministry of Power were established. The hydroelectric development may be undertaken by the Irrigation Department at the State level, but it increasingly came under the jurisdiction of the State Electricity Boards. At the Center, a Central Electricity Authority was constituted, and CW and PC became the Central Water Commission (CWC). Of late, the nomenclature of the Ministry of Irrigation at the Center has been changed to Ministry of Water Resources. However, at the State level, the concepts, practices, and administrative system remain as in the past, with total indifference to any sort of modernization.

A National Water Policy was formulated in 1989. Water is being looked upon as a national resource. A National Water Development Agency (NWDA) has been established as an autonomous body to study the possibility of an inter-river basin and thereby interstate coordination. However, there is considerable resistance from the States, and all the studies, particularly the hydrologic data, are classified. This is even more strictly enforced for the Ganges, Brahmaputra, and Meghna, as it is an international river basin. According to the Eighth 5-Year Plan (1992–1997), perspective plans have been prepared for the Ganga basin and the Brahmaputra and Bark basins, but they are all classified.

All activities related to irrigation were initiated in the Irrigation Departments, but over the course of time, some administrative changes have been made. Minor irrigation was reorganized as a high-priority program under the Ministry of Agriculture. In the Annual Plan of 1966–1969, it was given about one-third of the total allocation made for the agricultural program. However, the person-nel remained the same, transferred from the old irrigation departments. Again, while groundwater development was initially under the Irrigation Depart-ments, an Exploratory Tubewell Organization (ETO) was set up at the Center in 1954 to intensify efforts at deep exploration. This was reorganized in 1970 as the Central Ground Water Board, and groundwater development at the State level came under the jurisdiction of the Ministry of Agriculture.

Experience with intensive agricultural programs showed that irrigation, as was practiced in most of the command areas of canal projects, required modernization. A water utilization cell was set up at the Center in 1976. This was expanded to Command Area Development Programmes (CADP) in 1974–1975 and has been considerably expanded at the Center and the State even though Government experience is not very satisfactory.

While the concept of water management in India had remained canal irrigation and the engineers were essentially colonial bureaucrats, tremen-dous advancements in engineering of dams, power houses, and so on had taken place abroad. The U.S. Bureau of Reclamation (USBR) had attained considerable eminence, and the integrated planning of the Tennessee Valley (TV) has become well known. India was trying to develop indigenous self-sustained capability, and engineers were sent to USBR for training. Damodar Valley development was undertaken on lines of TV. There was commend-able commitment in the initial period of Independence, and emphasis on development of indigenous capability was made. The achievements in the early period, such as the Hirakud Dam, Bhakra Dam, and Rihand Dam, are commendable.[2]

However, as time passed, the commitment disappeared at a fast pace. The reason was the deterioration of the social system. The polity of Jawaharlal Nehru and other political leaders that ushered Independence disappeared in no time and was replaced with unprincipled and corrupt politicians. There was a rapid all-round social deterioration. The institutional revolution required to introduce modernization never took place. No effort was made to promote specialization in various facets of engineering such as planning and design, construction, and management. No effort again was made in promoting continuous education and collaborative activity between professional and academic institutions. The irrigation departments of the states expanded considerably, and even the earlier corps-de-spirit and discipline, two important contributions of the British military, were lost. Regular transfers were a central policy of the colonial system so that officers would perform their work objectively. While this was suitable for regular administration, this could not apply for specialized professional jobs. Even worse, occupancy in the highest offices was reduced to a few months, leading to increasing inefficiency and indifference. Corruption has also been an increasing trait. The political system shares the blame, but the profession is also responsible for the anarchic institutional and personnel state. In nomenclature, even the cosmetic change undertaken by the Central Government in replacing the name of the Ministry of Irrigation with Ministry of Water Resources was not undertaken by the States.

Mohile (2007) has described the government policies and programs in detail, but they remain pious declarations, changing names and organization structures, if any, without any institutional and technical modernization, because the characters remain the same—a corrupt polity and incompetent corrupt officialdom instead of a modernized, motivated scientific community.

10.4.4 Forests and Natural Habitat

The administrative and institutional arrangements that developed during the British period continued after Independence. Management of forests remains a provincial subject although coordination is provided by the Center through the formulation of policies by an Inspector General of Forests. Class I and Class II services continue, unlike the engineering fields. Interest has also been taken by several nongovernmental organizations.

10.4.5 Environment

Starting from the survey activities of BSI and Zoological Survey of India and, more importantly, forests policy management at the Central Government level, a Ministry of Environment and Forests was started by the Government of India. However, the change was only in name, with no specialization in the personnel.

10.5 Bangladesh

Bangladesh had the same history until Independence as India's and, after Independence, has faced even more difficult problems in institutional modernization because of an even more troubled history after Independence. Development embracing environmental management posed serious challenges, but Bangladesh had the same institutional antecedents. The indented administrative system displayed all the institutional trappings of a colonial bureaucracy that was elitist in character, centralized in structure, impersonal in behavior, amateurish in approach, and formalistic in operations. Because of historical reasons and its overdeveloped nature vis-a-vis other institutions in society, it was firmly entrenched within the social–political fabric. Its members were indoctrinated by techniques that were so alien in both letter and spirit as to create in them a peculiar temperament that would initially serve the interests of imperial and later internal colonial rulers. Generally, their social and economic background and upbringing made them contemptuous of indigenous politicians, and their operational style isolated them from the common man.

Attempts are being made to modernize the bureaucracy and make it responsible to people. Chronic political instability, economic mismanagement, social unrest, and, above all, ideological confusion on the part of the rulers make it very difficult (Huque, 1994). With this background, we may briefly discuss the institutional setup in the environmental sectors.

10.5.1 Land and Water

Management of land is limited to agricultural production. Soil conservation is still a neglected area. Water management has, however, been institutionalized, taking into account the recommendation made by foreign advisors. The East Pakistan Water and Power Development Authority (EPWAPDA) was established in 1959 to handle the planning and implementation of water and power development activities. The present Bangladesh Water Development Board (BWDB) was formed from EPWAPDA. It has the responsibility of flood forecasting, irrigation, and drainage with special regional O&M units.

With assistance from the World Bank, a Master Plan Organization (MPO) was established within the Ministry of Irrigation, Water Development, and Flood Control. The MPO produced the National Water Plan in 1986. It has been recommended in a UNDP-sponsored study by the Ministry of Planning that the MPO should be made part of the Ministry of Planning. Secondly, a new and high-powered authority called the National River Authority (NRA) should be established, distinct from BWDB for executing the flood policy.

At the educational and research level, the Bangladesh University of Engineering and Technology (BUET) is the only major engineering school in Bangladesh. It has a commendable background on hydraulic engineering and flood research, with collaboration with foreign agencies.

10.6 International Institutional Interaction

The developing countries lack technological capability and capital. Efforts are made by the countries to seek both in several ways. Donor countries and international organizations have also tried to help, but there have often been extraneous considerations and conditionalities. While international help has a positive facet, there also are negative aspects. The technology may not be appropriate. International support may engender dependency. We, thus, briefly discuss the international interaction so that appropriate policies may be formulated. We will first focus on India, and then we will consider the Bangladesh and Nepal scenes.

10.6.1 Indian Scene

The environment after Independence, particularly in India, was of euphoria and commitment. With a long history of colonial experience, attainment of technological modernization with self-reliance was a central policy issue in the new India. Backwardness in the field of natural resources development was well realized, and efforts were made toward modernization. Officers were sent abroad to learn modern developments, and experts were invited to guide the indigenous developments; new technological institutions were established with foreign support. A central issue, however, always was the development of indigenous self-sustaining capability. This was strongly emphasized by the Indian engineering leaders, led by Dr. A.N. Khosla. In view of the present negative attitudes, this may be brought out a bit, as the author has always strongly believed them, particularly with his experience with the capabilities of the so-called foreign experts.

The author had been involved in India's water resources development from the beginning of his career, from just before Independence, working under the guidance of the great Indian engineers, Dr. A.N. Khosla and Dr. A.C. Mitra. Since it was the beginning of the water resources development, some so-called experts were invited. Having worked with them for some time, the author realized that except for some practical real life experience, they hardly had any fundamental knowledge. The author therefore decided to go to the USA and study the subject at some leading university. This was in 1959. The University of Iowa, Iowa City, under the leadership of Dr. Hunter Rouse, was considered to be the leader in the area, and the author applied to it. Fortunately, it was also well known in the author's wife's area—education. The author was even luckier for having been awarded a Fulbright travel grant. A central experience is worth mentioning: while they had a strong base in fundamental areas, lack of practical experience was totally lacking—a problem that haunts engineering universally.

Indeed, Dr. A.N. Khosla was conscious of this aspect and, while establishing the Water Resources Development and Training Centre at the Thomason

College of Civil Engineering, later IIT Roorkee, he tried to integrate the two aspects as best as he could. However, this is a basic problem in engineering education, unlike the medical or even the legal areas. Fortunately, personally for the author, this has been mitigated by his equal association with real life as well as the academic area.

However, as the scale of development and need for financial support increased and, at the same time, the political environment changed, the implication of international interaction also changed. There has been an increasing shift toward neocolonialism and dependency. It was not that there was necessarily a conscious effort by the foreign governments and agencies to this end. Indeed, the responsibility and blame toward developing an increasing sense of dependency vest with the Indian technocrats. Several factors contributed to this change.

In operational terms, there was increasing involvement of the World Bank. As of June 1990, IDA and IBRD had given a total loan of $4588.93 million, about 13.65% and 12.96% of the total, respectively. After power, the two together accounted for the second largest share, 26.61%.

Loans require careful project preparation and evaluation by foreign consultants. This has a positive impact, but it also has another side. Gradually, the so-called foreign experts have an increasing say, not necessarily in a bad sense, but definitely with erosion of self-reliance. The low level of technological capability of the Indian professionals, who often are officers rather than specialists or even well-trained engineers, had much to contribute. There was also a lure of gains by senior bureaucrats through international assignments, which are often petty. Not that these attractions were not there in the earlier stages, but these were less, and the sense of self-respect was comparatively higher.

Besides the World Bank, United Nations, and country aids, another and newer activity was bilateral collaboration at the project level. A foreign government gives tied aid to a particular project in terms of finances, equipment, and technologists. There has not been any specific case in the GBM basin in India except for Russian support for the Tehri Dam, but with the collapse of the USSR, the proposed collaboration has also almost come to naught.

A new and recent interaction has been in the field of environment. Water resources development activities, particularly high dams, have come under attack by environmentalists in advanced countries. Indian environmentalists also often act as fundamentalists, without considering the entirely different conditions and priorities of the developing countries. They often get support from foreign agencies, explicit or implicit. The case of Tehri Dam is an example.

At the academic level, there has been some interaction in terms of visits by foreign academicians in the field of natural resources besides the sponsoring of Indian scientists under some program. The establishment of a Water Resources Development Training Centre (WRDTC) at Roorkee University has already been mentioned. However, the academic institutions also have the historicity of a colonial culture. The educational scene is the most depressing, as several eminent educationists have analyzed. The review by

Amratya Sen (1989) is one example. The academic institutions, including the much-venerated IITs, have to revolutionize to modernize themselves.[3]

10.6.2 Bangladesh Scene

Attitudes, circumstances, and reactions to international interaction in Bangladesh, formerly East Pakistan, are entirely different. East Pakistan did not have the historical mistrust toward international aid nor the urge for technological self-reliance as India has. It was a member of the Western military block and derived considerable financial support from foreign countries, particularly the USA and the World Bank. As we have seen earlier, East Pakistan and the later Bangladesh have also received heavy support, technical and financial, in the water resources area, which is a major component of its public foreign assistance. Since its independence in 1971 until the year 1980, Bangladesh has received a cumulative sum of about $20.8 billion in foreign aid (grants and loans) from donor countries and agencies. In 1990 alone, it received a sum of $1809 million in foreign aid, which constitutes about $16.5 per capita or about 8% of her 1990 GDP. More importantly, foreign aid constitutes an extremely high proportion (70% in 1990) of the investments of the country. (Corresponding figures for India are about 2% and 5.3% for 1982.) Besides implications on development, it casts serious doubts regarding the choice of appropriate technology and the establishment of indigenous self-sustained technological capability. It also has bearing on Bangladesh's independent decision-making capacity about international river water disputes.

10.6.3 Nepal Scene

The development of Nepal is even more dependent on foreign aid, and indigenous technological capability is little developed, while it requires the highest level and very wide development in water and environmental systems management. Foreign private companies exploit this scene.

10.7 Some Perspectives and Institutional and Cultural Revolution

It has been demonstrated that development and management of the water resources in India have to be revolutionized urgently. Considerable potential enhancement of water availability and of advancement exists from technolgical considerations. However, the management of water over the entire cycle—development, use, and management of the socioeconomic–environmental system—is a societal management issue, above and beyond the proposed technological revolutions.

The subject has been studied in the accompanying study, *India's Waters— Revolutionizing the Development and Management of Water* (Chaturvedi, 2011). The current state of water management in India has been brought out, the institutional and cultural factors which have led to this phenomenal neglect have been analyzed, and solutions have been offered.

A personal observation may be reproduced to emphasize the issue of "culture." The proposed Chaturvedi Water Power Machine will make dramatic improvements in the development of water. It has received encouraging response from the highest levels. It occurred to the author that because this will take time, we should start with some projects that can be undertaken immediately. He phoned one of his former students in Uttar Pradesh, who holds one of the senior-most positions. His response was, "Sir, have you not read today's papers about the murder of a young officer who refused to give his quota to the Chief Minister's birthday celebrations? The objective function here is making money for our political masters and us, not engineering."

10.8 Conclusion

Development of indigenous capability and a management system is the most important issue in the management of the environmental system, including water.[4] It is sadly lacking. Instead, the system has become totally corrupt and indifferent to scientific development.[5]

Notes

1. The author had close personal experience of the life and system as he grew up in small towns where his father, the Civil Surgeon, was posted. The bureaucracy was a tiny group of senior officials, living their own life and having an exclusive club. The author could feel it as he was being introduced to revolutionary books on one hand and playing tennis with the British officers in the club in the evening. This unique spirit of division of society has not left India even today, while democracy is buffeting at it. Unlike the Western society, where democracy had an indigenous growth, it is being grafted in India, with tremendous distortions.

2. The author had the privilege of starting his professional career under one of the distinguished engineers of the time, Dr. A.C. Mitra, and working for a long time under his guidance on many major projects

such as the Rihand dam, Ramgamganga Project, and then joining him and the other leading engineers of the country such as Dr. A.N. Khosla, Dr. M.R. Chopra, and Dr. Kanwar Sain as Member of the Board of Consultants.

3. The author had his higher engineering education in one of the leading universities in water resources area in the USA. The author has been interacting with U.S. universities, particularly Harvard, and European universities through collaborative research and visiting professorial assignments. In India, the author was the founding head of the Civil Engineering Department at the Indian Institute of Technology (IIT), Kanpur, and the Applied Mechanics Department, IIT Delhi. The author has particular interest in the field of Education, through collaborative study with his wife, Prof. (Mrs.) Vipula Chaturvedi, who specializes in the area of Higher Education and had been a student in the USA and had visiting assignments in U.S. universities at the same time, later working at Indian universities and, with the author, at U.S. universities.

4. The author and (Late) Prof. Vipula Chaturvedi have established a Vipula and Mahesh Chaturvedi Foundation to contribute to the advancement of education in India at the highest level. Under the new scheme of the IITs, contribution is being made to establish a Professorial Chair every year.

5. The author continues to work with the Indian Government agencies at the highest level and at U.S. universities, particularly Harvard, with which he has a long association, during summers as a visiting professor.

Part III

Proposed Revolutionary Policy

11

Current Scene and Official
Future Perceptions

11.1 Introduction

As emphasized at the outset, the GBM faces a tremendous development challenge in which the environment will have to be transformed on a large scale. This applies to all vectors of the environment but particularly to water, which is the focus of the study. Studies of future policies of the GBM basin, of which the largest component is India, have not been made so far. A recent study for the water resources of the entire country, as one unit, has been carried out by the National Commission of Integrated Water Resources Development Plan (NCIWRDP, 1999) to formulate policy decisions. It will be studied briefly, as the GBM constitutes the dominant river basin in India and the policy decisions have particular relevance to the GBM basin. The official studies are essentially technologically-oriented, and they do not bring out the real-life challenges of societal water needs and perspectives. It will therefore be presented first and then undertake the study of the NCIWRDP (1999) in their perspective. The focus is on India, but it will apply to all the riparian countries. The study is confined to the current perspective of technological development. It will be shown in Chapter 12 that an entirely revolutionary technology can be developed, which will transform the water scene.

11.2 Domestic Water Supply and Sanitation

India was, historically, characterized as a vast, amorphous, rural country with few towns and cities—which emerged, essentially, as seats of power—and a bit of trading. There was hardly any infrastructural development, including water supply and sanitation, be it in rural areas or urban areas. The condition, even at the time of Independence, had not changed much.

While the majority of the population, about 70%–80%, still lives in rural areas, a revolutionary change is taking place in terms of population explosion and habitation. Rapid migration of vast numbers of poor people to urban areas is taking place. It is estimated that about 50%, 600–700 million people, depending upon how high or low the population growth will be, will still be living in rural areas by 2050. The urbanization scene in the developing countries poses serious challenges in aspects such as the provision of infrastructural facilities, of which drinking water supply, sanitation, and wastewater management are important constituents. Provision of drinking water and sanitation at the rural level poses serious challenges in view of its neglect. We first briefly consider the subject, separately for rural and urban complexes, and then formulate an integrated policy for meeting the domestic demands.

11.2.1 Rural Scene

Review of the rural water supply in India brings out a distressing scene. The subject of providing drinking water supply and sanitation of the rural communities is routinely emphasized in the plan documents. Standards of water supply and targets for achievement are boldly stated. They remain unaccomplished in plan after plan but the ritual goes on.

Water is crucial for life. Providing clean water to meet the domestic needs and sanitation facilities is the first issue in terms of human development. A concept of water security as human security has been proposed under the United Nations Development Program, which integrates with the basic objectives of managing water (UNDP, 2006).

The requirements in terms of quantity are small, less than 5% of the total water withdrawals for both the urban and the rural sector. The technology of water supply, sanitation, and management of return flows is well developed. With the society in the GBM basin sitting on an ocean of groundwater or on the banks of major and minor rivers, it is paradoxical that even after half a century this basic need has been neglected. But the fact is that development is subordinate to the socioeconomic conditions, and this neglect confirms the fact. Two of the author's experiences may be worth narrating in this context.

The author was the Founder Head of the Department of Civil Engineering, Indian Institute of Technology (IIT), Kanpur, which was founded with the support of a consortium of U.S. universities. The campus was being developed at Kalyanpur, near Kanpur, a developing industrial city, over an area of about a thousand acres obtained from the villages. One of the author's American colleagues proposed that they should go and find out from the villagers what the villagers wanted, as that should be understood when formulating the educational policies. The author's group visited several adjoining villages. An American colleague had an interesting strategy. The Polaroid camera had just been developed then (circa 1962). A bunch of village children would surround the group as they reached the village. He would go to them uttering *"jadu, jadu"* (magic, magic), take their picture, and hand it over to them. With the rapport established, they would be led to the elders and the head of the Panchayat (the village group of five people and their chairman). He would again repeat the performance, and with the rapport established, the author would ask them as to what they wanted. The answer invariably was "water." The author would follow it up with the question, "water for drinking purposes?" And the answer invariably was "Oh no, for our fields. For drinking purposes, the girls go to the wells and fetch water. It keeps them occupied and out of mischief."

The author later joined IIT Delhi as the Founder Head of the Department of Applied Mechanics. Considering that rural development should be an important focus, he established an interdisciplinary Centre for Rural Development. The author's group visited a neighboring village, put the same question, and got the same answer, "water." "But domestic water supply has already been provided through pipelines?" "That is not much good. Now the girls have not much to do and go and spend time dancing in the forests. Our concern is only for water for the fields."

While the drinking water is of little concern, the issue of sanitation is beyond the wildest thinking of the rural community. The central issue, thus, is that the society has to appreciate the need for drinking water and sanitation and then see to it that this is satisfied.

Domestic water supply and sanitation demands do not require much water, and they can be well met if due attention is given to this vital issue. It has to be approached as a top priority issue, and if due attention is given to it, there is no reason why adequate and appropriate quality drinking water cannot be provided to the entire rural population of India. Technology of drinking water supply and sanitation is well known; financial requirements are small, and resources exist. The issue is only of commitment to achieve it. It is not the commitment of the state but of the people, with the state guiding and helping the people to develop and manage it for themselves.

11.2.2 Urban Scene

Although there was some urbanization in the past, the few urban centers had developed just in the context of a bit of trade and administration of large areas assigned to them. The components of the urban infrastructure—housing, transportation, energy, water, and sanitation—were woefully inadequate. The population of these urban centers has increased tremendously, with about 350 million people residing in them currently, but the infrastructure facilities have not been developed much. The state of water supply, sanitation, storm water drainage, and solid waste disposal, which are a set of integrated activities, is extremely poor. With the estimated urban population increase to about 650,970 million by 2050, depending on how high or low the population growth will be, urban development is one of the most difficult challenges faced by India, as by many other developing countries. This is further compounded in India because of the hydrologic–environmental conditions.

Although the water supply is not very large, the problem is compounded by three factors. First, development of the water supply, sanitation, and return flows is related to the totality of the urban infrastructure, which has to be planned as one composite whole. This becomes very difficult because of neglect in the past, which continues to the present. Second, this is further seriously compounded by the rapid and heavy onrush of poor people from the rural sector on account of population explosion and lack of job opportunities in the villages. The third serious problem arises because the low season river flows had been diverted for irrigation in the past, leaving the rivers almost bone dry.

Thus, action on three fronts is required. One, water resources policy has to be reoriented to give first priority to urban water supplies from habitat and environmental considerations. Thus, the low season flows have to be restored, and water can be diverted for irrigation only if there is surplus, above and beyond the urban and environmental requirements. Two, a long-term perspective of urban development, of which water and sanitation are one component, has to be worked out in the context of the likely pressure in urban–rural migration, so that appropriate arrangements and capacities are provided appropriately in advance. Even if it appears that there is some redundancy, it is better to err on the safe side because it is very difficult to expand the water and return flows infrastructure. The latest technology has to be adopted. The developments in Mexico City and Singapore in contrast to the Indian scene, to give two examples, are noteworthy.[1] Three, the challenge of rural influx to the urban areas has to be resolved. This means total spatial habitat–infrastructure planning has to be undertaken, in contrast to a limited focus on urban water supply.

All these are continuing long-term activities. Meanwhile, highest emphasis has to be laid on managing the urban system, as it is, as best as possible, while these three activities are pursued. In our view, much can be done if due attention is paid to the subject. Suggestions have been made that the metropolitan town water supply can be managed by rainwater harvesting, pricing, and so on. In our view, this is totally mistaken.

Water supply for the industrial sector also has to be provided in urban areas. It is considered separately, as it raises the serious issue of dealing with the water quality management of the return flows and their management at the source itself, in the first instance.

11.2.3 Human Habitat—Urban–Rural Complex

In view of the extremely serious challenge of urbanization, the subject may be given special attention even if there is some repetition, as it is an extremely difficult problem and has to be considered above and beyond the issue of water supply. Urbanization took place under conditions of socio-economic–political development in different parts of the world even before the industrial revolution. With the industrialization, a different trend of urbanization started in the developed world. It depended upon the socio-economic development of the society and on the available technologies of manufacturing, transportation, and communication. They also had the entire world at their disposal.

On the other hand, the urbanization of the developing societies followed in the context of the marginal expansion of the earlier urbanization under conditions of colonialism and the rapid development of these countries as frontier economies of the developed world. There was very little development of the infrastructure in these countries in the past.

As the transformation in the developing countries is taking place under the pressure of population explosion and economic development, a very serious problem of urbanization ensues. It has not been possible to ensure proper development of the civic facilities in the existing cities. In addition, a rapid migration of population from rural to urban areas is taking place, as the agricultural sector is not able to provide jobs and sustenance to the rapidly growing population. Cities also provide attractions even though the living conditions may be extremely depraved and employment opportunities limited.

The urbanization poses several problems. Industrialization is limited to a few large cities, which become the focal point of migration; the employment opportunities in these cities are very limited in contrast to the demands of the exploding population; and development of habitat and infrastructure is not commensurate with the demands of the exploding population. Civic management is a process of societal development, which also did not take place under the conditions of colonialism and neglect in these countries. Therefore, there are serious problems of availability of facilities such as housing, water supply, sanitation, transportation, and civic services, which deteriorate as the pressure continues. The concentrated impact of urban demands and wastes produces serious environmental problems at the city level and on a regional scale.

The approach to meeting the urbanization challenge has generally been reactive. A new anticipatory approach is required. Urbanization has to be considered in terms of spatially and comparatively more even growth as new means of transportation and communication are available, which were not existent when the urbanization in the developed countries took place and which are impacting the suburbanization of the mega cities in the developed countries. Thus, it requires a long-term perspective of development of the region, or even the country, as a whole, with the development of the rural environment as a continuum of the urban environment, dispersed industrialization, and appropriate development of the infrastructure on a futuristic basis instead of the current belated reactive approach. Emphasis on rural development, as part of the urban–rural continuum is an important component of the program. It is part of the basic national theme of eradicating the two Indias. A PURA (providing urban amenities to rural areas) program has been proposed by Kalam, the former President of India, in his capacity as a scientist himself (Gandhi, 2005). The ideas are still conceptual, and much work has to be done to detail them. For instance, it is an issue not only of providing the urban amenities but also of developing the urban culture in rural areas.

Concurrently, at the city level, it requires provision of the complex of the urban infrastructure and its efficient management. Management of urbanization is crucial for managing the environment at the regional level also because it will not be possible to manage the regional environment unless the concentrated wastes are properly managed. The problem of cleaning of the rivers in India, which essentially means completely transforming the current policy of water resources development, as discussed later, is a case in point. Thus, there is an urgent need for integrated environmental–societal systems management for meeting the basic needs of domestic water supply and sanitation of the people and the society.

11.3 Agriculture Demand and Agricultural Transformation

Besides its basic use as drinking water, water is equally important for livelihood, economic development, and environmental conservation. Water for agriculture, particularly in the environment of the developing countries in which India and China occupy the dominant position, is crucial. The quantity of water is dominantly large. For example, the current proportion of water in India for domestic (5% of total requirement), agriculture (83%), industries (5%), power (5%), inland navigation (1%), and ecology (0%) brings out a huge difference. Three aspects are brought out. First, management of water for agriculture determines the policy of development of water. Second, the development will have to be considered over the entire spectrum of water supply, its efficient use in the sector, and its return to the natural environment, thereby emphasizing consideration of the totality of the environmental system. Three, agricultural transformation is crucial for the transformation of the rural scene and management of water. A revolution will have to be brought about in the socioeconomic scene, that is, the rural scene, for efficient use of the environmental inputs—land and water. Thus, it again vindicates the need for integrated environmental–societal systems management.

The agricultural sector currently accounts for 83% of the water withdrawals and is estimated to account for about 68% even until 2050, according to the current official estimates (NCIWRD, 1999). It is important to point out that in these official estimates, little emphasis has been placed on the modernization of agriculture. Our point of departure is that we are proposing to revolutionize it with the highest urgency (Chaturvedi, 2011c). Some important issues that have bearing on the management of water are briefly discussed, as it is essential to consider the management of the total system for its modernization.

11.3.1 Agriculture and Food Scene

As one reviews the world agriculture scene, the differences in the two world systems glaringly stand out. World agriculture in fact is composed of two distinct types of farming: (1) the highly efficient agriculture of the developed countries, where substantial productive capacity and high output per worker permit a very small number of farmers to feed entire nations with comparatively much less use of the natural resources and (2) the inefficient and low productivity agriculture of developing countries where, in many instances, the agriculture sector can barely sustain the large farm population, let alone the burgeoning urban population, even at a minimum level of subsistence and even after exhausting the available natural resources. The gaps between the two are immense. The gaps not only are in terms of yields and productivity per worker but also relate to the agrarian structures. The agrarian structures are not only part of the production system but also a basic feature of the entire economic, social, and political organization of rural life. These are a result of the historical process of development. Thus, a very serious challenge ensues.

A common characteristic of agriculture in India, as in other developing countries, is the position of the family farm as the basic unit of production, with survival being the prime concern of the peasant and his family. As has been stated, "For the vast number of farm families, whose members constitute the main agricultural work force, agriculture is not merely an occupation and source of income; it is a way of life. This is particularly evident in traditional societies, where farmers are closely attached to their land and devote long, arduous days to its cultivation. Any change in farming methods perforce brings with it changes in the farmer's way of life. The innovation of biological and technical innovations must therefore be adapted not only to the natural and economic conditions, but also perhaps even more to the attitudes, values, and abilities of the mass of producers, who must understand the suggested changes, must be willing to accept them, and must be capable of carrying them out" (Weitz, 1971).

The peasant is a rural cultivator whose prime concern is survival. Subsistence defines his concept of life. He may strive to obtain his and his family's minimal needs by filling an inadequate piece of land that is his own or, more often, which is rented from or pawned to a landlord or moneylender, or by selling his labor for substandard wages to a commercial agriculture enterprise. Profits, which might come to him through the fortunes of weather or market, are windfalls, not perceived goals. Debt rather than profit is his normal fate, and therefore, his farming techniques are rationally scaled to his level of disposable capital: human and animal power rather than mechanized equipment, excrement rather than chemical fertilizers, and traditional crops and seeds rather than experimental cultivation.

No effective social security, unemployment insurance, or minimum wage law eases his plight. His every decision and act impinges directly upon his struggle for physical survival. In countries with a high proportion of peasantry, traditional food crops which a rural family can by itself convert readily into the daily fare for its grain- or tuber-based diet dominate the agriculture—corn in Mexico, rice in Indonesia, and soybeans in China. India is typical of peasant agriculture, with 75% of the cropped land devoted to food grains such as rice, wheat, millets, barley, and lentils. When these fail, a peasant is reduced to trading his bullock for a few bananas (Forland, 1974). These conditions are changing a bit with the "Green Revolution," but for the majority of the farmers, these continue to persist.

The agrarian structures are not only part of the production system but also a basic feature of the entire economic, social, and political organization of rural life. Briefly, the traditional Asian agriculture before European colonization was organized around the village community. Local chiefs and peasant families each provided goods and services—produce and labor from the peasants—to the chief in return for protection, rights to use community land, and the provision of public services. Decisions on the allocation, disposition, and use of the village's most valuable resource—land—belonged to the tribe or village community, either as a body or through its chief. Land could be redistributed among village members as a result of either population increase or natural calamities like droughts, floods, famines, war, or disease. Within the community, families had a basic right to cultivate land for their own use, and they could be evicted from their land only after a decision by the whole village community.

Colonial rule led to major changes in the traditional agrarian structure. It acted as an important catalyst to change both directly through its effects on property rights and indirectly through its effects on the pace of monetization on the indigenous economy and on the growth of population. In the area of property rights, European land tenure systems of private property ownership were both encouraged and reinforced by law. One of the major social consequences of the imposition of these systems was the breakdown of much of the earlier cohesion of village life with its often elaborate though informal structure of rights and obligations.

The creation of individual titles to land made possible the rise to power of another dubious "agent of change" in rural socioeconomic structure—the moneylender. Once private property came into effect, land became a negotiable asset that could be offered by peasants to the often unscrupulous moneylender as security for loans or, by default, could be forfeited and transformed. At the same time Asian agriculture was being transformed from a subsistence to a commercial orientation, both as a result of rising local demand in new towns and, more importantly, in response to external food demands of the colonial European powers. With this transition from subsistence to commercial production, the role of the moneylender changed drastically. By charging exorbitant interest rates or inducing peasants to secure larger credits that they could manage, moneylenders were often able to drive the peasant off their land. A new class of landless labor and bonded labor emerged, and, in general, the economic status of Asian peasant cultivation deteriorated steadily over time.

The final major force altering the traditional agrarian structure in Asia has been the rapid rate of population growth over the last 30 to 40 years, which continues today. When and where expansion in the cultivated land was a feasible alternative, population growth was reflected in the first instance in the cumulative subdivision and fragmentation of the acreages already under cultivation. Later, this process, in combination with the emergence of private property and the rise of commercial agriculture, and money lending, often contributed to the rise of large landowners, the demise of small peasant proprietors, and the increase of the landless. The ultimate impoverishment of the peasantry was the inevitable consequence of this process of fragmentation, economic vulnerability, and loss of land to rich and powerful landlords.

11.3.2 Yields

In terms of world food production, 94% of the rise in cereal production between 1970 and 1990 reflected an increase in yield per unit of land, and only 6% was due to an area increase. The increase in yield on account of several biological and technological factors and is often subsumed under the term "Green Revolution," including introduction of modern varieties of rice, wheat, and maize in combination with the more intensive use of inputs such as fertilizers and pesticides, and more intensive and judicious use of water. Indeed, scientific management of water—adequate, timely, and reliable—is critical even for the adoption and contribution of other factors of production. This is an important issue, which is often not considered in doing policy analysis of the subject.

Three factors are important regarding yield. One, the yields vary largely from country to country. Two, the yields are very low compared to yields in the developed countries. Three, the yields in India, which has the second largest area under cultivation, almost equal to that of China, are one of the lowest, less than half of those in China.

The potential of increase is brought out vividly, comparing the current yield of rice with yields in Japan, historically. The yields in India are still what the yields in Japan were in AD 1200 and are about one-fifth of the current yields in Japan. The trends and current yields in 93 developing countries are given in Table 11.1. The corresponding figures for India have also been added. The table clearly brings out the poor yields in India and the tremendous scope for improving them.

TABLE 11.1

Trends in Yields in 93 Developing Countries

Crop Type	Yield (kg/ha)			
	1961–1963	1969–1971	1979–1981	1990–1992
All cereals	1171	1461	1894	2466
Excluding China	1116	1271	1557	1951
China	1336	2070	3017	4329
Wheat	868	1153	1637	2364
Excluding China	964	1146	1460	1997
China	673	1169	2046	3208
Rice	1818	2218	2653	3459
Excluding China	1650	1855	2145	2790
China	2355	3281	4236	5722
Maize	1157	1456	1958	2531
Excluding China	1122	1291	1572	1837
China	1265	2005	3038	4545

Source: World Resources 1996–1997, p. 226.

11.3.3 Environmental Pressures and Impacts

Of all human activities, agriculture has the most severe impact on land and water. Increasing agricultural production implies further intensification of land and water use. More mineral fertilizers and, to a lesser extent, pesticides will be used in the future. Several adverse environmental impacts, such as degradation of land and water resources, soil erosion, soil nutrient mining, salinization of soils, desertification, water contamination, pesticide use implications, and global change issues, follow.

Two related issues stand out in this regard. First, what are the technological options for putting agriculture on a more ecologically sound pathway, while at the same time achieving higher productivity? Second, given the inevitability of some trade-off between the environment and development, what are the other actions required in minimizing them? In this context, first, it is necessary to describe and, where possible, to quantify the agricultural pressures on the environment that are implicit or explicit in the projections of the probable growth paths of the crop and livestock sectors. An analysis in an international context has been carried out by Alexendratos (1995) and FAO (1995) and has been summarized by Chaturvedi (2011a), to which reference may be made. Based on the study, we will briefly consider the modernization of agriculture and suggest further advances.

11.3.4 Modernization of Agriculture—Conceptual Issue

Review of the state of agriculture demonstrates that although advances in productivity have taken place in the developing countries since their independence, these are, with some exceptions, (i) modest and (ii) slow. There is a considerable scope for improvement, and it can be achieved much more rapidly. A comparison of yields in India, one of the largest developing countries in terms of land and people devoted to agriculture, shows that at least a threefold increase is possible, even in terms of the currently achieved yields in the developed countries (Table 11.1). A comparison with the rates of increase in China and the Korean Republic and those achieved in developed countries shows that these can be achieved at a much faster rate than what has been achieved in the past. The technology is well known. It is a matter of policies, institutions, involvement of people, and commitment to rapid development.

Making production more efficient has remarkable economic and environmental benefits. The role of irrigation is unique in this context, as it is the dominant input in the production function. Other inputs like fertilizers, high-yield variety seeds, and pesticides are used only if adequate, timely, and reliable water supplies are available. As we will elaborate later, there is hardly any extra requirement of water per unit of land. Indeed, there is a considerable scope in reducing unit water per unit land, as the irrigation practiced currently was inefficient by design. Some other factors of production such as fertilizers, pesticides, and energy will be required, but per unit of yield, costs will be considerably reduced.

It is also environmentally beneficial in many ways. First, for the same production, much less land and water will be required. The land can thus be restored to nature. Second, even when adopting current figures of soil degradation, it will be considerably minimized per unit of agricultural production. Third, the erosion and degradation are on account of poor agricultural and environmental practices. It is implicit in modernization of environmental practices that adverse environmental impacts on account of their development will be minimized. Some adverse environmental impacts, for example, those due to an increased fertilizer and pesticide use, will be increased. It is implicit in the concept of modernization that more attention is given to sustainability.

Technological modernization is an important ingredient of sustainable modern agriculture, but the technological pathway is not sufficient by itself. There is a wide range of policy and institutional measures required to provide the incentives needed for farmers, forest users, and fishermen to adopt sustainable technological and resource management practices.

The critical triangle of development goals has been considered to be (i) growth, (ii) sustainability, and (iii) poverty alleviation. A basic conceptual fixation, while examining the problems of developing countries, as we have emphasized throughout, has been to view them in a permanent state of developing countries vis-a-vis the developed countries. This concept has been particularly pervasive in the context of consideration of agriculture. Our first and most important submission is the jettisoning of this mental fixation. Modern agriculture has one definition—sustainable agriculture with highest output with factors of production. As far as poverty is concerned, the objective is eradication and not alleviation, joining the band of the developed countries on equal terms.

11.3.5 Technological Foundations for Sustained Agricultural Development

There are certain technological foundations for agricultural development. They are as follows: limiting land and water degradation, promoting integrated plant nutrition systems (IPNS), integrated pest management (IPM), and developing the potential of biotechnology.

11.4 Rural Transformation

The central issue in modernization of agriculture, as in the provision of the basic needs of water supply or any other facet of development, is rural transformation. Rural and urban divide is a social anachronism. However, while transformation is achieved, which is of the highest priority, much can be accomplished in scientific sustainable management of the environment, with particular focus on water.

11.5 Manufacturing and Industrialization

Manufacturing and industrialization are only at a rudimentary stage in India. Vast and rapid changes are going to take place. The quantity of water required in the manufacturing and industrialization activities will be small, but the return flows will require considerable quantities of river flows to attain acceptable environmental quality of water. For example, even with the current low level of manufacturing and industrialization, the return flows and human sanitation, as discharged in the rivers, have turned them into open sewers.

We look at provision of water to the manufacturing sector in a slightly different way than generally pursued. Our approach is to look at the ushering of modernization of industrialization and concurrent management of water.

One of the major challenges for developing countries is to expand and modernize the manufacturing sector. In other words, it means rapid industrialization. Industrialization has been a historical succession of periods of pervasive adoption of clusters of technologies and organizational innovations. Combined, they have enabled vastly rising industrial output, productivity, and incomes as well as reductions in the amount of time. Like any pervasive process of economic or social change, industrialization is driven by the diffusion of many individual but interrelated innovations. These are not only technical but also organizational and institutional, thus also transforming the social fabric. In fact, the term "industrial society" has come to describe a particular type of economic and social organization.

Industrialization has unpredictable environmental impacts stemming from effluents, new and old, coupled with an expanding scale of industrial activities. Industrialization has also created entirely new environmental impacts by virtue of introducing new materials such as DDT and CFCs. There are several generic strategies for responding to environmental impacts without a full understanding of all the details. Four such strategies are (i) dematerialization, (ii) material substitution, (iii) recycling, and (iv) waste mining.

A new approach has been that of the concept of industrial ecology. Industrial ecology provides a systems perspective based on its analogy to biological ecosystems. Changes are beginning to occur in the industrial activities with the understanding of industrial ecology, but there is a long way to go to make the technological activities sustainable even in the industrialized countries. The problem is entirely different in developing countries, even with a very low level of industrialization.

11.6 Conservation of the Environment and Water

Development of societies means increasing incorporation of the environment into culture. Until lately, environment was considered only in terms of providing resources. A concern about the quality of the environment followed, as there were numerous experiences of the abuse of the environment. There were also concerns about the limitations of the environment as a resource. According to the latest thinking, environment and the society have to be considered as one integrated whole, and the subject becomes one of environmental–social systems management of which management of water is an integral part (Bossel, 1998; Chaturvedi, 2011a).

11.7 Challenge of Managing the Waters—A Perspective

It is abundantly clear that very large scale water resource development will have to be undertaken as the society tries to pursue development. The Government of India constituted a National Commission on Integrated Water Resources Development Plan (NCIWRDP, 1999) to undertake a comprehensive study of the management of India's waters. Following them, we undertake the study of some of the important issues, with focus on the GBM basin.

A wide range of policy measures have been proposed by the NCIERDP (1999). Agriculture and rural developments have been given high priority. Raising the productivity of land and water has been emphasized. Rural infrastructure development has also been emphasized. Environmental degradation is noted. Monitorable targets include the following: increase of forest and tree cover to 25% by 2007 and 33% by 2012, all villages should have sustained access to potable drinking water within the plan period (2002–2007), and cleaning of all major polluted rivers by 2007 and other notified stretches by 2012. Policy suggestions regarding the modernization of all water-related activities have been made. They have all ended in failure. More than that, the NCIWRDP (1999) did not even consider the issues in the wider context of modernization of the management of water, as raised in Sections 11.2–11.6.

11.7.1 Review of Official Demand Estimates and Proposed Modifications

The issue is not merely to estimate the sectoral demands but to see how to manage them and satisfy them efficiently. The sector demands are therefore briefly examined.

11.7.1.1 Agriculture Sector

The agriculture sector will continue to have considerable importance in the economy for a long time to come. Increasingly and rapidly, the agricultural sector has to become the most modern, as it will not be possible to ensure livelihood, food security, and environmental sustainability if Third World agriculture continues. It has important implications on water demand and its service.

Therefore, urgent modernization of the entire spectrum of activities is required on a well-planned and organized basis as an interdisciplinary, interorganizational project. It is not an issue of appointing a national commission for some activity, say irrigation, at one time, which makes ad hoc recommendations and then lets the old colonial processes continue. Appropriate land use policy, cropping intensity, irrigation intensity, reduction in delta, and achievement of appropriate yields are vital matters, which will have an important bearing on water demand, and all have to be managed.

Focusing on yield, it is seen that, even without considering the modern advances in the field of biotechnology, such as the concepts of ever green and doubly green revolutions, a much higher yield than that adopted by NCIWRDP (1999) can be and must be achieved at the earliest possible time. It is considered that the adoption of a yield figure of 6000 kg/ha for cereals by 2050, which has already been achieved in most of the industrialized countries, in contrast to the currently recommended figure of 4000 kg by the NCIWRDP (1999), which has already been achieved in China, is logically imperative. Much higher figures will be justified (Conway, 1998). Implicit in the specification of the proposed figure is the modernization of agriculture over the entire sector. The supply of water is critical, and assured, adequate, and timely water supplies have to be provided. Water has to be applied with increasing efficiency in terms of productivity and environmental conservation. The issue is not just of providing an apology of irrigation on a highly subsidized basis for a Third World subsistence agriculture, as what current scene is. The central issue is embedding the development of water in the modernization of agriculture for rapid economic development and environmental conservation. Thus, for a preliminary exercise, a figure of at least 6000 kg by 2050 is adopted in our comparative study. All it means is that the issue is not only of providing irrigation but also of ensuring the quality of irrigation to provide adequate, timely, and reliable yields as part of the agricultural system modernization.

In addition, development of increasingly scientific methods of water use in agriculture is the topmost priority in water management. This will lead to considerably reduced delta and increased savings in water. However, in the first instance, for estimation purposes of development of water, we use the conventional delta figures adopted by the NCIWRDP (1999).

The subject of self-sufficiency or open agricultural policy has also been under discussion. We follow the current governmental policy of self-sufficiency, as it is a political policy matter.

We have developed two strategies for estimating water demand. In Strategy 1, all other assumptions are kept the same as those used by NCIWRDP, except that the yields of irrigated area by 2050 are specified to be 6.0 tonnes against NCIWRDP's figure of 4.0 tonnes. The water demand drops by 29%.

An arbitrary assumption is usually made about the share of groundwater and surface water in providing irrigation in estimating future water resources development (NCIWRD). In Strategy 2, we have dropped that assumption and put primacy on groundwater with an arbitrary limit of allocating 400 km³ against the estimated potential of 500 km³. (This should be further increased but, tentatively, a low figure has been adopted.) Groundwater can provide much more reliable supplies than public surface water supplies. Groundwater requirement is considerably less than surface water, the respective deltas being 0.49 and 0.61, and contributes to higher productivity. All other performance characteristics are maintained the same as before, and even then, only 56% of conventional water requirements are needed.

11.7.1.2 Domestic Water Demand

Domestic demand has the first priority. The total quantity is comparatively small. With increasing socioeconomic development, urbanization and industrialization are going to take place rapidly. From an environmental systems perspective, an issue of utmost importance is to ensure the appropriate development of these sectors, as water supplies and quality cannot be assured unless proper habitat development with appropriate spatial distribution, takes place. Thus, the issue is not merely that of adopting unit figures in these sectors but ensuring appropriate sectoral development.

Regarding urban water supplies, in our view, considering the future perspective of increasing socioeconomic development the hot climate of the country, and the fact that the infrastructure system has to be developed in a long-term perspective, reviewing the global water developments, a figure of 400 l per person per day (lpcd) appears to be reasonable. This may be contrasted with the figure of about 220–110 lpcd currently proposed and even more lowly figures recommended by some in terms of meeting the basic needs. Conjunctively, in view of the extreme scarcity of water, highest quality of return flows will also have to be specified. The important issue is not of the estimated figure but the fact that unless highest emphasis is accorded to habitat and infrastructure development, it will not be possible to achieve even the minimal environmental specifications for water. Secondly, obtaining proper domestic water supplies will require that reallocation be made from the agricultural sector to the domestic and environmental sectors. Thus, appropriate regional planning is extremely important if domestic and environmental requirements are to be met.

Considering rural water supply, resource scarcity is not the issue, as watershed management and rainwater harvesting can meet it. The issue is of mobilizing people and guiding them. For example, considering the GBM basin, people are sitting on an ocean of fresh water, which can be had just by digging some wells and lifting water. Even in the drought-prone areas, piped water supply can be provided as the first option. The subject, therefore, is related to the wider issue of rural development.

11.7.1.3 Energy

The energy sector is going to develop rapidly and shall have the second largest demand for water. There is also close interlinkage in terms of generation in view of the important role of hydroelectricity. Thus, water resources developmental planning has to be undertaken integrally with energy systems development. It is not a simplistic issue of hydrothermal balance, as generally emphasized currently.

11.7.1.4 Industrial Demand

Water requirements of industries have been studied in detail. There is considerable variation in unit figures depending on how conscious the society is regarding environmental management. Emphasis is now being laid on the subject in terms of the concepts of industrial ecology. The subject has not been given due emphasis in water resources management but is of great importance in view of the serious environmental challenges due to be faced as a result of serious problems of pollution, including toxic wastes, as industrialization takes place, and institutional arrangements will develop only with a certain inertia. We have, however, adopted the officially estimated figures (NCIWRDP, 1999) for the sector water demand, just for the sake of working out the overall water balance and estimating the minimum surface flows from environmental considerations. Highest priority has to be given to managing return flows in terms of quality and quantity in terms of industrial ecology.

11.7.1.5 Environmental Demands

Minimum flow and appropriate water quality specifications from environmental considerations have often been neglected, particularly in the developing countries. Specific guidelines are not available as of yet, but the issue is of great importance. Research is required on the subject, but the figure of 1%–2% of mean annual flow, as sometimes specified, is not correct, as this is arbitrary and does not specify the lean flow period requirements, which are the dominant determinants. Secondly, the figure has to be related to the return flow figures. However, we have continued with the official figures, just for the sake of illustration. Besides specifications for the terrestrial waters, consideration of water quality at the water's edge is conjunctively related to the subject.

11.7.2 Long-Term Perspective

11.7.2.1 Water Demand Estimates

The figures of water demand return flow and water availability on the basis of the official estimates and the proposed environmental systems concepts (under the caption Chaturvedi) are shown in Table 11.2 and Figure 11.1.

TABLE 11.2

Water Withdrawal and Requirement Projections

(a) Withdrawal (km³)

Sl. No.	Use	Year 1997–1998 Withdrawal	%	Commission (2050) Low	High	%	Chaturvedi (2050) Low	High Strategy 1	High Strategy 2	% Strategy 2
1	Irrigation	524	83	628	807	68	407	577	453	55
2	Domestic	30	5	90	111	9	90	111	111	13
3	Industries	30	5	81	81	7	81	81	81	10
4	Power	9	1	63	70	6	63	70	70	8
5	Inland navigation	0	0	15	15	1	15	15	15	2
6	Flood control	0	0	0	0	0	0	0	0	0
7	Ecology	0	0	20	20	2	20	20	20	2
8	Evaporation loss	36	0	76	76	7	76	76	76	10
9	Total	629	100	973	1180	100	752	950	826	100

(b) Return Flows—Year 2050 (km³)

		Irrigation Low	High	Domestic Low	High	Industries Low	High	Total Low	High
1	Groundwater	115	147	7	8	—	—	122	155
2	Surface water	13	17	36	47	40	40	91	104
3	Total	128	164	45	55	40	40	213	259

Source: Chaturvedi, M.C., *Water Policy*, 3, 297–320, 2001. With permission.

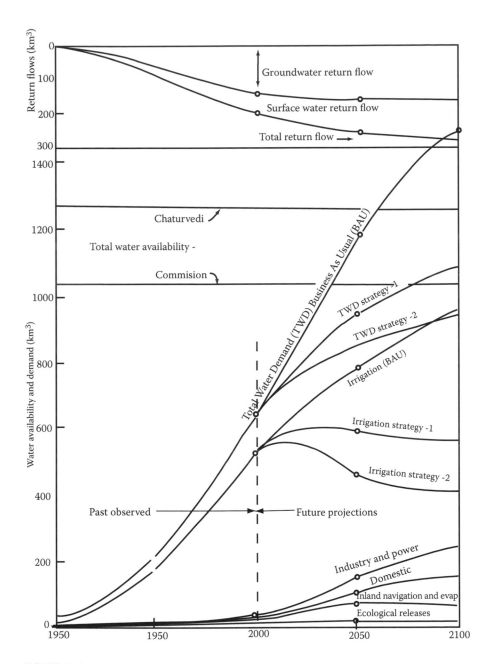

FIGURE 11.1
Development of India's waters—actual and future perspective.

It is seen that a significant decrease results. The demand estimates by Chaturvedi are still conservative, as the potential for decreasing demand by adopting better irrigation practices, which is extremely important from economic and environmental considerations, has not been taken into account.

Based on the aforesaid considerations, the total demand drops from 1180 km³ to 950 km³ and 826 km³, respectively, a change of about 19% and 30%. It will be seen that a major contribution toward achieving sustainability is made. The advances will be even more if the agricultural sector and other sectors are managed scientifically. These estimates are, however, only indicative, as the position varies considerably spatially and each river basin will have to be studied in detail.

11.7.2.2 Estimate of Water Availability

Estimates of water availability have been made as brought out in several official publications. We will present the official estimates and then try to make estimates on a more scientific basis, still confining ourselves to the accepted technological options.

11.7.2.3 Current Official Estimates

The NCIWRDP (1999) has made estimates of water availability (Table 11.3). The NCIWRDP estimates of water availability are based on the following criteria. The estimate of utilizable surface flows is based on the estimates of potential storage of surface flows and estimates of low season flows that can be developed by diversion. These have been worked out in detail for each river basin, further desegregated according to various tributaries and regions.

The utilizable groundwater estimates are on the basis of estimates of replenishable groundwater, as adopted by the NCIWRDP (1999) and as currently given by the Central Ground Water Board (CGWB). The CGWB has also made recommendations regarding the potential for additional recharge. These are also given in Table 11.3. They were not adopted by the NCIWRDP (1999).

It has been estimated by the NCIWRDP (1999) that a total of 1085.9 km³, comprised of 690.32 km³ of surface water and 342.43 km³ groundwater, could be developed on a sustainable basis. In addition, proposals have been made for interlinking the rivers, which is estimated to provide an additional 200–250 km³. Its feasibility, however, has yet to be established from technoeconomic–environmental–political considerations.

TABLE 11.3

Estimates of Water Availability

A. Regional

	Region	Commission			Chaturvedi		
		Surface	Ground	Total	Surface	Ground	Total
1	Himalayan region	320.00	185.00	505.00	320.00	316.50	630.50
2	Peninsular region	264.18	102.28	388.46	350.80	117.56	468.36
3	Rest	106.14	55.15	161.14	106.14	66.47	171.61
4	Total	690.31	324.43	1032.75	776.94	500.53	1276.47

B. National River Basinwise

Sl. No.	River Basin	Surface Water			Groundwater			Total Water	
		Mean	Utilizable Commission	Chaturvedi	Replenishable Commission	Artificial Recharge Chaturvedi	Total Chaturvedi	Commission	Chaturvedi
1	Indus	73.31	46.0	46.0	14.29	20.7	34.99	60.29	481.37
2	Ganga–Brahmaputra–Meghna basin								
2a	Ganga subbasin	525.02	250.0	250.0	136.47	94.9	231.37	386.47	65.62
2b	Brahmaputra	629.05	24.0	24.0	25.72	15.9	14.62	49.72	8.62
2c	Meghna	48.36	—	—	8.52	—	8.52	8.52	9.09
3	Subarnrekha	12.37	6.81	6.81	1.68	0.6	2.28	8.49	22.05
4	Brahamni–Baitarni	28.48	18.3	18.30	3.35	0.4	3.75	21.65	82.02
5	Mahanadi	66.88	49.99	66.88	13.64	1.5	15.14	63.63	149.82
6	Godavari	110.54	76.30	110.54	33.48	5.8	39.28	109.78	92.99

7	Krishna	69.81	58.00	69.81	19.88	3.3	23.18	77.88	10.99
8	Pennar	6.32	6.86	6.86	4.04	—	4.04	10.90	31.15
9	Cauvery	21.36	19.00	21.36	8.79	1.0	9.79	27.79	21.55
10	Tapi	14.88	14.50	14.88	6.67	—	6.67	21.17	56.42
11	Narmada	45.64	34.50	45.64	9.38	1.4	10.78	43.88	15.92
12	Mahi	11.02	3.10	11.02	3.5	1.4	4.9	6.60	7.61
13	Sabarmati	3.81	1.93	3.81	2.9	0.9	3.8	4.83	27.38
14	West-flowing rivers of Kutch and Saurashtra including Luni	15.1	14.98	14.98	9.1	3.3	12.4	24.08	53.86
15	West-flowing rivers south of Tapi	200.94	36.21	36.21	15.55	2.1	17.65	51.76	26.63
16	East-flowing river between Mahanadi and Pennar	22.52	13.11	13.11	12.82	0.6	13.52	25.93	33.58
17	East-flowing rivers between Pennar and south of Cauvery	16.46	16.73	16.73	12.65	4.2	16.85	29.38	—
18	Rivers draining into Bangladesh	8.57	NA	NA	—	—	—	—	—
19	Rives draining into Myanmar	22.43	NA	NA	—	—	—	—	1276.47
20	Total	1952.87	690.32	776.94	342.43	157.00	500.53	1032.75	—

Source: NCIWRDP, National Commission on Integrated Development Plan, Ministry of Water Resources, Government of India, 1999. With permission; Chaturvedi, M.C., *Water Policy*, 3, 297–320, 2001. With permission.

11.7.2.4 Author's Estimates

We have adopted the figures of surface water availability, tentatively, as proposed by the NCIWRDP (1999) for the Himalayan Rivers, although, as shown in terms of the novel technologies discussed in Chapter 13, the supplies can be highly increased. For the Peninsular Rivers, excellent opportunities of surface storage, both by large and small dams, exist. We believe that all annual flows could be economically stored and the figure for surface water availability will be equal to mean annual flows if carry-over storage is provided. This is, however, subject to more studies.

Implicit in these estimates is the bias of the historic approach of developing canal irrigation. In making the estimate of utilizable surface waters, a certain minimum of low season flows has to be left as free-flowing waters. Secondly, the estimates of utilizable groundwater have to be worked out with greater refinement, as the surface and groundwater constitute one whole, and the dynamics have to be carefully worked out.

There are also considerable shortcomings in the estimation of utilizable groundwater resources as noted by the NCIWRDP. For consistency, we have adopted the same figures of utilizable surface and groundwater used by the NCIWRDP, but allowance for reserving water for environmental demands should be made in development.

We would however, like to emphasize the availability of additional water through groundwater recharge. The CGWB has estimated that, in addition to the 342.43 km^3 of groundwater considered to be annually replenishable, 214 km^3 of surplus monsoon runoff could be stored in groundwater through conventional techniques, of which 157.0 km^3 is considered to be retrievable. These figures have not been adopted by the NCIWRDP on the plea that these need to be firmed up by detailed feasibility studies, and interbasin transfer could be considered if shortages arise. Our view is entirely different. We believe that, as studies have shown, the CGWB estimates are perfectly feasible, as discussed in Chapter 13.

We have not included the figures of augmenting water resources through the construction of reservoirs in upper neighboring states, following the NCIWRDP's (1999) approach, although we believe it is eminently feasible if an appropriate approach is adopted in developing collaborative development of the international rivers and should be given highest emphasis, because it has several other implications for sustainable development, such as hydroelectric development and flood control. We have, for the moment, also not adopted several other novel technological possibilities. Both these issues will be discussed in Chapter 13.

11.7.2.5 Estimate of Water Availability

The figures of utilizable surface and groundwater as estimated by the NCIWRDP and by us are given in Table 11.3. It will be seen that the surface water potential can be increased to 776.94 km^3 from 690.32 km^3, the ground-water potential can be increased from 342.43 km^3 to 500.53 km^3, and the total water availability is increased from the currently estimated 1032.75 km^3 to 1276.47 km^3—an increase of about 12.5%, 45%, and 24%, respectively, even through conventional technology, if scientifically developed. There are possibilities of further increase if some other options are found to be feasible, as discussed in Section 3.7. The formulations are just to obtain a general perspective of development, and details have to be worked out for each river basin. The perspective regarding some of them is briefly developed in the following sections.

11.7.2.6 Future Perspective

A perspective of development according to the official GOI perspectives is shown in Figure 11.1. The perspective, even with the modifications as proposed, confining ourselves to the currently developed technology, is also shown in Figure 11.1. It is seen that there is no need for the proposed inter-basin transfer according to the ILR proposal, as discussed later. The picture is even better transformed if the novel technologies and the environmental systems approach discussed in the following chapters are adopted.

11.7.2.7 Proposed Solutions

The NCIWRDP (1999) tried to work out a long-term perspective. Estimates of water demands until 2050 have been carried out by the NCIWRDP (1999). The estimates are given in Table 11.2. The key issue in the estimation was considered to be the agricultural sector demand, which is currently about 83% and is estimated to remain at about 68% of the total even until 2050. The major assumptions by the NCIWRDP (1999) in making the estimates of water demand for agriculture by the year 2050 were as follows: (1) the yield of the irrigated food crop by the year 2050 will be 4000 tonnes/ha; (2) the delta for surface and groundwater will continue to remain almost the same as at present, 0.61 and 0.49 m, respectively; (3) the cropping intensity will be 160; and (4) a certain specified proportion of supplies will be made from surface and groundwater sources, with corresponding delta figures.

Serious shortcomings can be noted in all the four assumptions made in projecting the long-term demand. One, the basis of the estimate of yield has not been explained, but perhaps it is on an extrapolative approach. Two, the cropping intensity estimate is also perhaps on an extrapolative basis. Three, estimates of delta are based on marginal improvement on the current figures. (In later estimates by the Central Water Commission [CWC], these estimated improvements in delta were considered to be unattainable, and revised enhanced figures of water demand were obtained.) Four, the proportion of supplies by surface and groundwater assumed and specified to remain constant is also not explained.

The central conclusion was that it was concluded that there will be a serious problem of shortages of available water nationally and even more so in several river basins. Accordingly, a plan for interlinking the rivers of the country has been proposed to meet the emerging shortages. Noting that there is considerable surplus in the northeast in the Brahmaputra basin and increasing shortage in the south and west, it was proposed that water from the east to the south and west may be transferred. There had been a proposal to transfer particularly from the north to the south, proposed by earlier Minister of Irrigation Dr. K.L. Rao himself, known as the National Water Grid. In principle, it considered that while it will not be economically possible to transfer Brahmaputra waters to the east and then to the south because of the difficulties of making the transfer on account of negotiating the numerous local rivers from the Himalayas. Enough waters are contributed to the Ganga near about Patna where three major tributaries, Ghagra, Gandak from the north, and Son from the south, join Ganga. It was proposed that water may be pumped up Son and then transferred to rivers flowing south, which originate near about the same location where Son arises.

The Ministry of Water Resources (formerly Ministry of Irrigation) and the Central Water Commission later formulated a National Perspective for optimum utilization of water resources in the country, which envisages interbasin transfer of water from surplus to deficit areas. A proposal called the Interlinking of Rivers (ILR) of India has been prepared by the National Water Development Agency (NWDA). It is being seriously followed by detailed investigations.

The rationale in developing the ILR proposal appears to be as follows. One, in view of the tremendous future need and shortages in some basins and surpluses in some, all the waters of India should be developed to meet the future demands. Two, transfer by pumping is expensive, while transferring by gravity is much more economical. Therefore, transfer by gravity should be adopted. Three, there is tremendous surplus in the Brahmaputra, and therefore, it should be transferred to the deficit states of the west and the south. The National Perspective Plan has two main components: (1) Peninsular Development and (2) Himalayan Rivers Development. These are shown in Figures 11.2 and 11.3.

The Himalayan Rivers Development envisages the construction of storages on the main Ganga and the Brahmaputra rivers and their principal tributaries in India, Bhutan, and Nepal so as to conserve monsoon flows for flood control, hydropower generation, and irrigation. In terms of the ILR proposal, a Brahmaputra–Ganga link will be constructed for augmenting dry weather flows of Ganga. Interlinking canal systems will be provided to transfer surplus flows of the Kosi, Gandak, and Ghagra to the west. Surplus flows available on account of the interlinking of the Brahmaputra system with the Ganga system are proposed to be transferred to the drought areas of Harayana, Rajasthan, and Gujarat. The scheme will also enable large areas in South Uttar Pradesh and South Bihar to obtain irrigation benefits from the Ganga with a moderate lift of less than 30 m. Further, all land in the Terai area of Nepal would also get irrigation apart from generation of about 30 million kW of hydropower in Nepal and India. The interlinking of the main Brahmaputra and its tributaries

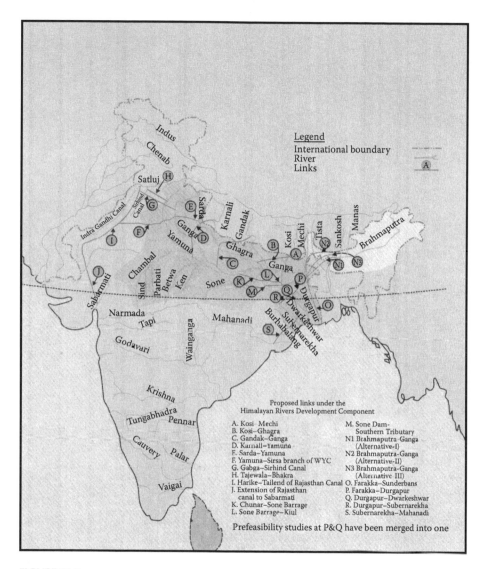

FIGURE 11.2
Himalayan Rivers Development component. (From NCIWRDP [National Commission of Integrated Water Resources Development Planning], 1999. Government of India, New Delhi.)

with Ganga and of Ganga with Mahanadi will enable transfer to the peninsular component. The Himalayan component would also provide the additional discharge for augmentation of flows at Farakka required inter alia to flush the Calcutta Port, the inland navigation facilities across the country and flood control in the Ganga and Brahmaputra.

The ILR has been studied in detail by a working group appointed by the NCIWRDP (1999) itself for this purpose. The working group concluded that

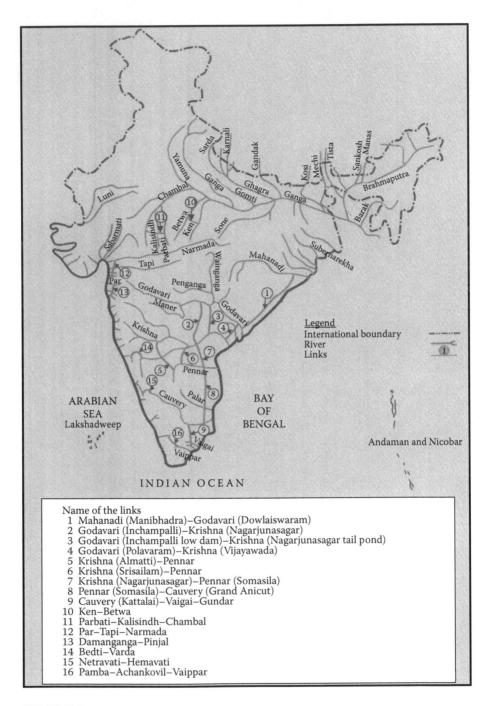

FIGURE 11.3
Peninsular River Development component. (From NCIWRDP [National Commission of Integrated Water Resources Development Planning], 1999. Government of India, New Delhi.)

the ILR proposal had serious basic shortcomings and was rejected. However, the NCIWRDP (1999) decided to accept the proposal to interlink Godavari with the southern states. In actual practice, the ILR proposal is being actively pursued. Detailed studies of various components are being carried out.

This leads to some speculative thinking as to how such unscientific actions are pursued. The proposal of the interlinking of Ganga with Yangtze Kiang by Cotton (1867), one of the most eminent engineers of the British period, who laid the foundation of the development of the water resources of India, comes to mind.

11.7.3 Latest Water Demand Estimates by GOI

We have restricted the discussion to the NCIWRDP (1999), as it represents the latest official position. However, one basic change may be mentioned, as it has important implications.

Estimates of long-term demand, as formulated after studying the NCIWRDP (1999), have been made by the GOI. It has been considered that the marginal efficiency over 50 years, as assumed in the usage in the NCIWRDP (1999), is not possible, and the current standards will continue! It is mentioned nowhere nowhere in the official documents, as we have discussed above, that our current efficiency is one of the lowest even in the 93 developing countries!

The issue is not of figures but of attitudes. This is the most important issue, and we will discuss it in Chapter 13. As emphasized, "Culture matters!"

11.8 Management of Environmental System

Our focus was management of water, as it is the most dynamic component of the environmental system—land, water, air, and the fauna and flora. But the entire system has to be considered integrally. Agriculture has a large-scale and complex impact on land, which is not merely clearing it of the natural fauna and flora. Urbanization also impacts the entire environmental system. The issue of managing air quality is dramatically brought out in terms of the serious climate change threat. We will, however, focus on water in Chapter 12 and consider it as a surrogate of the environmental system.

11.9 Climate Change

Humanity faces a formidable challenge of climate change. It is going to be particularly severe for the GBM basin. Besides the change in temperature, precipitation, pattern, and intensity, the GBM is threatened with two most serious disasters. One is the melting of the glaciers. This will create serious changes in river flows over a long time. Second is the threat of the rise in sea level, which is going to affect the delta region in a formidable manner. The essence of the message is the urgent need for the management of the environment, with particular perspective on climate change, in terms of mitigating it to the extent possible and facing it. Thus, it strongly reinforces the urgent need for a scientific management of the environment in the wider perspective of the environmental–societal management, which is the centerpiece of our message.

11.10 Conclusion

It follows from the perusal of the current scene that a serious challenge of managing the environment is faced in the GBM basin. The current development and future thinking have not addressed the challenge appropriately. A revolutionary change in the management of water will have to be introduced to meet this enormous challenge, efficiently, economically, and with the assurance of ensuring sustainability and fairness. Fortunately, advances have been made in the management of water, providing ideas, approaches, and tools to develop the path of transformation, known as environmental systems management (Chaturvedi, 2011a). We will have to change dramatically from the current colonial simplistic canal irrigation focus of managing water to stabilize sustenance agriculture to the modern concept of environmental systems management for sustainable development.

Note

1. The author is engaged in a study of Delhi's water supply and sanitation as discussed below. The Ministry of Science and Technology, Government of India, has sponsored a collaborative study of the problems of the metropolis by Indian and Mexican scientists. The first step in this context was an "Indo-Mexican Workshop on Water Management in Metropolis" in which some presentations of the present scene were given. Two aspects were particularly noteworthy—one, the excellent development in Mexico, and two, the poor perception and current development regarding Delhi. Three, the horrifying conditions in the slums of Delhi. This was brought out by

several NGOs in their excellent video presentations. Chan Yoon Kum, Assistant Chief Executive, Public Utilities Board, which is responsible for the water supply and sanitation, was a keynote speaker at a seminar entitled "An International Perspective on Environmental and Water Resources" organized by the Environmental and Water Resources Institute (EWRI) of ASCE and the Indian Institute of Technology, Kanpur. His talk brought out how the water infrastructure and environment have been transformed from being of a Third World standard to being one of the world's most efficient in a very short time and under severe resource constraints.

it. However, they considered that the dam was essential for the development of the country. This is important to be appreciated, particularly in the context of some renewed anti-dam movements in India of late. A brief reference is therefore made to it in Section 12.15.

The emphasis so far has been on the technological aspects. As it has been emphasized throughout, technology and social organization are the tools with which society transforms physical resources and human labor into distributed goods and services. Both have to be considered integrally. Development of the appropriate institutional framework and governance is one of the most important issues in the efficient and sustainable development of water. This has been neglected even though it is extremely important; the same goes for developing countries. The subject has been considered by Chaturvedi (2011c) in an Indian context to which reference is made.

A brief summary is presented in Section 12.15.

12.2 Development of Water—Historical Perspective

Water resources have been developed for a long time. Technologies were developed depending upon the needs and capabilities of the society and the physiographic–climatic–hydrologic conditions. For example, irrigation was an essential activity in the Mesopotamian, Indian, and Chinese environment, where civilizations developed in the earliest phases of human development. Dug wells, canals, tanks, and storages were developed. As development started taking place in Europe, mitigation of floods, navigation, and water supply and sanitation in cities became issues of concern. In science, fluid mechanics was an area of much interest, and so was pioneering mathematical analysis, to which mathematical leaders such as Gauss, Euler, Navier, to name a few, and even Newton made notable contributions. In terms of engineering sciences, hydraulics developed (Rouse and Ince, 1957). Hydroelectric development also soon became important as economic development gradually started. Developments on River Rhine were pioneering activities, which contributed much to the development of the society and the entire field of engineering science. As science and technology advanced, construction of large dams was increasingly undertaken. Navigation, flood control, and hydroelectric development became important in the USA as the development of the country started. As scientific advancement took place, focus, essentially, became on hydraulics in this context.

Advances in water resources development, with focus on irrigation, took place in a pioneering manner as developments were undertaken in British India at about the same time. Starting with the maintenance of the earlier inundation canals, a new era of canal irrigation began. Although considerable engineering developments were undertaken, not much organized scientific activity took place, as education and research had hardly developed in India.

Irrigation also became an important component of the multipurpose activities as development of the western United States was undertaken later. A scientific team even came to India to study the development being undertaken and reported the activities with much appreciation (Wilson, 1892). The science and technology of the subject was advanced considerably in the USA, as the large-scale developments were undertaken in the country and the increasing development of education and research took place. Close collaboration between educational institutions and profession followed. Increasingly, impressive dams, hydropower stations, and canals were constructed. These became the guiding examples as water resources development was undertaken in the developing countries later. There were attempts for these to be emulated in India as water resources development was undertaken after Independence, with Indian engineers going to the USA to study their development and U.S. engineers were invited to guide the Indian engineers.[1]

Conjunctively with the activities of development of water, advances were made in the science and art of the subject. Although the science of water has been developed for a long time, advances in the water field and, for that matter, over the entire spectrum of engineering made a quantum change after World War II (Chaturvedi and Chaturvedi, 1977; Chaturvedi, 1991). Two major changes were the introduction of science in engineering and the introduction of systems analysis in engineering, which, incidentally, found particular application in water resources (Mass et al., 1960; Hall and Dracup, 1970; Haimes, 1977; Loucks et al., 1981; Chaturvedi, 1987). Some of the important contributions were the integration of engineering and economics and scientific decision making. The increasing availability of computers considerably helped in these advances. Governing philosophies and principles for river basin planning, watershed management, and project design were developed. Particular advancement took place in the USA. These were codified and became the Principles and Standards for Planning Water and Related Land Resources. These principles, standards, and procedures were fashioned by combining the attributes of economic theory decision theory, and social choice theory that best suited the purposes of evaluating federal investments for water resources development and management. They have since been continually updated, and principles and guidelines have been developed. Procedures have been specified to evaluate, compare, and trade off the economic efficiency, technical performance, social acceptability, and environmental quality of numerous alternative measures for ameliorating the adverse affects of climatic variability. At the heart of the activity is engineering creativity to meet the challenge. It is considered that the governing philosophies and principles are comparable for most developed nations and international lending institutions.

It is relevant to note that most of the scientific advances in the field of water resources planning and management took place after the large-scale development of water resources had taken place in the industrialized countries and did not find much application in real life (Hall and Dracup, 1970;

Rogers and Fiering, 1986; L'vovick et al., 1990). Scientific advances were available for activities in India but were not readily adopted in practice, as water resources development is an official public sector activity and collaboration with the academic institutions is lacking, which hurts both.

12.3 Integrated Water Resources Management (IWRM)

Integrated development of water and hierarchical multiobjective developmental planning have been recognized as a central issue as water resources systems planning developed. Conceptual aspects of IWRM from geophysical, environmental, and socioeconomic considerations by computer-based mathematical models have continuously been advanced of late. Concern for environmental conservation and sustainability gave impetus to further advances in this context.

These ideas, often referred to as the Dublin Principles, have been increasingly emphasized (Young et al., 1994). The first of these principles is the "ecological principle," which requires that water be treated as a unitary resource within river basins with particular attention to ecosystems. The second is the "institutional principle," which recognizes that water management requires the involvement of governments, civil society, and the private sector, and that the principles of subsidiarity must be respected. The third principle is the "instrument principle," which requires that water be recognized as a scarce economic good and that greater use be made of user pays, polluter pays, and other market-friendly instruments.

Integrated freshwater resources management (IWRM) has been defined by the Global Water Partnership (GWP) as "a process that promotes the coordinated development and management of water, land, and related resources in order to maximize the resultant of vital economic and social welfare in an equitable manner without compromising the sustainability ecosystems." The conceptual issues following the WWC (2000) recommendations may be briefly stated.[2] A holistic systemic approach relying on IWRM must replace the current fragmentation in which water is managed. Such integration must cover all types of interrelated freshwater bodies, national and international, including both surface and groundwater and duly consider water quantity and quality aspects. The multisectoral nature of water resources development in the context of socioeconomic development must be recognized as well as the multi-interest utilization of water resources for water supply and sanitation, agriculture, industry, urban development, hydropower generation, inland fisheries, transportation, recreation, low and flat lands management, and other activities. Rational water utilization schemes for the development of surface and underground water supply sources and other potential sources have to be supported by concurrent water conservation and wastage minimization measures. Water does not understand jurisdictional differences, and trans-boundary development should be undertaken with due cooperation, taking into account the interests of all riparian states concerned. Water is affected by everything, and water affects everything and everyone. A graphic comb structure to depict IWRM has been developed with cross-sectoral integration, depicting (1) water for people, (2) water for food, (3) water for energy, (4) water for nature, and (5) water for other uses (industry, navigation) as "uses" and (1) enabling environmental policies, laws, (2) institutional framework, (3) management mechanisms, and (4) political economy as "framework."

Participatory institutional mechanisms must be put in place to involve all sectors of society in decision making. The old model of this is "government's business" must be replaced by a model in which stakeholders participate at all levels. Experience shows that partnerships between governments and with governments are playing a vital role in creating the enabling environment and in providing technical and enforcement support.

Freshwater must be recognized as a scarce commodity and managed accordingly. Full cost pricing of water services with equity will be needed to promote conservation and to attract large investments that are needed. "Polluter pays" principles must be enforced and mechanisms must be found whereby those who use water inefficiently have incentives to desist and transfer that water to higher-valued uses, including environmental purposes.

Water is a fundamental need. All human beings, including the poor and the marginalized, must have access to water. This aspect has particularly been emphasized recently under the United Nations Development Program in Human Development Report 2006 (UNDP, 2006). A concept of human security looking beyond narrow perceptions of national security was introduced under the Human Development Program by the United Nations in 1994, with a vision of security rooted in the lives of people. Water security is an integral part of this broader concept of human security. In broad terms, water security is about ensuring that every person has reliable access to enough safe water at an affordable price to lead a healthy, dignified, and productive life, while maintaining the ecological systems that provide water and also depend on water. The concept brings out an orientation to policy making and management, which must be taken into account.

Ensuring access for all, with full-cost pricing of water, will often require financial assistance to help poor communities develop their own water supply. Subsidies should be delivered directly to people, not to service organizations, in a manner that is transparent and well targeted. It is essential to separate the welfare task (the task of the government) from the business task (which service organizations should be asked to do). The public goods nature of the needs of future generations, of nature, and of people outside of the political and administrative units where the decisions are being made must be recognized.

Incentives for resources mobilization and technology change are needed. Considerable resources will be needed to develop and conserve water and much effort will be required to generate these resources.

Political will is needed. Difficult decisions and complex tradeoffs can be minimized by seeking win–win solutions, but they will not be eliminated. A technically and scientifically informed, participatory, transparent process of decision making at all levels (from the community to the river basin) must be put in place as the action arm of IWRM.

Governments, both national and local, are the sovereign and legitimate decision-making bodies. They will be the key actors to make all this happen—not by undertaking the tasks themselves but by setting the enabling framework for local community-based action and for properly motivated and regulated private sector.

Behavioral change is needed by all. We are all involved in the management of water, by what we choose to do or not do. Unless human behavior changes dramatically, technological solutions will be set at naught. Public awareness, education, identification and dissemination of best practices, and incentives for action are all parts of sustainable management of water.

12.4 Societal–Environmental Systems Issues and Management

The concept of IWRM brings out several important issues but also misses many, particularly, for example, the issue of integrating it with the societal activities and sustainability (Chaturvedi, 2001; Savenje and Hoekstra, 2002). Several conceptual advances have been considered to be necessary for considering management of the environment for sustainable development. Taking note of the current scene, it is considered that attempts at sustainable development of the environment have been pursued within narrow disciplinary or sectoral frameworks that ignore the strong linkages and interactions that exist between environment and human activities across many different issues, sectors, and scales. In addition, prevailing systems for decision making tend to separate economic, social, and environmental factors at the policy planning and management levels. It is important that the decision-making process is improved or restructured so that considerations of socioeconomic and environmental issues are fully integrated (Agenda 21; Chaturvedi, 2011a). In this context, certain conceptual clarifications are required in the first instance. These may be considered under the following heads for convenience.

12.4.1 Society and Environment

The society is embedded in the environment, and, in the first instance, environmental management depends upon the concepts, values, activities, and capabilities of the society. The most important issues are ethical and conceptual. Society and environment are one entity, and perturbation of the environment has far-reaching implications for the society. Thus, the very concept of exploitation of natural resources or nature, as in the past, is fundamentally mistaken. Nature cannot be exploited without jeopardizing not only the sustainability of the society but also the purpose of the society. At the same time, environment has to be developed to make it more appropriate for societal survival and development because the natural state may not be conducive. Thus, as has been said, society has to be freed from environment, but environment also has to be made free of the society (Asubel and Langford, 1997).

There are bound to be limits in any physical system, and even with the most efficient technology, there will be limits to perturbation of the environment. Therefore, the developmental objectives of the society have a determining impact on sustainable development of the environment. For instance, if consumerism, fun, and pleasure leading to so-called affluenza become the dominant determinants of development, sustainable development becomes doubtful. It is also questionable if this will contribute to peace and happiness. On the other hand, if mere survival continues to remain the occupation of the society, as in the underdeveloped societies historically, social development cannot take place. Thus, as in all fields, there is a need to find the middle path, as Buddha taught a long time back.

Society is a complex dynamic system, or to be more precise, it is a conglomeration of a very large number of complex dynamic subsystems, organizations, and individuals that are interconnected in a myriad of ways. Several approaches have been developed for understanding social change. In terms of environmental management, the focus of interest in terms of societal interaction is on the economic sector, but, as has been emphasized of late, culture and institutions have central importance. Institutional development, including ideological commitments, and awareness generation, therefore, become a central issue for sustainable development.

Much work has been done in the area of economic development, with which the issue of environmental development is closely linked. A new approach called "comprehensive development framework" has been developed of late in an attempt to operationalize a holistic approach to development (World Bank, 2000). The proposed new approach is based on four areas of development—structural, human, physical, and sectoral—in which structural elements relate to government and the administrative system. Environmental development could be considered in terms of these component activities.

In another approach, for long-range integrated planning, a concept of socioecological system has been proposed (Bossel, 1998). The socioecological system is considered to be composed of economic, social, and environmental subsystems and their interlinkages. In a similar formulation based on systems thinking, the societal system has been considered under six subsystems: (1) economic system, (2) social system, (3) individual system, (4) government system, (5) infrastructural system, and (6) environment and resource system (Figure 12.1). Each of these subsystems represents a component of "capital" that is vital to the development of the total system. We thus consider the interlinkage of the environmental system with each of the subsystems.

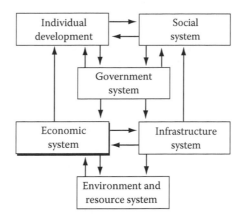

FIGURE 12.1
Societal system. (From Bossel, H., *Earth at Cross Roads*, Cambridge University Press, Cambridge, UK, 1998. With permission.)

12.4.2 Economic System

Human pressure on the environment arises principally through the economic–environmental system interaction. The economic exchange system is comprised of productive, consumptive, and accumulative activities, which create flows of goods and services between the system components. The environmental system consists of natural, man-made, and social components. Natural resources from the environmental system are used in the production activities and wastes or pollutants that go back to the environment, often considered in terms of externalities. The first point that follows, as has been well emphasized, is that the activities have to be considered over the entire life cycle and the externalities have to be fully internalized. This needs to be carried out not merely by economic valuations, as currently proposed, but in the process of environmental management activities or technological activity themselves (Marglin, 1962).

Second, the economic system needs to be made as efficient as possible over the entire spectrum of activities of the sector, so that the factors of production are used most productively with minimal environmental system perturbation. For example, the yields in the agricultural sector in the developing countries are almost one-third of those achieved in the industrialized countries just because all the factors of production are not used appropriately or the activities are not undertaken efficiently. Therefore, if the sectoral activity is made more efficient, the use of resources and environmental degradation can be minimized by an order of magnitude. Similar examples can be given from the energy and the industrial sector, bringing out the possibility of improvement of the system by several orders of magnitude. The issue is that environmental management must not be considered in isolation, developing it to meet the demands, but the entire socioeconomic system has to be considered integrally as brought out in the industrial ecology (Richards and Pearson, 1998) and land-use wiring (Rayner et al., 1994) concepts. Management of the demands is particularly important. This can be accomplished through (1) adoption of appropriate end-use technologies, (2) pricing mechanisms, and (3) generation of value systems.

12.4.3 Environmental System

The environment is a living organism, composed of biotic and biotic components, including humans, in synergetic interaction. It is not a mere assemblage of the constituent components. The physical world—land, water, and air—evolved over geologic time and continues to be shaped by geophysical–chemical–biological processes. Emphasis has increasingly been laid on comprehensive environmental impact analysis of proposed activities, but at the planning stage, the focus generally is in terms of the constituent components in isolation and that too principally in terms of a resource development perspective. It is important to consider the environment in totality and integrally in the ecosystem perspective as part of socioeconomic activities over the entire life cycle from the outset, albeit breaking it down in terms of the components for detailed consideration. This is particularly important for the fluid elements, particularly water, which provide the dynamic interlinkage through geochemical–biological processes. For example, water has to be considered as a holistic vector in the context of the hydrologic cycle in terms of hydrosolidarity thinking, rather than considering it in isolation only in terms of precipitation, surface, and groundwater. Land has to be considered integrally with it. Indeed, since land has the dominant demand and land mediates dominantly to modify the precipitation, land becomes the dominant component, and development of water is indeed development and management of land and water integrally.

Reference may also be made in this context of the usual conceptual conflict between development and conservation of the environment. As has been forcefully brought out, development and conservation are two sides of the same coin. The term "management" is therefore used as emphasizing embracement of the two activities integrally (International Union of Conservation of Nature [IUCN], 1991). This is particularly important in the context of the developing countries, where it has been amply demonstrated that economic development, which requires environmental development, leads to environmental conservation and improvement. Poverty degrades and pollutes the environment, as there is little understanding of environmental implications of the activities, often taken under desperate conditions of survival, besides poor capability in terms of productive use of the resources. Third World agriculture is a glaring example of inefficient use of land and water.

12.4.4 Environment and Infrastructure

Next to the usual economic activities of the agricultural and manufacturing sector, infrastructure—the man-made environment—has a significant impact on the natural environment. There are several components of infrastructure. The particularly relevant ones are (1) setlements and cities; (2) water, sanitation, and waste management; (3) transportation; and (4) energy.

Infrastructure develops as part of the historical development process. The development in the past was gradual, and environmental pressure increased slowly as the different ecological transitions took place. Even during the industrial revolution, urbanization and the development of the transport system were a slow evolutionary processes. The conquest of the new worlds and colonization of foreign countries provided tremendous new space for settlements for the expoloding population of the currently industrialized countries. These benefits of time and space are not available to the currently developing world, and instead of following the current gradual evolutionary approach, habitat and infrastructural development has to be urgently undertaken creatively. It is crucial for economic development, environmental conservation, and even for providing the basic amenities like drinking water. For example, the pollution of rivers in the developing countries cannot be managed unless sanitation and drainage are provided properly in the human settlements. Even drinking water cannot be provided unless ubanization is properly undertaken, in which the Third World rural–urban migration creates a serious problem. This can be done only if rural development is emphasized as part of spatial habitat and new habitats are developed urgently at a large scale as part of socioeconomic development. Conjunctively, the modern facilities, even drinking water, cannot be provided to the rural communities unless appropriate rural develoment is undertaken. Similarly, the air quality and minimization of adverse climatic impacts can be managed only if the transportation sector, besides the overall energy sector, is properly developed.

Large-scale development of settlements, the infrastructural system, and the energy system has to be undertaken in the developing countries. This will require complex long-term systems planning so that the environmental–economic impacts are optimal. For example, in the transportation system, different intermodal choices have to be considered. City, regional, and interregional interlinkages in the intertemporal context have to be considered. Similarly, for the energy system, a careful long-term analysis is required in the context of different energy supply possibilities in the intertemporal context. The planning of urban settlements itself introduces the consideration of proper division of space, locally and regionally, so as to generate community and environmental relationships. The central issue is that these choices become irreversible, and therefore, very careful planning is required. A review of the current activities in the field shows that adequate attention has not been paid to this feature.

12.4.5 Social System and Environment

Traditionally, capital has been considered in terms of natural capital, physical or produced capital, and human capital. Together they constitute the wealth of nations and constitute the basis of economic development and growth. It has now been recognized that they overlook the way the economic actors interact and organize themselves to generate growth and development. The missing link is the social capital (Putnam, 1993; Cornea, 1994). While the importance of the social capital has been well accepted, its concept and specification have not been fully agreed upon. The most encompassing view of social capital includes the social and political environment that enables norms to develop and shape social structure. Research is underway to specify the social system characteristics as they impinge on development and, correspondingly, environmental management, but the fact remains that environmental management must include consideration of impact on social system development and vice versa. For example, watershed development can best be undertaken if the local village community manages it. Similarly, the community has to be fully involved in the urban system development. A central point to be emphasized is that one of the most damaging implications of the Indian colonial history has been the throttling of the spirit of development in the people and the society. The most important challenge for attaining sustainable development is to establish, vitalize, and institutionalize this spirit.

12.4.6 Human System and Environment

Human capital has been recognized to be one of the most important components of capital. Emphasis has increasingly been laid on improving the human capital. It is much more than mere literacy or even education. As has been widely emphasized by noble laureate Amritya Sen (1999), human and social capital development in terms of entitlement of people is the central and primary activity for socioeconomic development. It is equally critical for sustainable development. This involves not only development of appropriate skills but perceptions about sustainable management of the environment in a wide sense, and attitudinal developments in terms of adoption of an ideology of attaining the rightful place in the human community (Yamumara, 1996; North, 1996; Harrison and Huntington, 2000).

12.4.7 Government System

Societal processes are regulated through organizational processes, essentially government and administration. They enable and constrain much of what happens in other sectors. Government has to define and enforce the rules regulating human action and interaction within the social system. It has to provide social and coordinating services that the economic system cannot provide. It has to finance the developmental activities and the services in case of public sector activities, through appropriate means. It has to resolve conflicts. In the context of the environmental system, it has to regulate interaction of the society with it. Thus, along with the technoeconomic development, the issue of evolving an approriate environmental developmental framework becomes very important (World Bank, 1993, 2000).

12.5 Societal–Environmental Systems Management (SESM) with Focus on Water

Large-scale environmental development and conservation, termed "societal–environmental systems management" (SESM), to emphasize the integrated nature of these activities, will be required as the developing countries gradually achieve their rightful place in the human community. It will have to be undertaken much more efficiently than in the past to ensure sustainability. Considerable advances have taken place in the field recently, and building on them, the concepts and approach for SESM, has been proposed (2001, 2011a). Thus, SESM builds on the modern advances considering (i) environment as a socioecological system, (ii) sustainable development of the society, (iii) the environmental vectors integrally and conjunctively in the context of the multifarious demands and objectives, and (iv) the environmental processes integrally with the resource transformation adopting a holistic and integrative approach over the entire life cycle. Several other important issues, though not emphasized currently, also have to be considered. Environmental transformation is only one part of the developmental activities. The meaning, objectives, and strategies of development will have to be carefully examined. Secondly, development is a societal activity. Attitudes, institutions, and capabilities have to be developed continuously.

Focusing on water, IWRM has to be undertaken considering it as part of the socioecological or environmental system. There is a need to introduce increasing creativity and sensitivity. Water resources development has to be undertaken adopting a holistic systemic approach. Water, along with other environmental vectors, particularly land, shall be considered integrally in terms of the ecosystem development and in the context of the multifarious sector demands and objectives for rapid economic development. Concurrently, the economic sector activities have to be undertaken with increased efficiency. Water resources development in terms of demand and return flows has to be considered in this dynamic context. Both the environment and economic sector modernization are intimately and circularly interlinked. Agriculture and water sector interlinkage is a good example. Third World agriculture, with its poor yields and little concern to environmental degradation, is not an efficient use of environmental resources and amenity services. At the same time, unless adequate, timely, and reliable water supplies are provided, in contrast to the current extensive irrigation, yields cannot be improved. Similarly, the land use has to be given primary focus. Development of habitat, in the past, was a secondary activity, with the population being dominantly rural. As population increases and economic development is vigorously undertaken, development of infrastructure and other sector activities of industrialization and services will have to be undertaken at a rapid rate. Urbanization is a concurrent process. It puts a lots of concentrated pressure on the environment, and it is important that the activities are organized so that pressure on land is spread out as evenly as possible. Our focus is on water, but the interlinkages with the ecosystem and the economy have to be duly considered, and development of water has to be considered in the context of the development of the entire socioeconomic system.

Conceptually, development of water has been considered as shown in Figure 12.2. One important point that is being emphasized is that modernization of the sector activities must also be considered integrally. For example, as stated earlier, it will not be possible to keep on providing water if the current Third World agriculture continues. Again, it will not be possible to maintain the ecological integrity of the mountain areas if historic subsistence agriculture continues. Instead, it is imperative that, instead, horticulture is practiced in the hilly terrain. Similarly, it will not be possible to provide water to meet urban demands and maintain clean rivers if the current process of Third World urbanization continues. From this perspective, water resources development may be better perceived as shown in Figure 12.3 and is carried out integrally with other activities in terms of obtaining an overall perspective. First and foremost, a developmental trajectory to bring the societies in the mainstream of human endeavor should be worked out. Policies, supported by appropriate analytical models, have to be developed for each sector and interrelated. Detailed development of the water sector has to be carried out as emphasized in the approach proposed in the context of IWRM with further advancements that all the environmental components have to be considered integrally in terms of an ecosystem. Conjunctive use of land and water in terms of watershed is particularly important and pragmatic. Both these activities have to be undertaken iteratively to evolve an integrated environmental systems management policy.

Development has to be considered at different scales spatially and integrally. Thus, at one end is the watershed or a characteristic ecosystem. At the other end is the river basin, which again has to be considered as a macro ecosystem. Integration will be at several levels hierarchically. Again, there will be a number of interrelated activities. Thus, a portfolio of activities has to be creatively developed and considered integrally. Besides spatial interlinkage, there is an intertemporal interlinkage as development and impacts occur over a period of time.

It has to be understood that there are all sorts of uncertainties and differences. For example, there are multiobjectives, and valuations also differ. Management approaches differ. It is therefore important that a transparent participatory decision-making approach is employed. In view of all these considerations, the approach of policy making, besides creativity, is

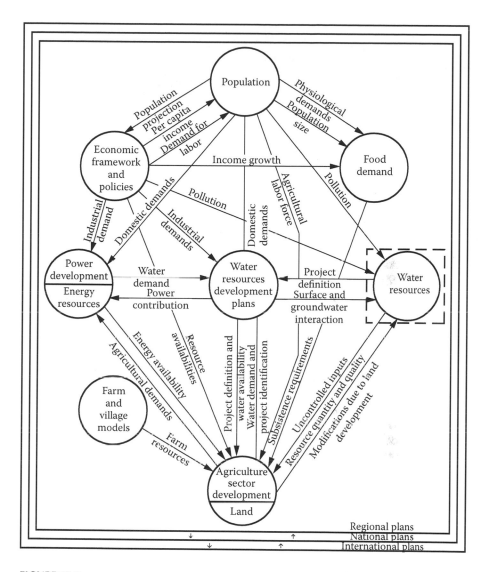

FIGURE 12.2
Resources and policies interactions in water resources development. (From Chaturvedi, M.C., *Water Resources System Planning and Management*, Tata McGraw-Hill, New Delhi, 1987. With permission.)

important. Systems planning, backed by approaches that help make decision making participatory and transparent, is extremely important. Climate change is an important issue and must be considered concurrently.

It has to be emphasized that the proposed formulation is only a conceptual framework of the study. Creative engineering and management are the cornerstone of scientific development.

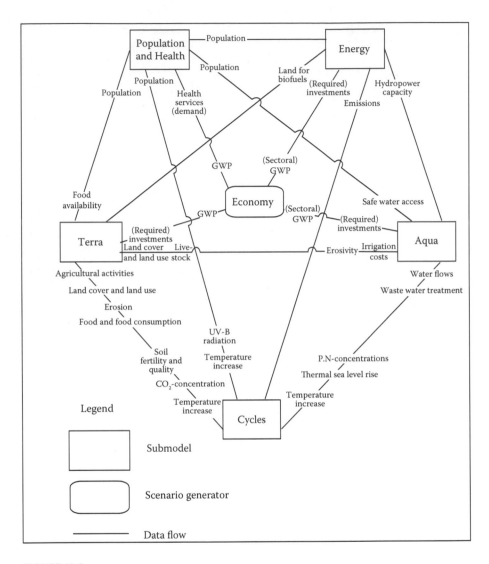

FIGURE 12.3
Regional environmental system study perspective. (From Chaturvedi, M.C., *Environmental Systems Management for Sustainable Development*, 2011. With permission.)

12.6 Societal–Environmental Systems Management Principle Implications for Management of Water

As emphasized earlier, technology and social institutions are the two tools used to manage the environment. The challenges posed in the policy formulations and in overcoming the shortcomings being experienced in the sector require urgent management of water resources on sound policies and strengthening of the institutional arrangements.

The former requires a comprehensive analytical framework that takes into account the various concepts raised in the context of the environmental systems approach. Water resources should be managed in the context of a national water strategy that reflects the nation's social–economic and environmental objectives and is based on an assessment of the country's water resources. The assessment should include a realistic forecast of the demand for water, based on projected population growth and economic development and a consideration of options for managing demand and supplies, taking into account existing investments and those likely to occur in the private sector. The strategy should spell out priorities for providing water services: establish policies on water rights, water pricing and cost recovery, public investment, and the role of the public sector in water development. The framework should facilitate the consideration of relationships between the ecosystem and socioeconomic activities in river basins. In essence, this comprehensive approach breaks down very complex problems in a river basin into more manageable elements to achieve coherent cross-sector water management. Alternative public investment option and patterns of water management can be considered and evaluated, especially in light of supply and land use interdependencies and of the major social and environmental consequences that characterize water resources sector investment or reallocation.

Complexity of the analysis will vary with the circumstances. In the first instance, it is often possible to clarify priorities and to take account of key interdependence, using a relatively uncomplicated framework. Once a suitable overall framework has been established, more detailed analytical approaches have to be undertaken.

The analytical framework provides the underpinnings for formulating public policy. Regulations governing pollution, health standards, and environmental protection address water quality interdependencies among uses. Appropriate pricing and charging systems (and water markets where feasible) that provide the correct signals must be established so that decentralized decision making can provide the allocation of resources. Within a decentralized system, adequate charges would also endorse entities with operational and financial autonomy to provide reliable and sustainable services. In general, the move could be made toward generalizing the delivery of water services, to the extent possible, and toward using appropriate pricing to create the precondition for an efficient, sustainable delivery of water resources.

The conventional approach is to work out the long-term water demand and try to plan the development of the water resources accordingly. This approach, however, needs considerable changes. In view of the importance of the activities from economic and environmental considerations, it is necessary that systems studies with multiobjective tradeoffs based on watershed, river basin, and national level be carried out. Coupling them with integrated intertemporal analysis has to be undertaken from sustainability considerations.

Following these concepts, water management may be considered as a "comb," in which the teeth are the water-using sectors and the handle is the resource itself, defined by the location, quantity, and quality (World Bank, 2000). There are two components of the developmental activities: (1) those that focus on the water using sectors and (2) those that focus on water resources management and the interlinkage between resource use and service management. The former embraces the technological activities of the water demand sectors. This is well covered in the technological literature and does not need to be repeated except when emphasizing that there are tremendous opportunity and urgency in managing the demand sectors efficiently, leading to a phenomenal decrease in water use and its consequent debasement. This has been totally neglected.

The latter, called "water management strategy," means addressing the following: (1) the development and management of infrastructure, for annual and multi-year flow regulation, for flood and droughts, multipurpose storage, and for water quality and source protection; (2) the institutional framework, including the definition and establishment of laws, rights, and licenses; the responsibilities of different sectors; and standards for water quality and service provision (especially to the poor) for the environment, for land use management, and for construction and management of infrastructure, which affects the quantity and quality of water resources, including international, national, and resource management agencies, river basin and aquifer agencies, and watershed management institutions; (3) the management instruments, including regulatory arrangements, financial instruments, standards and plans, mechanisms for effective participation of stakeholders, and knowledge and information systems that increase transparency, motivate effective water allocation, use, and conservation, and secure maintenance and physical sustainability of water resource systems; (4) and the political economy of water management and reform, in which there is particular emphasis on the distribution of benefits and costs and on the incentives, which encourage or constrain the more productive and sustainable resource use.

12.7 Systems Planning

The complexity and scale of environmental management require that the systems approach be followed such that systems planning is the first step. This is a subject by itself, but briefly, we need to emphasize that, systematically and iteratively, we have to define the objectives, goals, development indicators and evaluation criteria, and development policies and plans and evaluate them objectively through integrated systems modeling. It needs to be emphasized that systems planning is not esoteric mathematical modeling but organized scientific creative engineering, supported by computer-aided studies, because often this aspect is lost, and it degenerates into academic modeling acrobatics (Chaturvedi, 1987).

One of the first tasks in the context of systems planning is development of the system dynamics framework. Much work has been done recently in this area although advances are considered to be necessary. The usual approach has been in the context of a Pressure–State–Impact–Response (PSIR) framework (Figure 12.4) (Rotmans and de Vries, 1997). Pressure represents social, economic, and ecological forces underlying the pressures on the human and environmental system. State represents the physical, chemical, and biological changes in the state of the biosphere, as well as changes in human population and resource and capital stocks. Impact represents the social, economic, and ecological impacts as a result of human and/or natural disturbance. Response represents human intervention in response to ecological and societal impacts. The PSIR framework can be conceived of as a dynamic cycle.

The PSIR framework is a valuable concept, but it puts more emphasis on the evaluative aspects of the activity. For sustainable development, a helpful approach may be conjunctive use of guidance through some orienters in terms of basic characteristics of sustainability, in the first instance. Suggestions have been made for establishing some basic principles of orienters and these have been developed for each subsystem (Bossel, 1998). These could be helpful for the initial system architecture, following which the PSIR framework may be used for more detailed analysis through modeling. Secondly, instead of adopting a passive approach, as the term response indicates, a creative anticipatory formulation may be more appropriate.

Systems analysis has been found to be very helpful in comprehensive critical and creative analysis and planning of ever-increasing complex techno-economic–environmental systems. It has found particular application in the area of water resources as discussed earlier. The approach taken for large river basins is first described to bring out various issues that need to be investigated and the approach that needs to be adopted for creative development,

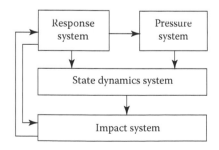

FIGURE 12.4
The Pressure–State–Impact–Response (PSIR) framework. (From Chaturvedi, M.C., *Environmental Systems Management for Sustainable Development*, 2011. With permission.)

supported by computer-oriented mathematical models for undertaking thought experiments.

Water resource development is undertaken for meeting the societal demands in the context of an increasing population and a developing economy. Some interactions in this context were shown schematically in Figure 12.1. The developmental policy formulation will require an analysis of several aspects of the socioeconomic system and the physical system through a number of studies. An overview of relevant issues and models is shown in Figure 12.5. Their modeling morphology is shown in Figure 12.6. A scheme

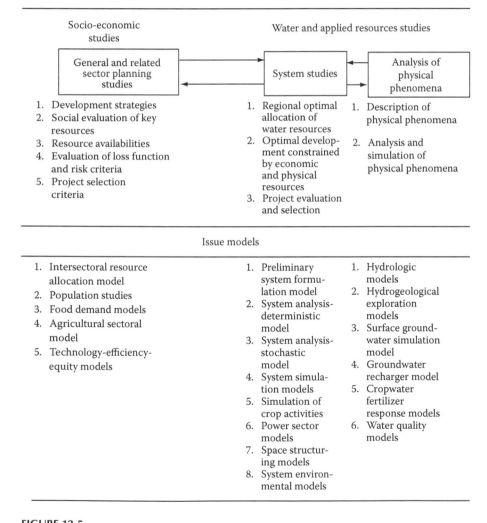

FIGURE 12.5
Relevant issues and models. (From Chaturvedi, M.C., *Water Resources System Planning and Management*, Tata McGraw-Hill, New Delhi, 1987. With permission.)

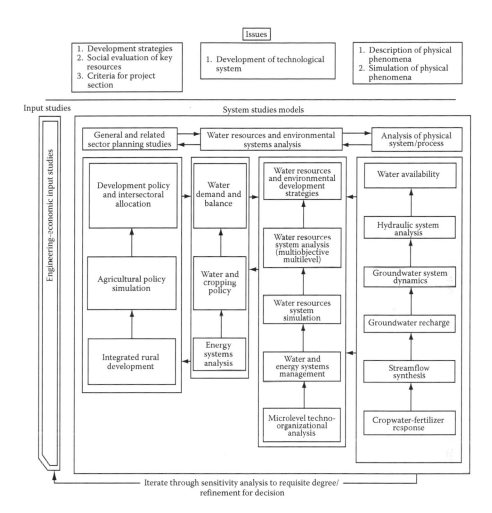

FIGURE 12.6

Issues and systems analysis modeling morphology. (From Chaturvedi, M.C., *Water Resources System Planning and Management*, Tata McGraw-Hill, New Delhi, 1987. With permission.)

of study at the national, river basin, and project levels is shown in Figure 12.7. It also shows the relevant sectional and economic studies, which can also be illustrated in terms of the PSIR approach as shown in Figure 12.3 at any level. These are only indicative, and specific studies will depend upon the river basin to be studied. The framework has been applied to the developmental planning of the Ganges–Brahmaputra basin, as shown in Figure 12.8, as an applied-research study. Other approaches have also been developed, as those for the Maumee River basin, the Netherlands, and Zambezi River, to name a few (Haimes, 1977; Chaturvedi, 1987; Hoekstra, 1998).

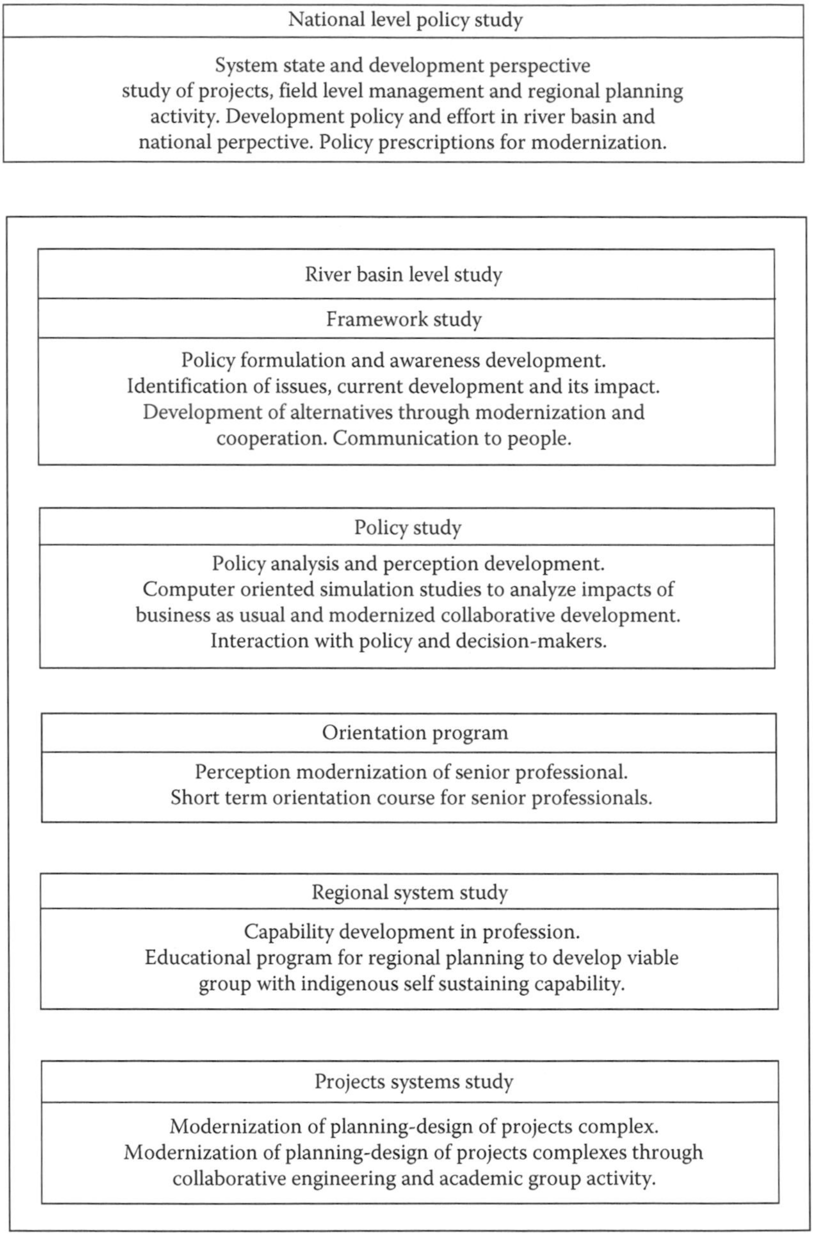

National level policy study
System state and development perspective study of projects, field level management and regional planning activity. Development policy and effort in river basin and national perpective. Policy prescriptions for modernization.

River basin level study
Framework study
Policy formulation and awareness development. Identification of issues, current development and its impact. Development of alternatives through modernization and cooperation. Communication to people.

Policy study
Policy analysis and perception development. Computer oriented simulation studies to analyze impacts of business as usual and modernized collaborative development. Interaction with policy and decision-makers.

Orientation program
Perception modernization of senior professional. Short term orientation course for senior professionals.

Regional system study
Capability development in profession. Educational program for regional planning to develop viable group with indigenous self sustaining capability.

Projects systems study
Modernization of planning-design of projects complex. Modernization of planning-design of projects complexes through collaborative engineering and academic group activity.

FIGURE 12.7
Schemes of studies for environmental systems management at different spatial scales. (From Chaturvedi, M.C., *Water Policy*, 3, 297–320, 2001.)

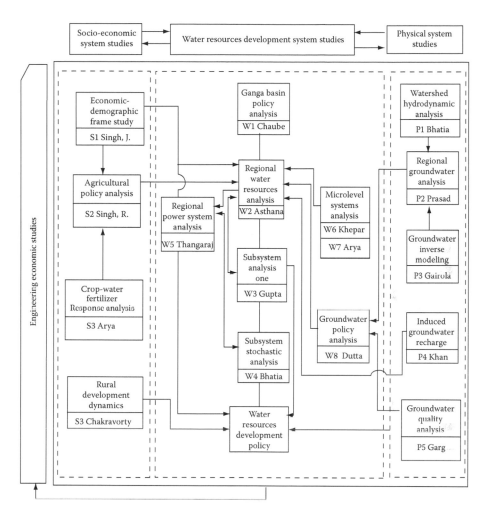

FIGURE 12.8
Regional water resources systems planning scheme. (From Chaturvedi, M.C., *Water Resources System Planning and Management,* Tata McGraw-Hill, New Delhi, 1987. With permission.)

12.8 Scenario Analysis

Sustainability implicitly underscores long-term perspective. Consideration of interplay and dynamic evolution of social, economic, and natural systems over a long time frame is required for this purpose. Addressing all concerns in a single integrated model particularly to explore the various possibilities is difficult in an integrated model, and scientific scenario analysis has emerged as a valuable way of developing policies for environmental management in the first instance (Raskin et al., 1998). The objective of scenario analysis is not to predict what the long-term outcomes are most likely going to be but to explore the future and develop an approach for interactive development of the policies. Integrated assessment models to explore the environmental management in greater detail could follow scenario studies (Rotman et al., 1997).

Scenario studies have been undertaken in the context of sustainable global development. Similarly integrated assessment modeling has been carried out in the context of climate change or global environmental change. Considering the context of the developing countries, where development of the land–water system is particularly important, it will be appropriate to enlarge the integrated system modeling currently developed in the context of climatic studies in terms of regional studies in which the land–water system could be examined in detail while maintaining the integrity of the total system. The choice of modeling approach also requires some consideration. The tool generally used in integrated systems modeling is system dynamics, which provides an excellent way of maintaining system integrity as well as temporal continuity. Optimization models could integrally supplement it so that system development could be reflected and that it does not remain a mere business-as-usual study (Hoekstra, 1998; Kothari, 2000; Hasan, 2003).

12.9 Some Advances in Modeling Technology

The field of system planning for sustainable management of water is being continuously advanced. Attempts have been made to examine the various issues and challenges raised by the concept of sustainability applied to water resources system design and management, review the guidelines that have been suggested for achieving a greater degree of sustainability, and outline some approaches for measuring and modeling sustainability (Loucks and Gladwell, 1999).

Reference may be made particularly to modeling technologies that have been developed of late in this context, as these need to be applied to real life studies. Two subjects are of particular importance: (1) generalized water resources models and (2) decision support systems.

Generalized water resources models mean that the computer model is designed for application to a range of problems dealing with systems of various configurations and locations. They cover watershed runoff; river hydraulics; river and reservoir water quality; reservoir/river system operation; and groundwater, water distribution system hydraulics, and demand forecasting.

Activities and comprehensive modeling of complex, multicomponent water resources systems have been considerably enhanced by developments that permit various mathematical objects to be linked to databases and operated through computer graphic user interfaces. Mathematical optimization and simulation modeling techniques, often together with geographical information system capabilities are now widely incorporated within numerous interactive software programs developed to study water resource planning and management issues.

Repeated use of computer-based models and databases under different assumptions is considerably facilitated through the use of interaction menu-driven graphic-based user interfaces. Such interfaces give users interactive control over data entry editing and display and over a particular operations being performed by the computer. This interactive and display capability facilitates user understanding of both the input and the output. They are typically called "decision support systems" (DSSs) (Loucks and Gladwell, 1999; Chaturvedi, 2007a). DSSs are only interactive computer-based information providers. They, like underlying data management components, do not make decisions. DSSs for integrated river basin management have been developed by focusing on real-life problems.

Another effort is to develop DSSs to improve the capability of predicting a variety of economic, hydrologic water quality, and ecosystem health indicators associated with alternative land and water management and use decision. The DSSs will be building upon the object-oriented, data-driven, simulation models already developed. Examples are numerous. This object-oriented modeling approach will be expanded to include GIS-based objects that define the temporal and spatial distribution of water quantity and its constituents (e.g., sediment and nutrients) in the runoff from watershed lands. It will be linked to demographic/economic models for predicting the probabilities of possible changes in land use over time driven by economic growth and various land and water use policy decisions.

The overall structure of the DSSs will include four interdependent components combined within a synthesis component. Module I predicts the probable changes in local land use that are likely to occur over time in response to economic and population growth projections as well as in response to current or proposed land use and water management decisions. These changes may impact surface and groundwater quality and hence the aquatic ecosystem themselves. Modules 2 and 3 simulate the rainfall runoff and water flows, temperatures, oxygen concentrations, and sediment and nutrient loading over time. Module 4 defines the relationships between environmental parameter values (gradients) and ecological performance and integrity. Module 5 is the interacting interface for data input, editing, and analysis for controlling the operation of each of the four component modules and for creating various displays of input data and results on a microcomputer.

The ongoing pursuit of evolutionary algorithm-based but relatively simple-to-use ecological models discussed above is one among many others. It is considered that the integration of individual ecosystem communities within the real-world dynamics of large ensembles of communities, effectively accounting for adaptation, change, and evolution (e.g., from one species group to another), remains a significant research challenge.

Whatever modeling technology is developed and implemented to study a particular water resources system, it cannot address sustainability issues unless it addresses and simulates the variables of concern to those who will be affected by its management. It must focus on what is culturally or socially important. The value of predictive technology lies not only in its economic viability and its technical soundness but also in its adaptation to the local institutional and cultural environment. This becomes particularly important when addressing the problems of the developing countries. As has been emphasized, it is important that while addressing the Third World problem a collaborative study is undertaken, because it is not possible for outsiders to understand the cultural and institutional nuances (Bower and Hufschmidt, 1981).

12.10 Study of Changes in Land Use and Land Cover

Land is normally defined as a physical entity in terms of topography and spatial nature; a broader integrative view also includes natural resources—the soils, minerals, water, and biota that the land comprises. Land use/cover has been recognized as a critical factor mediating between socioeconomic, political, and cultural behavior and global environmental changes. Emphasis is increasingly laid on increasing the move, toward more effective and efficient use of land and its natural resources. Integrated physical and land-use planning and management are eminently practical ways to achieve this. The essence of the integrated approach finds expression in the coordination of the sector planning and management activities concerned with various aspects of land use and land resources. Water is a part of land, and integrated planning of land or water implicitly denotes integrated planning of the two and is one of the basic aspects of environmental systems development and management.

The integration is done in several ways. Water is delivered through precipitation and is transformed into various components of the hydrologic cycle as it is received on land. The topographical features, land cover/use, and soil, and subsoil characteristics determine the water availability and geochemical–biological characteristics. Watershed management is thus a central issue in water resources development and management. Similarly, the demand for water is overwhelmingly dependent on requirements for agriculture. Thus, the land cover is also dependent on development of water and is determined by it.

Land cover has increasingly been changed from the natural state to land use from agricultural considerations. Forests have been drastically cleared with the current cover of only about 23% of the country's geographical area.

The forest cover falls far behind the international one-third optimal acreage. Moreover, the quality is poor. Forestland with a crown density of 40% covers only 11.5% of India's total area. Open forestland with crown density covers 7.6% of the country. Further, forests are shared very unevenly by different states. Efforts are being made for reforestation. An important contribution can be made through efficient water resources development and management. As will be argued later, if productivity could be increased by efficient land use and inputs, in which water is critical, it may be possible to give land and water back to nature. Thus, development and management of water are synonymous with integrated land and water use.

It has been usual to plan the land cover from agronomic, water availability, economic, and environmental considerations, taking into account the local, regional, and national interlinkages. This is being expanded to take into account the international linkages. However, it is being emphasized that considerable change in land use/land cover will take place as economic development and population increase takes place, and it is important to manage it from sustainable development considerations. The subject is particularly important for the developing countries and requires careful study, especially in the context of development and management of water. The proposed approach is still a research scheme, but it merits attention, as it provides an organized approach for looking at an important issue for development and management of water.

12.11 Management for Climate Change

Anthropogenically induced change in global climate would have major impacts on water supplies and demands and its management in a wider perspective. It is also closely related to, besides other factors, change in land use and cover, which are related to water use in a cyclic manner. It is therefore necessary to understand it, quantify it, and develop appropriate ways to incorporate the implications of this possible change in the development and management of water. At the outset, it is abundantly clear that efforts to restore natural land cover are eminently important, in which efficient water use has an important role.

Studies have been made to incorporate ways of incorporating the subject of impact of climatic change in water resources planning. There are considerable uncertainties regarding climate change. The timings and magnitudes of such changes are uncertain; even the directions of change in the availability of water are uncertain, especially at the scales relevant to water resources planning. Although very substantial surprises are possible, the climate-related uncertainties are not expected to be qualitatively different from those stemming from changes in other factors that have traditionally played a central role in water planning and project evaluation, such as population, income, technology, and social values. It is generally considered that the methods of sensitivity analysis, scenario planning, and decision analysis may help us in studying the implications of climate change and are generally appropriate for planning and project evaluation under the prospect of climate change, although new applications and extensions of some criteria may be warranted. These include nonstationarity. However, it is important to keep the perspective in view and develop measures that can be used to adapt to it in whatever way possible. In our view, the issue is very complex and important, and much more attention is needed than what is currently bestowed on it.

Two other important issues have to be noted. First, land cover/use change has important bearing on climate change. Development of water for agriculture has important bearing on land cover/use. For example, if productivity of land can be increased through appropriate irrigation and corresponding use of other inputs and modernization of agriculture, land can be given back to nature. This will contribute to the mitigation of climate change. Thus, while detailed study of implications of climate change is undertaken, efforts should be made to adopt policies that contribute to its minimization.

Second, climate change is a very serious matter. It is extremely complicated. Besides other impacts, it will have serious implications on water resources. As brought out in the third assessment report of the Intergovernmental Panel on Climate Change (IPCC, 2001), "Climate change will lead to an intensification of the global hydrologic cycle and can have major impacts on regional water resources, affecting both ground and surface water supply. Changes in the total amount of precipitation and its frequency and intensity directly affect the magnitude and timing of runoff and the intensity of floods and droughts; however, at present specific regional effects are uncertain."

The current Government of India's approach to implications of climate change is brought out in Appendix 1. In our judgment, it has not received adequate consideration. With the revolutionary changes that we are proposing, we will be able to meet the climate change challenge much better, but the subject needs careful consideration. The approach to planning may not change, but the plans will change substantially.

12.12 Sustainable Development—Third World Perspective and Challenge

The subject of development has been extensively discussed since the former colonized countries achieved independence and the era of cold war set in. The issue of economic development has been particularly discussed. The issues of purpose of development and meaning of wealth continue to be discussed.

Of late, the emphasis has been on sustainable development. The genius of the idea of sustainable development lies in its attempt to reconcile the real conflicts between the economy and the environment and between the present and the future. Much has been written on the subject, but it remains vague, as it is a conceptual issue as yet (BSD, 1999; Chaturvedi, 2011a).

At the outset, a clear distinction arises between the First and the Third World perception. For the former, the emphasis has to be on an urgent transition to sustainability, as their environmental impact is not sustainable. For the latter, the emphasis has to be on rapid socioeconomic development, with due concern for sustainability of the environment. Confining to the current meanings of development for the sake of focusing on the main issues, a curious paradox emerges. Equity, spatially and temporally, among the human community and partnership with nature are implicit in the concept of sustainability. Therefore, the operative principle to contribute to sustainability and justice becomes an endeavor by the Third World to achieve the rightful place in the global community, at the earliest. This has also been officially declared in India's tenth 5-year plan (GOI, 2002). As a corollary, with about 20% of the global population in China and India each, it means a leadership role for them in all aspects of human endeavor, particularly in science and technology, while paradoxically they are two of the poorest countries. It is not a matter of chauvinism but a logical conclusion (Yamumara, 1996). As has been pointed out, the most important issue in achieving development is commitment to achieve it (Landes, 2000). Eradication of poverty is part of this rapid transformation, but it is a necessary and not sufficient condition. There should be no misconception that the current state of disparities will take long to be corrected, and tremendous dedicated effort by the developing countries will be required. However, it is essential for achieving the transition to sustainability. It may also be emphasized that the initial and boundary conditions of the developing countries in all facets of a socioeconomic system are different from those of the current industrialized countries, and they will have to evolve their own developmental trajectory. Thus, the latest technology, adopting a leapfrogging approach with creativity, has to be developed in all sectors instead of adopting an incremental approach. In the water sector, this has been emphasized by Chaturvedi for a long time (Chaturvedi, 1976; Chaturvedi and Rogers, 1985). In contrast, it may be noted that in much of the literature on the subject, the general assumption is that the Third World will keep on peeping at modernity from the threshold, content with mitigation of hunger, and the currently industrialized world has to help them in achieving this task. The Millennium Developmental Perspective is a sad testimony to this concept.

A point that has been emphasized in studies and needs to be particularly stated is that any successful quest for sustainability will necessarily be a collective, uncertain, and adaptive endeavor in which society's discovering of where it wants to go and how it might try to get there will be inextricably intertwined (BSD, 1999). The development therefore needs to be understood in terms of a journey, which has to be adaptively navigated. We have therefore put considerable emphasis on systems modeling to help us in this endeavor.

12.13 Environment, Development, and Participation

Considerable criticism of development of water resources has recently been meted out by the so-called environmentalists. There is considerable confusion about the subject and acrimony leading to adverse impacts on both environment and development. It is unfortunate, as development and environmental conservation are essentially two sides of the same coin, as emphasized by the leading environmental agencies (IUCN, 1991). The subject may be briefly discussed, as it is important for rapid development to resolve the differences. Differences will, however, persist, and this is one more reason to back up the studies with transparent participatory models that are being developed, as discussed earlier.

Environment is often understood to refer simply to the state of "nature," e.g., the extent of forest cover, the depth of the groundwater table, the number of living species, and so on. It is assumed that this pre-existing nature will stay intact unless impurities and pollutants are added to it and is, therefore, best protected if we interfere with it as little as possible. The common elements of the criticism are the assumption that technology and development are essentially on a collision path in which the common man is the worst sufferer both in terms of adverse impacts of environmental change and denial of benefits of development. This understanding is, however, misleading and defective for three distinct reasons as brought out by Dreze and Sen (2002).

First, even though the human perspective on the environment is not the only one, the impact of environment on human lives must inter alia be among the relevant considerations in assessing the richness of the environment. In understanding why the eradication of smallpox is not viewed as an impoverishment of nature in the way, say, the destruction of lovely forests would be, the connection with human lives has to be brought into the explanation. Given that inescapable connection, the assessment of the environment needs to be dependent on many other features of the human lives that are directly—and often constructively—dependent on the process of development.

It is, therefore, not surprising that the environment sustainability has typically been defined in terms of the prospects of preserving and enhancing the quality of human lives. The Brundtland report (WCED, 1987) defined "sustainable development" as "development that meets the needs of the present without compromising the ability of future generations to meet their own needs." We can broaden the formulation of sustainability to include what we value (and feel committed to), rather than only what we enjoy as part of our living standards. There is, thus, an inescapable human connection in assessing environment, in the light of our values and priorities. What we value and the reasons that we have for valuing one thing or another are not independent of the process of development. Valuation is, among other things, a developmental process.

Second, while human activities that accompany the process of development may have destructive consequences, it is also within human power to enhance and improve the environment in which we live. Our power to intervene and to do it with effectiveness and reasoning may be substantially enhanced by the process of development itself. Development is empowering, and that power can be used to preserve and enrich the environment, and not just to decimate it.

Third, even though it is tempting to think of the environment exclusively in terms of pre-existing natural conditions, our idea of environment does, in fact, encompass aspects of human creation. For example, the purification of water is a means of improving the environment. Elimination of epidemics or other so-called acts of God is another good illustration of the created component of benign nature.

It is also perhaps worth remarking here that the development is not a new process, and in examining the effects of development on the environment, we must not presume that the entire mechanism has only just been initiated. Some of the strong environmental effects of changes that have occurred over the last thousand years are strong enough on their own. Consider, for example, that the air pollution in urban India is a major problem today, which can be seen as a by-product of modern development. However, it is interesting to note in this context that even the hapless commuters in the most congested parts of Delhi breathe a healthier air today than the average village woman, who is exposed to staggering levels of "indoor pollution" from kitchen smoke (CSE, 1999). Many other examples of environmental improvement with the advancement of technology and development can be given (Isabel and Langford, 1997). While the rural environment may seem predevelopmental and evokes romanticism about charms of "unspoiled nature," it can, in fact, incorporate a heavy dose of man-made pollution of traditional varieties.

These are some of the reasons for questioning the view of the environment as (1) composed of pre-existing nature, (2) valued for its own sake, and (3) can only worsen—not improve—through development. We have to move away from that limited vision, and that re-examination makes the environmental issue inseparably linked with the demands of development. Not only must the assessment of development be inclusive of environmental concerns we must also take note of the various ways in which the process of development may influence the nature of environment and the values that are invoked in assessing it. This recognition does not, in any way, change the basic fact that the process of economic development can also have very destructive environmental consequences, sometimes even swamping the constructive perspective. However, it is important to see the relationship between development and environment in an adequately broad way, taking note of the constructive as well as destructive possibilities. With poverty as the worst source of human misery and the worst polluter, development has a very positive qualification.

Similarly, the issue of benefits going to the poor segment of the society, first and foremost, is also the central issue in environmental systems management. However, it is true that this has not happened as desired. It does not mean that developmental activities should be identified as being anti-poor. The same is equally true in respect of participation of all the stakeholders in the planning process.

12.14 Dams—A Scientific Perspective

In view of the temporal and spatial mismatch of water availability and demand, attempts at storage of water have been made for a long time. Tanks and storage structures have been developed over time. River flows have been tried to be stored through the construction of dams. With increasing advances in technology, the size and sophistication of dams have been increasing continually. They impacted on human populations in the catchments area, and therefore, there have also been protests.

Dams were the most adored technological activities in water resources development, called the "new temples" in India, until recently. Dams have been constructed since time immemorial to store water and change the temporal availability. As water resources development was undertaken in Europe for hydroelectric generation, dams of increasing capacity were constructed. The environmental and developmental conflicts started with the classical difference of opinion on the Hitchiti developments in California as water resources development in western USA started. A compromise between development and conservation was reached, and numerous high multipurpose dams were constructed. As this phase almost came to a close in the USA with the construction of most of the feasible major dams, the era of dam construction started in the developing countries as they became independent, considering the differing hydrologic–climatic characteristics of the south, where 90% of the rainfall takes place during the three or four months of monsoons, concentrated in a few days and hours, and with long periods of hot summer and low flows or zero flows over the year, as most of the rivers are nonperennial. Adverse impacts of dams were known, but they also have very positive social and environmental impacts. Accordingly, following the earlier developments in the USA, with irrigation, hydropower, and flood mitigation as their urgent need, multipurpose dams were given the highest priority. Guidance was sought and provided by U.S. agencies in this endeavor, as they developed indigenous capability. There were short-comings, both technological and in caring for people adversely affected by the construction of the dams, but there was no antagonism, and they were hailed as the new temples of development. Comparative developmental experience of regions, with and without dams, such as Punjab and Bihar in India, vividly demonstrated the beneficial effects of dams. It is the considered opinion of all knowledgeable people in the developing countries that dams are essential for sustainable development. There are no alternatives, as dams are an integral part of a portfolio of activities for water resources development.

As the era of environmentalism started, technology came into disrepute. Dams were obvious symbols, and they became the select targets. Much has been written about their adverse impacts. Environmental activism was also undertaken. The epicenter of all activity is the developed community. Sardar Sarovar, which is one of the most well-designed dam and project in the context of development of Narmada River, became a whipping item of all environmental activists. The World Bank had extended financial support to the project, but it withdrew its support in the context of the activist's propaganda.

Debunking of dams has remained a favorite pastime of environmental activists. Since they are vocal constituencies of the World Bank, a Commission was appointed, largely at the initiative of the World Bank and the IUCN, to evolve a framework for decision making about dams. It was pompously called the World Commission on Dams (WCD). The report was released in November 2000.

Reactions to the report have varied widely. The international engineering organizations, the International Commission on Large Dams (ICOLD), the International Commission of Irrigation and Drainage (ICID), and the International Hydropower Association (IHA) have all rejected the report. Two leading water resources developing countries, India and China, have also rejected the report. Many environmental NGOs have hailed the report.

The report represents considerable work regarding water resources development focusing on dams. As the Commission noted, the key decisions are not about dams but about options for water and energy development. The report is essentially about improving the way such decisions are made.

12.14.1 Performance of Large Dams

A performance of several large dams was studied. It was found that, generally, they did not perform up to expectations. They had undesirable environmental impacts, which were not taken care of. Since the environmental and social costs of large dams have been poorly accounted for in economic terms, the true profitability of these schemes remains elusive. Perhaps of most significance is the fact that social groups bearing the social and environmental costs and risks of large dams, especially the poor, vulnerable, and future generations, are not the same groups that receive the water and electricity services, nor the social and economic benefits from them.

12.14.2 Options for Water and Electricity Services

It was considered that today, a wide range of options for delivering water and electricity services exists. Many of the non-dam options today—including demand-side management, supply efficiency, and new supply options—can improve or expand water and energy services and meet evolving development needs in all segments of society. There is considerable scope for improving performance of dam projects and other options.

12.14.3 Decision Making, Planning, and Institutional Arrangements

According to the Commission, the decision to build a dam is influenced by many variables beyond immediate technical considerations. These could be focal points for interests and aspirations of a large number of powerful people. Opportunities of corruption further distort decision making. Involvement from the civic society varied but was limited.

12.14.4 Core Values for Decision Making

Improving development outcomes in the future requires a substantially expanded basis for deciding on proposed water and energy development projects that reflects a full knowledge and understanding of the benefits and risks of large dam projects. It also requires introducing new voices, perspectives, and criteria into decision making, as well as processes that will build consensus around the decisions reached. This will improve the development effectiveness of future decisions.

The Commission grouped the core values that informed its understanding of the issues under five principal headings: (1) equity, (2) efficiency, (3) participatory decision making, (4) sustainability, and (5) accountability. These five are more than simply issues—these are the values for advancing an improved decision-making processes. The debate about dams is a debate about the very meaning, purpose, and pathways for achieving development.

12.14.5 Rights, Risks, and Negotiated Outcomes

Reconciling competing needs and entitlements is the single most important factor in understanding the conflicts associated with development projects and programs—particularly large interventions such as dams. An approach based on the recognition of rights and assessment of risks can lay the basis for greatly improved and significantly more legitimate decision making on water and energy projects. Only decision-making processes based on the pursuit of negotiated outcomes, conducted in an open and transparent manner, and inclusive of all legitimate actors involved in the issues, are likely to resolve the complex issues surrounding water, dams, and development.

12.14.6 Recommendations for New Policy Framework

The Commission put the controversy surrounding dams within a broader normative framework. This framework, within which the dam's debate resides, builds upon international recognition of human rights, the right of development, and the right of a healthy environment.

Within this framework, the Commission developed seven strategic priorities and related policy principles. It translated these priorities and principles into a set of corresponding criteria and guidelines for key decision points in the planning and project cycle. Together, they provide guidance on translating this framework into practice. They help us move from a traditional, top-down, technology-focused approach to advocate significant innovations in assessing options, managing existing dams—including processes for assessing repatriations and environmental restoration—gaining public acceptance, and negotiating and sharing benefits.

The WCD policy framework is shown in Figure 9.12. The WCD seven strategic priorities are shown in Figure 9.13. The seven strategic priorities, each supported by a set of policy principles, provide a principled and practical way forward for decision making. These are (1) gaining public acceptance, (2) comprehensive options assessment, (3) addressing existing dams, (4) sustaining rivers and livelihood, (5) recognizing entitlements and sharing benefits, (6) ensuring compliance, and (7) sharing rivers for peace, development, and security. These guidelines add to existing decision-support instruments and should be incorporated by governments, professional organizations, financing agencies, civil society, and others as they continue to improve their own relevant guidelines and policies over time (Figure 9.14).

12.14.7 From Policy to Practice—Planning and Project Cycle

The Commission's recommendations can best be implemented by focusing on the key stages in decision making on projects that influence the final outcome and where compliance with regulatory requirements can be verified.

12.14.8 Conclusion

The Commission considered that the report is a milestone in the evolution of dams as a development option. Three main contributions were considered. One, comprehensive global review of the essential aspects of dams and their contribution to development was carried out. Two, the center of gravity in the dam's debate was shifted to one on investing in options assessment, evaluating opportunities to improve performance and address legacies of existing dams, and achieving an equitable sharing of benefits in sustainable water resources development. Three, it was demonstrated that the future for water and energy resources development lies within participatory decision making, using a rights-and-risks approach that will raise the importance of the social and environmental dimensions of dams to a level once reserved for the economic dimension.

The Commission, as rightly claimed, shifts the debate on a more scientific evaluation of development of dams, which includes socioeconomic–environmental aspects. These are generally neglected, particularly the social ones. Secondly, it emphasizes participatory and transparent development, which is essential and generally absent in developing countries. However, the Commission has neglected two important issues. It is now well known that planning and design of dams or water resources development in the past and even now are generally not undertaken according to modern scientific advancements even in the industrialized countries, much less in the developing countries. The Commission displays a bewildering ignorance of the matter and has not considered this subject at all. As we have emphasized, considerable advancement has to be undertaken in terms of environmental systems management of water resources, and dams have to be considered as an integral part of the activities. Thirdly, in many situations in the developing countries, there are no alternatives for dams. There is also a difference of opinion about the overall contribution of dams. They have, in the conditions of many developing countries, very positive socioeconomic–environmental impacts, which could be considerably enhanced if scientific development was undertaken. They are the centerpieces of the portfolio of the development of water resources projects, and there is no alternative to them, at least in South Asia. The subject of dams in India has been discussed by two Indian scientists, to which reference is made (Bhatia et al., 2006; Rangachari, 2006).

12.15 World's First Anti-Dam Movement and India's Societal Response

The world's first anti-dam movement took place in India in 1920. It was directed against the Tata Company's proposed dam at Mulshi village, in the Pune (Poona) district of what was then Bombay. The societal response, in which the leading personalities of Indian society, including Mahatma Gandhi, a champion of the poor in India, were involved, brings out a scientific perspective that is valid and important to be noted regarding the role of dams in society, particularly under the Indian conditions.

The Tata Hydroelectric Power Company proposed to construct a dam for generating hydroelectric power on the confluence of the Mula and Nila rivers in the Mulshi Peta tract of the Pune district. The electricity was to be supplied to textile mills and other factories and to the railways in Bombay city. It was estimated that some 48 villages will be submerged by the reservoir. By the end of 1919, the "Mawalas" (as the residents of the Maval region, of which Mulshi Peta is a part, were known) realized that their land and houses would go under water if the project went through. The peasants and moneylenders of these villages came together with leaders of the Indian National Congress and went on *satyagrah* to fight the scheme. The struggle, which began in 1920, continued until 1924.

The anti-dam proponents contacted the national leaders fighting for the Independence of India, including Mahatma Gandhi, who was a leading champion of the poor. Initially they supported the anti-dam movement, but when the full facts of the benefits of the dam were brought out to them, particularly those regarding the economic development of the country and employment generation as well as those for the poor and marginalized people, they supported the construction of the dam. The subject has been well discussed by Vora (2009) to which reference is made.

12.16 Institutional Issues—Culture Matters

A fundamental difference between the development of India and the West is that in India, it was undertaken by kings, followed by the British, whereas people in the West started ruling themselves almost from the beginning of their development from about the seventeenth century. Even after their achievement of independence in 1947, people in India were governed and essentially ruled by a bureaucracy, which is the successor to the British bureaucracy. The "culture" has not changed. As the eminent economic historian Landes emphasizes, "If we learn anything from the history of economic development, it is that culture makes all the difference" (Landes, 1999, p. 516). We will demonstrate that the development of India's waters, in which the development of the GBM is crucial, continues with the colonial concepts and technology. We will show that it can be and must be revolutionized. The central issue is development and establishment of a new culture—India as an equal component of humanity. Paradoxically, in terms of ends and means, it means a leading position for India in the human community.

There is a wider issue in terms of policy perspective. We have to bear in mind, as Mahatma Gandhi emphasized, the Truth. And as he emphasized, the Truth is that India is backward and poor. The concurrent Truth, which Nehru emphasized, which we brought out at the outset, is that India has to achieve her due place of equality in humanity. By a paradox of ends and means, it means global leadership. This is our central perspective, while considering the development of the GBM basin, which represents almost about 50% of the Indian people. And this has to be done by us ourselves.

12.17 Summary

Development of water resources has a long history, and advances in science and technology have continuously been made. A recent advance has been to incorporate the challenge of sustainable development, which is a critical challenge. It has dominant focus on the management of the environment. Emphasis on sustainable development of the environment has accordingly been much emphasized of late. Management of water in this context is crucial. This has led to the concept of IWRM, which is accepted to be the guiding principle of management of water.

Further recent advances have been made in the context of sustainable development. It is considered that the environment is generally considered in isolation from the society. In contrast, integrated management of the two with the perspective of sustainability has been suggested. The concept of SESM has emerged in this context. Management of water is thus further advanced in this context. This has been discussed in detail by Chaturvedi (2011a). The conceptual issues and the approach to implement these ideas have also been briefly presented.

With the ever-increasing complexity and scale of problems, uncertainties, and the pervasive role of technology, the need has been felt for creative, comprehensive, and critical planning. New concepts, approaches, and techniques have been developed in recent times to meet this challenge. A new science—systems analysis and planning—has been developed for handling complex problems. Much work in this context has been made in the area of water resources, and further advances are taking place.

An important aspect of sustainable development is developing a long-term perspective. Advances have been made in the area of scenario analysis and integrated systems modeling. This is particularly important for the management of water.

According to the environmental systems perspective, development and management of water are to be undertaken in terms of integrated development of the socioeconomic–environmental system with focus on water. From a geophysical–chemical–biological perspective, development and management of water are synonymous with the development and management of the ecosystem. In this context, the concept of land embraces water, and land use/cover management is crucial.

Anthropogenic climate change is now considered an established fact, and estimates of climate change are being made. Estimates of impact on water are also being made. Approaches to estimate these impacts and take preventive and adaptive measures have been developed. It is extremely important that water resources management is undertaken with due concern for the impending climate change. Unfortunately, the subject has not been given due recognition so far in terms of its likely precarious impacts and the urgency of dealing with it.

The emphasis so far has been on technological aspects. As we have emphasized throughout, technology and social organization are the tools with which society transforms physical resources and human labor into distributed goods and services. Both have to be considered integrally. Development of the appropriate institutional framework and governance is one of the most important issues in efficient and sustainable development of water. Unfortunately, due emphasis has not been paid to it.

There has been considerable controversy about development being antithetical to environmental conservation. Indeed, the two are two sides of the same coin. In this context, considerable controversy has been generated against dams. It is demonstrated that considering the challenges in India, dams are essential for sustainable development.

While considering the development of water, it is important to remember that water is in many ways different from the "created sectors" such as power, telecommunications, and transport. It can best be developed and managed in terms of the proposed SESM approach. However, the challenge can be best understood as brought out by Boulding (1964), an eminent resource economist, in an ode to water:

> Water is far from a simple commodity
> Water's is a sociological oddity
> Water's pasture for science to forage in
> Water's mark of our dubious origin
> Water's link with distant futurity
> Water's symbol of ritual purity
> Water is politics, water's religion
> Water is just about anyone's pigeon
> Water is frightening, water's endearing
> Water is lot more than mere engineering
> Water is tragical, water is comical
> Water is far from the Pure Economical.

Notes

1. The author started his career on Nayar dam in 1947. It was a very high dam on a tributary of Ganga. Dr. A.C. Mitra and Shri S.K. Jain, the Superintending Engineer and the Executive Engineer, went to the USA for about six months and Dr. Norman, from the U.S. Bureau of Reclamation, guided us. It was the study of the calculation of the backwater from Posey's book on Open Channels which led the author to Iowa to work under Posey.
2. Lenton (2011) has brought out the latest position, briefly, and may be used as reference.

Appendix 1: The National Water Mission

One of the key missions of the National Action Plan on Climate Change aims at the "conservation of water, minimizing wastage and ensuring its more equitable distribution both across and within states through integrated water resources development and management."

The mission has five goals:

- Comprehensive water database in public domain and assessment of impact of climate change on water resources — Collect comprehensive data on water resources, develop water resources information system by 2011, make information available in the public domain, assess the impacts of climate change on the country's water resources by 2012. Scientific data collection includes additional hydrometeorological data, wetland inventory, reassessment of basinwise water situations, and finally, using this data to predict the impacts

- Promote citizen and state action for water conservation, augmentation and preservation — Includes expeditious implementation of irrigation projects, minor irrigation schemes, groundwater development, mapping flood-affected areas, capacity-building and awareness

- Focused attention on overexploited areas — Intensive rainwater harvesting and groundwater recharge programs, pursuing enactment of groundwater regulation and management bill

- Increasing water use efficiency by 20% — Both on the demand side and the supply side, particularly in the agriculture and commercial sectors. Guidelines for incentivizing recycled water, water neutral and water-positive technologies, improving efficiency of urban water supply systems, benchmark studies for urban water use, water efficiency indices for urban areas, manuals for mandatory water audits in drinking water, irrigation and urban systems, promoting water-efficient techniques including sprinkler and drip irrigation systems

- Promote basin-level integrated water resources management — Basin-level management strategies, review of National Water Policy in order to ensure integrated water resources management, appropriate entitlement and appropriate pricing, review of State Water Policy

- Mission goals including collecting and reviewing data for an integrated information system have been in the pipeline ever since the National Water Policy (NWP) was revised in 2002. Water being a State subject under the Indian Constitution, the NWP and the policies under the Water Mission will have to address policy and project implementation at the state level.

- The mission hopes to revisit the NWP in consultation with States, in order to ensure basin-level management strategies that can deal with changes in rainfall and river flows due to climate change.

Soil Erosion and Its Management

Soil erosion is a serious issue, and its management is important for environmental conservation. The problem is very serious in the GBM basin, in the Himalayan region. Recent studies in the Nepal region have indicated that annual soil loss rates are the highest (up to 56 tonnes/ha/yr) in the areas with rainfed cultivation, which is directly related to the sloping nature of the terraces. The lowest soil losses (less than 1 tonne/ha/yr) are recorded under dense forest. In the degraded forest, the soil loss varies from 1 to 9 tonnes/ha/yr, and in the grazing lands, it is estimated at 8 tonnes/ha/yr. The rice fields seem to trap the sediments brought from up-slope. Similar results and suggestions have been given in other studies in the region (Gardner and Gerrard, 2003).

13

Revolutionizing the Environmental Management of the GBM Basin

13.1 Introduction

The physiographic–hydrologic characteristics of the GBM basin offer opportunities for certain novel technologies which can revolutionize the modernization of the development of water resources of the GBM basin and of India, leading to a revolutionizing of the management of the environment. The management approach has also to be modernized in this context. This is the central theme of the book, and the proposed revolution is discussed briefly.

Environmental processes do not understand man-made boundaries. Scientific considerations dictate that the management of the GBM basin environment should be considered in totality. The technological considerations and sociopolitical considerations reinforce this basic scientific postulate. However, since this is not possible at the moment, we first focus on the region within India, which is the dominant part and which determines the developmental policy for the entire GBM basin. Part of the environment of the two riparian countries, Nepal and Bangladesh, contains features of the Indian environment and, thus, enables India to appreciate the entire basin challenge. An important perspective of the novel technologies that have been developed is that they consider the development of the GBM in an integrated physiographic perspective, particularly the development of Ganga. This means that when we have to consider the development of Ganga, even in terms of India, we consider the entire physiographic region. We will consider integrated development of the GBM basin in Chapter 14, taking due account of the political perspectives. India will have to play the lead role.

Development of the environment has to be in accordance with the developmental goals of the society. The development objectives, as specified by the State, are therefore discussed first, in Section 13.2.

We start with management of water, which is the dynamic component of the environment. The novel technologies are discussed in Section 13.3. They have been well investigated and proved. Contributions of the unique proposed revolutionary technologies in the development and management of

India's waters are specifically brought out in Section 13.4. An estimate of the tremendous increase in water availability on the basis of the proposed novel technologies is given in Section 13.5.

The GBM has three major river basins, Ganga, Brahmaputra, and Meghna, followed by the Delta, which integrates the three. Each has a different environmental characteristic. Each is therefore considered individually first in Sections 13.5.1–13.5.4, respectively. Integrated development policy follows in Section 13.6.

While considering the constituent river basins, development has been considered taking into account all aspects of the physical environment. However, an overview of the basic specific issues of environmental conservation, hydroelectric development, flood management, and navigation, is given in Sections 13.7.1–13.7.4, respectively.

Groundwater is usually considered only as a resource. It is part of the land–water dynamics, and the issue of management of the chemical cycles becomes necessary, particularly in the context of the large-scale development. The challenge of managing arsenic, which is particularly noticeable in the delta region but is being identified all over the basin, is a serious challenge and is a pointer of the need for a new perspective in managing groundwater. The geologic history of the basin also poses the groundwater challenge in a new light. The new perspective required for management of groundwater is briefly discussed in Section 13.8.

The central concern has been to develop policies in the environmental systems perspective with focus on water as part of the physical environment in the context of the societal activities. The societal–environmental systems management is further reviewed in Section 13.9 in this context.

Some basic perspectives of management of water of the GBM basin are presented in Section 13.10. They represent the core of the thinking on the subject.

These ideas presented have to be developed through detailed scientific analysis in terms of systems analysis. Considerable work has been done in the context of the GBM basin (Chaturvedi, 1987). Further advances have taken place in terms of environmental systems planning backed by systems dynamic modeling (Chaturvedi, 2011a). This is briefly presented in Section 13.11. A perspective of societal–environmental systems management is given in Section 13.12.

Above and beyond the revolution in the engineering is the needed revolution in institutions and management. The historicity of the colonial–bureaucratic system was discussed in Chapter 10. This has to be revolutionized, and new management policy and practices have to be introduced. This has been discussed in Chaturvedi (2011c) but is briefly presented in Section 13.13. The new management is particularly applicable to the GBM basin.

Modernization of the spectrum of activities is presented in Section 13.14. Besides development of the management system, changes have to be made in the capabilities of the development personnel. Currently, it is entirely engineering oriented. There has to be diverse professional specialization capable of working as a team for the development and management of the environmental system.

Of late, the central emphasis has been laid on societal consciousness and institutionality. As has been said, "Culture matters." This has been discussed in Chaturvedi (2011a) but is briefly presented in Section 13.15.

A brief conclusion follows in Section 13.16.

13.2 Development Objective

India was one of the leading areas of well-being, with the current Bangladesh area being one of the most prosperous, which attracted the East India Company (Spear, 1978). Since then, the world has been divided essentially into two—the rich and the poor—almost in the shape of a champagne glass, as discussed in Chapter 1. India is at the bottom, currently accounting for about 17% of the humanity but soon to account for about 20%, with Bangladesh being even worse off. A sacred commitment was made by the people, at the first moment of Independence, as declared by Nehru, the first prime minister, in his famous Tryst with Destiny speech in the Parliament: "this ancient land attain(s) her rightful place in the world" (Nehru, 1947). He also declared that "the Ganga above all, is the river of India which has held India's heart captive and drawn uncounted millions to her banks since the dawn of history. The story of the Ganga from her source to the sea, from old times to new, is the story of India's civilization and culture." (Nehru, 1947).

These views have been reconfirmed in the 5-year plan of the Government of India. For instance, the tenth plan's (GOI, 2002) guidelines stated that (1) the objective is to bring the people in the mainstream of human endeavor at the most rapid rate and to redress the poverty and backwardness simultaneously, ensuring ecological security; (2) the society is very backward and the people are very poor, and there is a very long way to go to come in the mainstream of human endeavor; (3) the issue, however, is not merely to mitigate poverty but to attain our due position in the human community; and (4) this has to be achieved by our own efforts.

The importance of the challenge is presented by noting that the GBM basin is currently the home of about 650 million people, which is estimated to stabilize at about a billion people by 2050, more than the population of the entire Western world.[1] A revolution in the societal system has to be brought about to achieve these ends. Sustainable transformation of the environment at a large scale will also be required. Management of water, which is crucial for life, economy, and environment, is of utmost importance in this context. It is a big challenge. It will be our central focus of attention in the environmental systems perspective.

13.3 Some Novel Technologies for Development of GBM Waters

Given the climatic characteristics, with precipitation taking place essentially during the monsoons and over a few days, it is essential that water is stored. The conventional technology comes in the form of storage projects, which must be developed to the fullest extent. However, in view of the physiographic characteristics of the Himalayas and the GBM basin, storage opportunities are very limited. The physiographic and hydrologic characteristics of the Indo-Ganga basin, however, offer the potential of developing the water resources through some novel technologies. These richly add to the resource and energy potential and will contribute significantly to the comprehensive development of the new India, Nepal, and Bangladesh. They will also promote collaborative international development of the GBM basin by the three political entities. They are:

1. Chaturvedi Water Power Machine
2. Chaturvedi Water Machine
3. Development of Artesian Aquifers in GBM basin
4. Development of Static Ground Water Potential, and
5. Integrated Water and Energy System Management

13.3.1 Chaturvedi Water Power Machine

The Indo–Gangetic basin has, geologically, one of the world's highest and youngest mountain chains in the north and one of the oldest highlands in the south. In between is a narrow alluvial plain, about 1500 m in depth, formed essentially by the erosion of the Himalayas. The potential of the storage projects in the Himalayas is therefore very limited. It is not possible to even store the average yearly runoff, the storage being limited to just about 30% in some storage projects. Development of the storage projects is comparatively expensive, and storage per unit of height is comparatively low on account of the physiographic features. Thus, besides the low storage potential, the cost per unit of water stored is comparatively high.[2]

On the other hand, storage potential in the Central Highlands is one of the highest in the world, on account of its geologic history. Moreover, the physiography and excellent geologic foundations make construction very convenient and economical. Thus, they provide opportunities for large and economical storage of water.[3] Storage figures of some reservoirs on the Himalayan and Central India river storages, shown in Table 13.1, bring out the dramatic difference.

The runoff is highly variable over the year, as shown in Figure 13.1 (Figure 2.8 reproduced). It further reduces the potential for storing the water and generating hydroenergy, which currently is normally generated on a 90% reliability basis.

TABLE 13.1

Salient Characteristics of Storage Projects in Himalayan and Vindhyan Region

Sl. No.	River	Dam	Height (m)	Storage (million m³) Gross	Live
1	Ganges	Tehri	260.5	3540	2615
2	Rihand (Son)	Rihand	93.5	10,800	8967
3	Chambal	Ranapratap	54.0	2900	1573

Source: Central Water Power Commission, Storages in River Basins of India, Government of India, New Delhi, 1997.

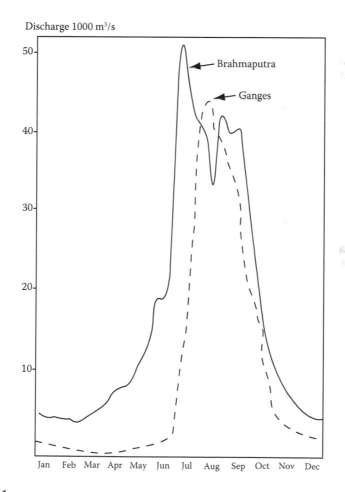

FIGURE 13.1
The annual hydrograph of Ganges and Brahmaputra (Figure 2.8 reproduced). (From Bruijzneel, L.A., with C.N. Bremmer, "Highland–Lowland Interactions in the Ganges–Brahmaputra River Basin: A Review of Published Literature," ICIMOD Occasional Paper No. II, Kathmandu, 1989. With permission.)

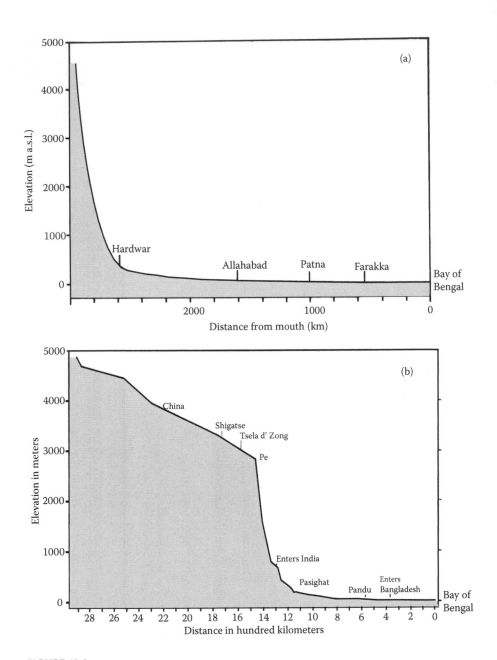

FIGURE 13.2
(a) Approximate longitudinal profile of Ganga. (b) Approximate longitudinal profile of Brahmaputra (Figure 2.10 reproduced). (From Bruijzneel, L.A., with C.N. Bremmer, "Highland–Lowland Interactions in the Ganges–Brahmaputra River Basin: A Review of Published Literature," ICIMOD Occasional Paper No. II, Kathmandu, 1989. With permission.)

These characteristics offer the feasibility of a novel technology of pumped storage on an intertemporal interspatial basis with vast economic and enviromental benefits (Chaturvedi, 1998). It is explained briefly as follows.

It may be noted that water, as it flows, is not only matter but also represents energy. Besides the energy generated on storage reservoirs, there is a huge potential of run-of-river schemes as the longitudinal sections of Ganga and Brahmputra bring out (Figure 13.2) (Figure 2.10 reproduced). Currently, the development of the run-of-river schemes is undertaken on a 90% reliability basis, in the pespective of providing firm energy. Even this has been stopped, according to a recent government notification, in view of the environmental requirements of the low flows.

If we look at Figure 13.1, it will be seen that there is tremendous energy more than that is provided by the low flows at 90% reliability. Thus, considerably more energy can be generated on run-of-river hydroelectric schemes, above and beyond the conventional development of 90% reliability, as currently designed. Thus, while there is tremendous energy potential in the Himalayas, the possibility of development of firm energy, according to the current technology, is comparatively very limited, and even this is not feasible in view of mainaining the environmental integrity.

According to current international technology, hydroelectric energy is generated essentially on a diurnal pumped storage basis. They work in conjunction with the thermal projects, providing the peaking energy, and the water so released is stored at the downstream end, which is pumped up during off load hours of the thermal system. The turbines are reversible turbines and pumps. The Orville Project in the California hydrosystem is a classic example (Hall and Dracup, 1970). Integrated hydroelectric generation in Sweden and thermal generation in Germany are based on this principle.[4]

The attractive economic feasibility of the pumped storage scheme is evident from the well-known fact that the price of peaking power is several times higher than that of base load power. While the profitability varies from scheme to scheme, depending upon topographic, geologic, and hydrologic characteristics, the figure from Orville Dam, constructed as a pumped storage scheme in the California system, will give some examples. As stated by Hall and Dracup (1970), "Pumped storage is exceedingly valuable as peaking power is almost thrice the power of off-peak power. Secondly and more importantly, the big return from pump-back is the ability to increase the total amount of energy that can actively be sold as firm energy. Thus, an additional unit of pump back, which can actively be sold as firm energy in the constraining month, would be worth 600 times the nominal energy price differential to a 50-year power contract." Pumped energy profitability ratios will vary from scheme to scheme, but in the Californian case, it has been considered to be about nine times more profitable (Hall, 1980).[5]

The storage projects in India are currently not being designed even on the basis of pumped storage on a diurnal basis.[6] Secondly, the storage projects are generally designed for irrigation supplies, with a cushion for managing power variation at different times of reservoir levels and inflows. Further, they are generally not being designed for conjunctive flood management. Since the capacity of storages in the Himalayan region is limited, compared to total inflows, much hydroenergy potential is thus wasted, which could be used to generate secondary energy.

The Himalayan system has a large potential for run-of-river hydroelectric projects, unlike U.S. or European systems, in view of the physiographic conditions. It is also very variable over the year on account of the hydrologic characteristics. Much of the energy is lost because the projects are designed on the basis of 90% reliability over the year, as brought out in Figure 13.1.

There is a characteristic of run-of-river scheme that makes generation of hydroelectric energy attractive. The major cost of the hydroelectric system, almost 75%, is on a hydraulic conveyance system. The capital cost increases with the size of the tunnel, but the capitalized cost of the head loss decreases with the size of the tunnel for the same discharge. The diversion structure is of a fixed cost, irrespective of the generation capacity. Therefore, increase in cost on account of increased generation capacity is not substantial. It therefore becomes very attractive to utilize the secondary energy if it could be converted to firm energy.[7] This also applies to the storage projects.

In contrast to the Himalayas, the Central Highlands have one of the world's best storage sites.[8] It is proposed that the secondary energy of the Himalayan rivers during monsoon months, generated at economical levels in terms of peaking energy, be conserved by pumping up water to be stored in the Southern Highlands. It could be brought back as releases are made later, as part of pumped storage in the system, on an annual basis. The limit up to which the hydrosecondary energy over the year can be thus transformed depends upon economical analysis over the year. Similarly, in conjunction, the thermal energy of the hydrothermal system can be transformed for peaking purposes on a diurnal basis. The scheme will also lead to enhanced postmonsoon low river flows in the basin or for interbasin transfer.[9] The scheme has been called the Chaturvedi Water Power Machine (Chaturvedi, 1973, 1998).

Considering the economics of the run-of-river schemes and the storage characteristics of the Central Highland projects, it appears that the proposed scheme will be very attractive even after allowing for the additional expenses of two different components in contrast with the usual diurnal pumped storage schemes. Irrigation benefits are additional. It will also avoid the adverse and environmental consequences of the Himalayan run-of-river schemes designed according to the conventional practice of diverting all the low flows, as undertaken currently.

The water from the storage projects in the Central Highlands can be released on the GBM basin or the peninsular region side. It can thus be extended also for interbasin transfer if the economics so recommends.[10]

The scheme is not limited only to the Ganga. The secondary energy of the Brahmaputra on an intertemporal basis and the thermal complex of the region on a daily basis can be converted into peaking energy as found economical.

The full potential of the scheme, beyond the principles discussed above, can be determined by undertaking an integrated water–power systems study of the total region on a long-term basis. According to the judgment of the author, who has designed several major projects, it will revolutionize the power systems development of India (Chaturvedi, 1989).

It should be emphasized that when we speak of Ganga or Brahmaputra, we mean all the Himalayan rivers of the Ganga and Brahmaputra system. Further, as we are proposing to store the hydroenergy, we explore economic generation of the entire river system, down to the smallest unit.

This introduces another aspect for consideration. One of the problems in generating hydroenergy on a run-of-river basis is dealing with sediments, as they are detrimental to the turbine blades. The current practice, as they are dealing only with the very low flows, is to provide stilling basins on the intake end of the hydrogeneration system. Our perspective is entirely different. We are talking of energy generation of the entire river system, starting with the smallest economic unit, round the year. Thus, sediment management is an important issue from the context of hydroenergy generation. This will also have important beneficial implications for the storage projects.

This puts sediment management in an entirely new and independent perspective. Four perspectives may be brought out. One, sediment control is a valuable environmental management component in its own right. Two, it can be undertaken by people in a very diversified, economical manner, bringing them into the picture. Three, it will be important in all the proposed run-of-river schemes, and since we are dealing with round-the-year sediment control, sediment management from the beginning in the entire catchment of the river system, as brought out above, gets an important position from this perspective. Four, the cost of managing sediment is a pittance, compared to the economic value of the loss in the storage system, as emphasized from the author's real-life experience of designing the Ramganga storage project. To conclude, soil erosion is extremely important in an environmental system management activity, and that too in the Himalayan region. The Chaturvedi Water Power Machine puts special emphasis on it.

13.3.2 Chaturvedi Ganga Water Machine

The climatic conditions and the rich alluvial plains of the GBM basin offer the potential for excellent agricultural development, but irrigation in both seasons, monsoon (kharif) and winter (rabi), is necessary. Given the hydrologic conditions, irrigation should be in terms of assured, adequate, and timely water supplies. It should therefore be under the control of the farmer.

As discussed earlier, the present irrigation system was developed in the colonial times with the objective of stabilizing sustenance agriculture and minimizing famines. The techology consisted entirely of diverting the low flows of the Himalayan rivers after the monsoons. It has been extended with the construction of storage projects on the major Himalayan rivers, Ganga and Ramganga (with the possibility of storage in Yamuna in the process of development). The perspective, policy, and technology continue to be unchanged, except that some additional water has been provided by putting up storage dams and that the irrigated area has been enlarged. Irrigation is essentially provided only during rabi and that, too, in terms of three waterings to stabilize Third World agriculture. The waters are delivered through a network of canals taking off from the perennial rivers and diverting almost all the low season flows. They are operated by the Government. In addition, some diversion canals have been constructed of late to utilize the monsoon flows in Ganga, but they too follow the old perspective. The major difference is the introduction of tube wells, which have been developed mostly on the initiative of the farmers of late.

As stated above, the storage potential in the Himalayas is very limited on account of the steep slopes. Further, it has not even been possible to develop the potential fully on account of the disputes among the international riparians, Nepal, India, and Bangladesh. The storage of monsoon precipitation and runoff over the alluvial plains, which is very large, is not considered to be storable, apart from the normal groundwater recharge in natural course, on account of the flat alluvial plains, which are heavily populated. Since the precipitation is intense, for a brief duration, most of it flows away, and even the recharge is limited.

Thus, three serious disadvantages have emerged. First, irrigation is very limited. Irrigation was provided initially only during the winter (rabi) period because of the temporary nature of headworks. Kharif irrigation was also provided after permanent headworks were constructed but was limited because initially, headworks were not constructed from consideration of utilizing flows round the year. They are, however, being extended also to provide Kharif irrigation, but the potential developed so far is small. Second, the irrigation continues to be an apology of irrigation in terms of the historic objective of stabilizing primitive agriculture and mitigating famines in view of the scarcity of resources and lack of the people's involvement in managing the waters according to their perspectives. Third, with the increasing urbanization and industrialization, a calamitous degradation of river waters and the problem of providing drinking water to the urban areas have emerged because the rivers become almost bone dry as all the low season flows are diverted for irrigation as they debouch on the plains from the mountains.

In view of the highly variable availability of water over the year, it is necessary to store it. Our concept is that in view of the excellent alluvial formations of great depths, it should be possible to store the monsoon waters on the alluvial plains if sufficient time is made available for recharging and the groundwater is appropriately lowered in advance. This has been confirmed by several analytical studies (Revelle and Herman, 1972; Laksminarayana and Revelle, 1975; Chaturvedi et al., 1975; Srivastava, 1976; Chaturvedi and Srivastava, 1979; Chaturvedi, 1987).[11] There is considerable potential for the large catchment in the GBM basin as the rivers, big and small, debouch from the Himalayas or are established in the sub-Himalayan reach, where this storage could take place.

A scheme called the Chaturvedi Ganges Water Machine has been developed on this principle (Chaturvedi, 1981, 1987).[12] It is postulated that for irrigation in the rabi (postmonsoon) period, groundwater in the private sectors shall be the core of the water supplies, in contrast with the current policy of putting primary emphasis on public canal water supplies. Its development will be at the village level, individually or collectively. These supplies can be supported by surface flows duly enhanced by releases from the storages, in which the first priority has to be assuring minimum flows from environmental considerations and meeting the demand for urban drinking water supplies. It may be noted that considerable low flows will have to be conserved for this purpose with growing urbanization and industrialization.

According to the proposed policy, the development will be undertaken so that the groundwater is lowered to a precalculated level. The proposal that groundwater be made the dominant irrigation water supply source has the implication that its supplies are considerably enhanced. This is ensured by working out the level down to which it has to be lowered, in the context of Rabi supplies, and assuring that it is recharged during the monsoon period while increasing Kharif irrigation and local level recharge generally. The feasibility has been confirmed by the studies.

Second, instead of the current neglect of the Kharif (monsoon period) irrigation, we put primary emphasis on it. There is little potential for storing rainwater in the plains in the major rivers and major tributaries on account of afflux. However, storage barrages to divert it and to maintain river flows up to certain levels, from considerations of minimum annual afflux, can be developed. They already exist on the main tributaries in the context of the current provision of Rabi (winter period) and Kharif irrigation. However, they have to be extended, and the development has to be undertaken in the context of developing them as storage barrages. All these storage barrages shall be built in conjunction with bridges, which will also have to be constructed in the context of developing the transportation system and will therefore be very economical. Indeed, the river should be considered to be developed in the context of the transportation considerations also through navigational facilities. Two, storages have to be developed on minor streams. Three, an extensive watershed development has to be undertaken. They can be integrated with the Kharif irrigation system. It is important to emphasize that the two, transportation infrastructure and the water storage and diversion system, should be developed conjunctively as the environmental systems philosophy emphasizes. (Rubber dams are recommended for storage on rivulets because of the problems of flood discharge. This, however, is a secondary issue.) Four, another novel technology is that instead of the conventional tube wells, reversible tube wells should be used, so that rainwater, duly collected at the farm level, is directly recharged, if found feasible. Further supplies can be assured from storages in the hill region and groundwater without minimizing their potential for Rabi (winter period) irrigation. We can thus provide assured water supplies for Kharif and Rabi agriculture, almost doubling the irrigation intensity and increasing the yields in both seasons several times.

The feasibility of all these proposals has been established. All this requires sophisticated analysis and design. The Chaturvedi Ganga Water machine is way above the current populist broadcast of rainwater harvesting.

The Central Ground Water Board (CGWB) prepared a "National Perspective Plan for Recharge to Groundwater by Utilizing Surplus Monsoon Run-off" (CGWB, 1996). It was estimated that it is possible to store 21.4 mha.m in groundwater, out of which 16.05 mha.m can be utilized. This was not accepted by the National Commission for Integrated Water Resources Development Plan (NCIWRDP, 1999) on the basis that the technology has not been proved as yet. A "Master Plan of Artificial Recharge" has been prepared since then (CGWB, 2002). It is in the process of examination by the appropriate sources and processes. Our proposed scheme is different from these proposals, considering the subject in a more advanced and dynamic manner.[13]

13.3.3 Development of Artesian Aquifers in GBM Basin

During the course of investigations in the GBM basin by a team of Indian and foreign experts in the context of oil and further investigations that came later, it was well established that the region is well served with artesian waters, which could profitably be used for water supplies (Jones, 1985). The subject was well publicized by the World Bank.[14] However, for some reasons, it is not mentioned at all now. It appears that the subject should be pursued, as these are very valuable resources and energy sources.

13.3.4 Development of Static Groundwater Potential

Besides the dynamic groundwater resource, there is a very large static fresh groundwater resource in the Indo-Ganges–Brahmaputra basin. The figures of the two, static and dynamic, for the three river basins are 1338.2/26.49, 7825.3/170.99, and 917.2/26.5 km^3, respectively (NCIWRDP, 1999). A dynamic groundwater development policy could be adopted, which may overcome the problems being encountered on account of the delays in the storage projects. The idea is not to use the static reserves but only contribute to the development of the optimal surface and groundwater development in view of the different temporal trajectories of development.

13.3.5 Integrated Water and Energy Systems Management

Conjunctive surface water and groundwater use is well known. Similarly, conjunctive thermal energy and hydroenergy advantages are known. We propose the conjunctive use of all four sources, which will lead to large economies of production and yield. The region offers some unique and rich conjunctive surface water, groundwater, and energy management opportunities in view of some significant characteristics. The relative surface water and groundwater potential is almost of the same order of magnitude and spatially and temporally very varied. The hydropotential is very large, and thermal potential is also of a large order of magnitude and temporally very complementary. The pumped storage has valuable and large intertemporal and interspatial potential in contrast to the conventional diurnal characteristic of limited potential. Considering all these factors together, significant advantages can be obtained if integrated management of water and energy is undertaken. It was in this context that the use of static groundwater was mentioned.

A schematic perspective has been developed as shown in Figure 13.3, but it has to be developed in detail. Planning is complex, but modern advances

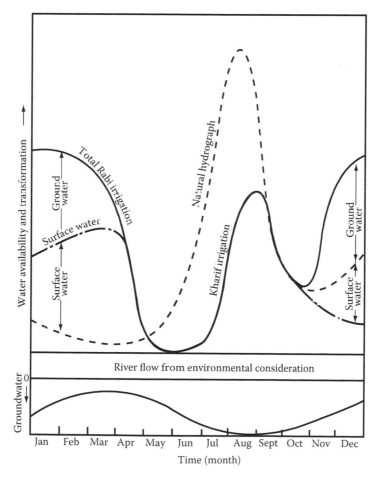

FIGURE 13.3
Conjunctive surface–groundwater energy management.

in systems planning make it workable. Some studies have been undertaken, as will be discussed later.

13.4 Emphasizing Some Unique Contributions of the Proposed Revolutionary Technologies

Some unique contributions of the proposed revolutionary technologies may be specifically presented.

13.4.1 Chaturvedi Water Power Machine

The proposed technology aims to store and repeat store as much potential of the Himalayan waters as economically possible on an annual pumped storage basis. It is in stark contrast of the present hydroelectric technology, which converts only the energy of the low flows on a 90% availability basis. This makes a world of difference in the energy potential generated and even more in terms of its financial value. Secondly, it emphasizes soil conservation as a very important and valuable component of the environmental management, in conjunction with the management of water.

13.4.2 Chaturvedi Water Machine

The proposed technology enables the storing of all preciptiation on the GBM basin and its use over the year, at the discretion of the farmer, placing waters in his hands. It differs from the conventional groundwater storage proposed in vague pronouncements by some water enthusiasts, including the CGWB, in that it emphasizes very scientific management of the monsoon precipitation in the alluvial plains. It is assisted by a novel technology, the Chaturvedi Reversible Pump, which has been patented by IIT Delhi on behalf of Prof. M.C. Chaturvedi.

13.4.3 Development of Artesian Aquifers in GBM Basin

We are mentioning the possibility of this technology, which was proposed by some international researchers working in India, together with their Indian counterparts, and has also been acknowledged by the Indian researchers and the Ministry in conferences to which the author was also invited. However, it has yet to be established.

13.4.4 Development of Static Groundwater Potential

This is mentioned only to show that the feasibility exists. However, detailed system studies are required to consider its development in terms of the long-term policy of development of the GBM waters.

13.4.5 Integrated Water and Energy Systems Management

We believe that the policy of development of India's water and energy has to be reconsidered by taking into account the possibility of integrated water and energy development, in view of the unique hydrologic characteristics.

13.5 Estimate of Water Availability Revolution

A brief assessment of water availability, in view of the proposed advances, may be presented to give an idea of the magnitude of increase. The estimates are tentative because the projects have to be developed in terms of the economic viability. However, it is considered that the proposed estimates are realistic.

According to the current official estimates, mean surface flow, utilizable surface flow and replenishable groundwater, and total utilizable water are 1952.87, 690.32, 342.43, and 1032.75 km^3, respectively. It was argued that the utilizable surface flows of the peninsular rivers could be increased to their full mean value without any novel technology, increasing the total of utilizable surface flows to 776.94 km^3 (Chaturvedi, 2001). Similarly, the utilizable groundwater was estimated at 500.43 km^3, in accordance with the CGWB recommendation, making the total utilizable water resources 1276.47 km^3, in contrast to the current official estimate of 1032.75 km^3, according to the earlier estimate called Chaturvedi 1.

We are vastly increasing these estimates through the proposed novel technologies. We are arguing that through these proposed novel technologies, almost all of the 1275.46 km^3 of surface flows of the rivers of India, excluding the Brahmaputra and Meghna, can be utilized. Adding these two rivers' estimated utilizable figure of 24.00 km^3, we obtain a figure of 1299.46 km^3 for utilizable surface waters, in contrast to the current official estimate of 690.32 km^3, an almost 1.9-fold increase from the current estimates! In addition, there is an increase in groundwater availability through the proposed Chaturvedi Ganga Water Machine. This will increase the groundwater availability to levels much above the current total CGWB-estimated availability of 500.43 km^3, but since estimates are difficult to make, we have limited ourselves to the CGWB figure of 500.43 km^3. We come to a total figure of 1799.89 km^3 for utilizable waters. Thus, there is an increase of about 88% in utilizable surface waters, 46% in groundwater, and 74% in total utilizable waters. The transfer to the peninsular region of the estimated currently unutilized surface waters of Ganga, 275.02 km^3, represents a vast increase from the current estimated value of the utilizable waters of the peninsular rivers Godavari, Krishna, Pennar, Cauvery, Tapi, Narmada, Mahi, and Sabarmati, and west-flowing rivers Kutch and Saurashrta, including Luni, to which these could be added, estimated at 229.17 km^3. It was shown that

TABLE 13.2

Estimates of Water Availability (km³)

| Sl. No. | Agency | Surface Water | | Ground | Total | Peninsular[a] Surface Water |
		Mean	Utilizable			
1	NCIWEDP	1952.87	690.32	342.43	1032.75	229.17
2	Chaturvedi 1[b]	1952.87	776.94	500.43	1276.47	298.90
3	Chaturvedi 2[c]	1952.87	1299.46	500.53[d]	1799.89	573.92
4	Increase (%)[e]	—	88	46	74	150

[a] Peninsular rivers considered are Godavari, Krishna, Pennar, Cauvery, Tapi, Narmada, Mahi, Sabarmati, and west-flowing rivers of Kutch and Saurashrta, including Luni, to which the Ganga waters could be added.
[b] Chaturvedi 1 estimates relate to increase based on conventional advances.
[c] Chaturvedi 2 estimates relate to increase through novel technologies.
[d] As stated, the estimates are tentative.
[e] The increase estimate is on the basis of the novel technologies.

this could be increased to 298.90 km³, even by means of some conventional advances (Chaturvedi, 2001). Further adding the figure of 275.02, we get 573.92 km³, an increase of about 150%. The scene is summarized in Table 13.2.

We are also arguing that the figures of utilization, particularly in the agricultural sector, which currently amounts to about 80% of the utilized waters, can be substantially reduced. Thus, a revolution can truly be brought about in the management of India's waters. The central point, however, is creativity, modernization, and, above all, implementation. It must, however, be clarified at the outset that the achievement of these figures depends, first and foremost, on the state of the social system and is not a matter of technology alone.

13.6 Overview of GBM Development

The GBM has three major river basins, Ganga, Brahmaputra, and Meghna. The Delta integrates the three. Each has a different environmental characteristic. Each is therefore considered specifically first in Sections 13.5.1–13.5.4, respectively. The possible technological options of development are proposed, and a portfolio of developmental activities and their trajectories is identified as a preanalytical vision.

13.6.1 Ganga

Ganga flows in a narrow basin flanked by the world's highest mountains, the Himalayas on one side and the world's oldest peninsular uplands on the other, with a low, almost unnoticeable watershed separating it from the another major river basin in the Indo-Gangetic plains, the Indus basin. Ganga is joined by several major tributaries from the north as well as from the south, each of which has to be considered in detail independently. The precipitation also increases significantly as one moves downstream. Therefore, consideration of Ganga will be in terms of the three geophysical regions: (1) the Himalayan region, (2) the alluvial plains, and (3) the Southern Highland region.

While there may be different technological options for the development of the physical environment in different regions, there are some common priorities all over the basin. People or the *am admi* (common man), as the saying goes, come first. Drinking water supply, provision of appropriate housing, and jobs for the people are the first priority. It should be well recognized that development is a long arduous journey. Tremendous development of the physical environment and the infrastructure has to be undertaken. A long-term plan of development focusing on the eradication of poverty and on rapid economic development, is thus essential for all regions.

13.6.1.1 Himalayan Region

The consideration of the Himalayan region, as of all the ecosystems, proceeds from the twin perspective of bringing the people in the mainstream of human development and environmental conservation. In both respects, mountain regions in general and the Himalayan region in particular pose a formidable challenge (Agenda 21). However, the spectacular transformation of Switzerland and Sweden, currently the most prosperous regions of the world, offers valuable guidance and inspiration (Pfister and Messerli, 1990; Hagerstrand and Lohm, 1990).

Mountains are an important source of water, energy, and biological diversity. Furthermore, they are a source of such key resources as minerals, forest products, and agricultural products, and recreation. As a major ecosystem representing the complex and interrelated ecology of our planet, the mountain environment, in which the Himalayas occupy an important position, is essential to the survival of the global environment. They are a storehouse of biological diversity and endangered species. They are, however, highly vulnerable to human and ecological pressures. They are susceptible to accelerated soil erosion, landslides, and rapid loss of habitat and genetic diversity. On the human side, the poverty of the developing countries is most acute in the mountain areas, as it was in the currently developed world historically. They urgently require socioeconomic development of the people and a new perspective of environmental management.

It is well recognized that urgent research is required for understanding the complex issues of sustainable development of the mountain ecosystems (Agenda 21). Certain tasks, however, are well understood for urgent action. It is important that scientific land-use planning be undertaken for both arable and non-arable land in mountain-fed watershed areas to prevent soil erosion, increase biomass production, increase the value of produce on the arable lands, and maintain ecological balance. It may be necessary to completely change the historic agricultural practices where people had developed terraces and *kuhls* (small irrigation channels) with tremendous labor over centuries for subsistence agriculture, and yet, as population increased, they could not even secure a livelihood. This was the historic experience in many mountain areas such as the Alps. People drifted to areas where they could make a livelihood in view of lack of any skills, such as military activities for men and prostitution for women.[15]

Modernization of society, skill-based industrialization, and hydroelectric development is a crucial activity for the socioeconomic–environmental development of these mountain regions as the history of Switzerland and Sweden demonstrates. Development of multipurpose storage projects and run-of-river hydroelectric projects is thus of extreme importance in this region. It is instructive to note that it contributed significantly to the development of Switzerland and Scandinavian countries in many ways, which provide an illustrative examples of the dynamics of mountain region development and environmental management. It can be undertaken even more effectively in the GBM basin, in contrast to the usual conventional approach, through an integrated land–water energy development of the Himalayan region, the alluvial plains, and the Southern Highland through a novel technology called the Chaturvedi Water Power Machine, as discussed in Section 13.3. In short, the environmental characteristics can be very profitably developed, and contribution to environmental conservation can be made by scientific development of the Himalayan region's water resources, particularly in collaboration with development of the water resources of the rest of the basin.

However, it is important that the development be undertaken in the perspective of transformation of the society and the environment, rather than in terms of an ad hoc development of hydroelectric projects merely from the perspective of generating energy, as is currently being undertaken. It may be pointed out that with the rich hydroelectric potential of the region, in contrast to the comparatively limited potential of the Alps, there is much higher potential for the development of the region if undertaken scientifically. Indeed, the hydroelectric energy is what oil is to the Middle Eastern countries, except that it is even more valuable, as it is everlasting, and environmentally benign.

The development of these projects has to be integrated with the infrastructure development and soil conservation on a well-planned-out basis. A very large-scale infrastructural development will be required, which will cause much land wasting, and can severely reduce the storage potential. The region is inherently prone to very high soil erosion and sedimentation in the lower areas. Soil conservation constitutes an important and priority activity, which is not only environmentally valuable but also highly remunerative.[16] Projects undertaken recently for soil conservation, such as Sukhomajri in the foothills of Himalayas, which is highly vulnerable to soil erosion, have demonstrated that soil erosion can be controlled and that the land can be made very productive. The essence of the success of the project was employing a scientific approach and the initiative and cooperation of the people. It can therefore be stated that the soil erosion and mass land wasting can effectively be managed. The land use may be in terms of horticulture and forestry. Groundwater development is not possible, but water can generally be harnessed for all agricultural and habitat needs through storages.

It must also be stated that while there is tremendous potential as brought out from the example of the currently most developed mountain regions, there is also the tremendous challenge of high population, abysmal poverty, and limited time. The development has to be planned and undertaken very scientifically, earnestly, and urgently. The central issue is that the perspective is the development of the people and the environment and not of the projects. The hydroelectric development will take time, and the immediate tasks are to manage the environment, particularly water, in the context of a well-planned trajectory of socioeconomic development, which itself has to be increasingly modernized.

13.6.1.2 Alluvial Plains

The Indo-Gangetic plains have been considered the world's richest agricultural estate (Ruttan, 1997). The region can be transformed into a most prosperous region, as the example of California has shown. Scientific development and management of water, as in California, have a key role in this revolution, but water resources development has to be undertaken in terms of integrated socioeconomic–environmental management as the environmental systems approach envisages.

Social transformation and industrialization are again the core of development, but agriculture will continue to have an important role for employment and economic growth. However, it has to be modernized rapidly. The key to its transformation is the provision of adequate, timely, and reliable supplies of water at the command of the farmer. This can be assured all over the basin, from the arid west to the flooded east. Our focus becomes the people and the ecosystem. We start at the village/watershed level at one end of the scale and the state/river basin at the other. Noting that water is made available as a diffused input in the hydrologic system, we start with watershed management by the villagers, individually and collectively, to store the water and mitigate land/soil erosion, while increasingly leapfrogging to the latest sustainable agriculture using the latest water development and application in terms of agricultural ecology according to the concepts of the environmental system. The historic and continuing focus continues to be, as brought out in the NCIWRDP (1999), on the Rabi (postmonsoon–winter) season, with almost total neglect of the Kharif (monsoon) season, trying to stabilize Third World agriculture. We have an entirely developmental and technological perspective. We propose to develop the most advanced, productive, and sustainable agriculture round the year, with assured water supplies through the Chaturvedi

Ganga Water Machine. The associated scientific integrated surface groundwater and power development further enhances the technoeconomic potential.

13.6.1.3 Southern Highland Region

The region is characteristically and significantly different, physiographically and hydrologically, from both the Himalayan region and the alluvial plains. The rivers are not snow fed, and therefore, low season flows are negligible. Again, the groundwater availability is much lower in view of the rock formations. However, they have an excellent storage facility. Therefore, a much higher proportion of total flows, up to all the surface runoff, can be stored, and by carry-over storage, currently estimated availabilities can be increased substantially. Usually, concentration has been on the development of dams on major tributaries. Conjunctively, storages may also be developed at the watershed level in small streams to store all the precipitation, which has popularly been called "rainwater harvesting." It must be, however, emphasized that this aspect should not be overblown, as is currently being done by some. People's participation is important, but the activities must be undertaken under the supervision of competent persons.[17] This region can also be planned to the highest levels of socioeconomic–environmental development as in the rest of the two subbasins.

13.6.1.4 Ganga Overview

The scale and manner of land–water system management in the Ganga basin will vary in the northern and southern and the western and eastern rivers but are interrelated because of the river basin integrity. Integration with the other major components, Brahmaputra and Meghna, is also important in view of the land–water integrity, particularly at the delta level. Again, integrated water and energy development has to be specifically considered in view of the high hydropotential. Integration at the national level also has to be considered in view of the interlinkage of agricultural production and possible interbasin water transfer, if any.

We have concentrated on the development of water resources for meeting the agricultural demands and developing the hydropotential. We have emphasized integrated land–water management in terms of watershed management for mitigating soil erosion and enhancing groundwater recharge. Flood mitigation is also extremely important, and it acquires increasing importance as one moves eastward. The complex subject of water management has many facets, and all have to be attended to for sustainable development. For example, river geomorphology has to be integrally considered. In this context, reference may be made to the management of River Kosi. It is the last major tributary of River Ganges, joining it from the north. As extremely high monsoon flows take place, during the monsoons, it meanders hundreds of kilometers. The area is intensely populated. It has ravaged the plains of north Bihar, in the lower Ganga basin, and it is called the "Plight of Bihar." Its management poses a formidable challenge.

13.6.2 Brahmaputra

Brahmaputra takes a big U-turn to enter India, flowing almost west and takes an abrupt southward course at the end of the confining southern mountains, to meet the Ganges at the confluence of the Delta. The Brahmaputra basin in India is a long narrow valley of irregular shape, surrounded by mountains on all sides except the flow channel of the river Brahmaputra, with narrow valleys of several tributaries and with high discharges and severe floods. The Brahmaputra basin is very rich in water availability and hydropotential. Brahmaputra alone accounts for about 30% of India's total runoff and constitutes a mighty storehouse of power. Little development has taken place.

The developmental perspective is a mix of the two regions of the Ganga— the Himalayan region and the alluvial region. Development of storage projects and careful management of the drainage for managing the floods are extremely important for the region. The guiding objective of storage projects are flood mitigation and hydroelectric generation and not irrigation. Land is a very scarce resource, and in agriculture, high return land use should be considered. The focus has to be on rapid socioeconomic development providing avenues of employment rather than merely focusing on agriculture.

13.6.3 Meghna

Meghna is a small river defined by low mountains on all sides but which experiences one of the world's heaviest rainfalls. It meets the Ganga as it approaches its end. The Meghna region requires an even more carefully developed management of infrastructure, floods, drainage, hydroelectric development, and land conservation and development. The central issue, again and again, is organized social, economic, and environmental systems development.

13.6.4 Delta

The GBM basin delta is the tail end of the world's second largest river system. It is the world's largest and wettest, and it is also, over most of the year, an arid delta. It is a unique ecosystem. It is a land of very fertile soil, with numerous rivers and abundant groundwater. It once had very rich fauna and flora, and even now, it is still somehow quite rich in this respect. It receives annually, on average, about 1073 km^3 of surface water inflow and 203 km^3 of rainfall, enough to submerge each unit of land under more than 9 m of water. During each monsoon season of June through September, it is gripped with very heavy floods. During the dry months, November through May, major river flows decrease drastically. Fresh water becomes scarce for use in agriculture, fisheries, navigation, industries, and drinking or domestic purposes.

The rhythm of the annual water cycle, severe annual flooding during the monsoons, and drought during the dry season, has fashioned the environment and the life of the people. Over a long period of time, people had evolved a way of life with the vagaries of environment, enjoying the gift and calamities of abundance. Until the industrial revolution, it was one of the world's most prosperous regions, which attracted the British trading East India Company. Its exploitation started by the mid-eighteenth century, and it became the springboard of the establishment of the British Empire in India. With the establishment of the British rule, peace and law, and order were established, but development did not take place. One tragedy of development, however, was population explosion, which started taking place by the early twentieth century. Currently, the region is one of the world's poorest and most densely populated regions. It can be described as a super saturated solution of land, people, and water. It poses another facet of one of the most difficult problems of integrated socioeconomic development, natural resources development, and environmental management, just as Nepal and India pose different facets.

The three major facets of water resources development are (1) flood management, (2) adequate water supplies for people and the economy, and (3) management of the environment. Unfortunately, there has been bitter Indo–Bangladesh dispute about the sharing of the river waters, adversely affecting the development in both countries.

Management of the Delta region, which is principally in Bangladesh, must be considered integrally with the totality of the development of the GBM basin, irrespective of the political divisions. Its development poses difficult challenges and has been considered in some detail in Chapter 9. Detailed studies have been carried out, and it will be demonstrated in Chapter 14 that these can be adequately managed if an integrated development of the river basin is undertaken. However, a new revolutionary perspective has to be adopted.

13.7 Integrated Development of GBM Basin

We have considered the different regions just to keep the specific perspectives in view, but the GBM region, even in India, has to be considered as one unit. Kharif irrigation in the alluvial plains is almost an independent activity, but management of floods has to be duly considered in totality at the same time. The storages will develop over time to manage and mitigate the floods, but they will never be controlled.

The Himalayas are vulnerable to soil erosion of the highest order, which is supplanted by erosion in the alluvial plains, leading to meandering all over but particularly in the delta. It has serious environmental implications and has to be mitigated.

Development of storages will take time. Their operation has to be well planned from diverse considerations—environmental conservation, irrigation, energy generation, and flood mitigation, all over the basin conjunctively. All the regions of the Ganga basin are integrated in terms of the novel technologies. The GBM basin may be integrated even with the peninsular region if interbasin, in any form, is found to be attractive.

These developments will take place over time. The conclusion is that development cannot be undertaken on an ad hoc basis but has to be organized through the modern advances of systems planning, as will be discussed later.

13.8 Some Specific Issues

The development of the environmental system was considered in the context of the socioeconomic conditions under the differing physical environment conditions of the basin. Some specific issues of the development of the physical environment, with particular reference to water, may be discussed to ensure that they have been properly accounted for.

13.8.1 Environmental Conservation

Traditionally, environmental conservation has been undertaken as an afterthought in water resources development. Our point of departure is that it should be considered conjunctively, with it as one of the primary objectives. Thus, besides conducting environmental impact assessment of the developmental activities and trying to mitigate the adverse affects, we may examine as to how to implement environmental conservation and manage it as one of the multiobjectives. For example, maintenance of certain minimum low flows in rivers from environmental consideration, and drinking water and return flow management requirements are the first priority. Any diversion can be made only if these are maintained. One of the priority functions of the multipurpose dams may therefore be to release water from ecological considerations in the low flow season from these considerations, if needed. Similarly, groundwater is not merely another source but a dynamic system whose quality has to be carefully considered through what is called a study of the chemical cycles.

Urbanization and infrastructural development is going to be an enormous challenge, and therefore, the issue is on how to undertake it so that environmental conservation is ensured. Same considerations apply to industrialization and agricultural development. Therefore, we emphasized the consideration of these sectoral activities themselves in this context at the outset in terms of integrated regional development, modernization of the agricultural activities considered as a system, industrial ecology, scientific management, and so on, as briefly referred to in Chapter 12. Thus, an important issue in water resources development is that of undertaking rapid sustainable development itself.

13.8.2 Climate Change

Climate change is an important factor, which has to be taken into consideration in managing the water resources of the region. A study entitled "Development and Climate Change in Nepal: Focus on Water Resources and Hydropower" has been undertaken by Shardul Agrawala, Vivian Raksakulthai, Maarten van Aalst, Peter Larsen, Joel Smith, and John Reynolds under an OECD project (2002). The main findings are as follows.

This report presents the integrated case study for Nepal carried out under an OECD project on development and climate change. Recent climate trends and climate change scenarios for Nepal are assessed, and key sectoral impacts are identified and ranked along multiple indicators to establish priorities for adaptation.

Analysis of recent climatic trends reveals a significant warming trend in recent decades, which has been even more pronounced at higher altitudes. Climate change scenarios for Nepal across multiple general circulation models meanwhile show considerable convergence on continued warming, with country-averaged mean temperature increases of 1.2°C and 3°C projected by 2050 and 2100. Warming trends have already had significant impacts on the Nepal Himalayas—most significantly in terms of glacier retreat and significant increases in the size and volume of glacial lakes, making them more prone to glacial lake outburst flooding (GLOF). Continued glacier retreat can also reduce dry season flows fed by glacier melt, while there is moderate confidence across climate models that the monsoon might intensify under climate change. This contributes to enhanced variability of river flows. A subjective ranking of key impacts and vulnerabilities in Nepal identifies water resources and hydropower as being of the highest priority in terms of certainty, urgency, and severity of impact, as well as the importance of the resource being affected.

The in-depth analysis of water resources in Nepal identifies two critical impacts of climate change—GLOFs and variability of river runoff—both of which pose significant impacts not only on hydropower but also on rural livelihood and agriculture. A preliminary discussion on prioritization of adaptation responses highlights some potential for both synergies and conflict with development priorities. Microhydro, for example, serves multiple rural development objectives and could also help diversify GLOF hazards.

On the other hand, storage hydro might conflict with development and environmental objectives but might be a potential adaptation response to increased variability in stream flow and reduced dry season flows, which are anticipated under climate change. Further, while addressing one impact of climate change (low flow), dams could potentially exacerbate vulnerability to another potential impact (GLOFs), as the breach of a dam following a GLOF might result in a second flooding event. Finally, the in-depth analysis also highlights a transboundary or regional dimension to certain impacts, highlighting the need for regional coordinated strategies to cope with such impacts of climate change.

A study of climate change, focusing on India, has been undertaken by Gosain et al. (2006).

13.8.3 Hydroelectric Development

A hydroelectric potential of 150,000 MW at 60% factor has been estimated in the Himalayas, of which only about 1.5% has been developed so far. Of the total, 50% is estimated to be through run-of-river schemes, as conventionally designed on 90% water availability, with the rest being through storage projects. This has been tried to be developed with the highest emphasis after Independence, but the developments could not be undertaken expeditiously as planned due to international bickering and have further slackened on account of adverse publicity of the environmental impacts of the multipurpose dams. The appropriate development has also not taken place on account of certain technical shortcomings, as will be discussed later. We consider all these issues.

Dams are absolutely essential for sustainable development in South Asia. This is discussed in the context of the development of the Indus basin in detail through a study of the Bhakra Dam (Rangachari, 2006; Chaturvedi, 2008c). There is, however, a need to make several advancements in the field. For example, hydroelectric development is part of energy systems development, and the capacity and economic returns will be much higher, as hydro will be essentially for peaking purposes and the load factor will be only about 20%. This also brings out the need for emphasizing pumped storage schemes, which has been neglected so far. Secondly, conjunctive surface water, groundwater, and energy planning is required, as briefly pointed out earlier and will be discussed in detail later. This is much beyond the conventional conjunctive surface and groundwater development, even which is not being undertaken currently. Even the capacities in run-of-river schemes are not developed adequately, as these have to be in terms of cost of alternative energy modal choice, considered in terms of the total system intertemporal development, rather in terms of 90% reliability of the individual unit as adopted currently. The proposed novel technologies enhance the potential enormously. Scientific development can best be undertaken through collaborative development of Nepal and India, to the immense benefit of both.

Development of hydroelectricity has to be considered, besides integration with the development of water, with scientific study of development of energy. Two sets of studies are required, as in the case of water as discussed in Sections 13.11 and 13.12. One is integration of hydroelectricity with the energy sector. Second is integration of the energy sector with the socioeconomic system, which will also bring in integration with the water sector. It may be pointed out that there is a tremendous scope for modernization, as in the water sector.

13.8.4 Management of Floods

Floods are severe in the Himalayan region, becoming increasingly furious as one moves east. Storage reservoirs are the conventional mitigating devices and are trying to be developed. However, it is necessary that flood mitigation is studied comprehensively and that warning systems, land management, and integrated flood management as part of comprehensive water management, which has been rather neglected, are undertaken with greater concern and urgency. The scene is going to become even more uncertain with the likely climatic changes. Therefore, scientific management and mitigation become increasingly important.

13.8.5 Navigation

Navigation was of primary importance in water resources management in Europe and particularly in Britain, at the time when British hegemony was being established in India. Navigation was therefore provided in the design of canals when first developed. The Ganges Canal, which was the first one to be developed, provided for navigation. Cotton, the leading canal engineer of the East India Company, was most enthusiastic about navigation and proposed the interlinking of Indian rivers from this consideration. In view of his stature, some private companies even started action on his suggestions, which, however, turned into a disaster, and the policy of Government that all future water resources development should be in the public sector evolved from this experience. (He even proposed the interlinking of Ganga and Yangtzekiang [Cotton, 1867].) Railways became the natural choice for transportation, although from political considerations, and sight of navigation was lost.

Transportation has to be developed in a big way as development of India takes place. Roads, railways, and water can all be considered to be developed integrally as a system. Management of waterways has to be considered not only from the perspective of irrigation but from that of their management in terms of the huge annual variation in flows, leading to floods and submergence of large areas in the alluvial plains. They have therefore developed comprehensively, and navigation may contribute to their economical development and environmental management. The central issue is creative development instead of being tied to the perspective of rivers only for providing Third World irrigation.

13.9 Groundwater Management

Groundwater is usually considered merely as a resource. Its development is planned on the basis of the estimated annual recharge. With some recent experiences of its quality, such as the arsenic problem in Bangladesh, some concern has also been expressed about this aspect. However, the conventional perspective of an unknown resource to be extracted to a safe extent persists.

Groundwater is a component of the dynamic hydrologic cycle, dependent on the hydrologic input and transformed in a complex manner by the geophysical–chemical characteristics of the land system. It has to be planned in a much more scientific manner than the current approach, with particular concern for the chemical cycles, as they are called. The implication of the arsenic problem of Bangladesh is only one of the manifestations.

Widespread presence of arsenic in Bangladesh raises very serious challenges for management of the groundwater and water resources in general. Arsenic has since been detected widely in the GBM basin. Widespread presence in Bihar and even in the western parts of the GBM has been reported. The subject has been discussed in more detail in Chapter 9.

As pointed out at the outset, the possibility of immense tectonic disturbances leading to the disappearance of Saraswati appears to be existing. If so, it would have created complex interactions in the subterranean regions, which could have serious implications on the groundwater quality of the region, which has very low hydrologic inputs. The salinity problems of the Indus basin led to modernization of the management of water resources. With the problem of arsenic, as experienced in Pakistan recently, consciousness of scientific management of groundwater is reinforced. The central point is that groundwater is an important and complex resource and that its management has to be undertaken much more scientifically than it currently is. The implication of River Saraswati is that the drainage of the entire groundwater field will have to be studied in totality because if proper subterranean drainage is not developed, geologically, serious groundwater quality problems may arise. These become important because water resources, including groundwater, are being developed at such a large scale. Detailed groundwater finite element modeling has been carried out for some regions of the GBM basin, as shown in Figure 13.4, but the studies have to be vastly expanded (Chaturvedi, 1987; Prasad, 1981). The central point, as emphasized in the environmental systems perspective (Figure 13.3), is that the chemical cycles have to be duly and integrally studied.

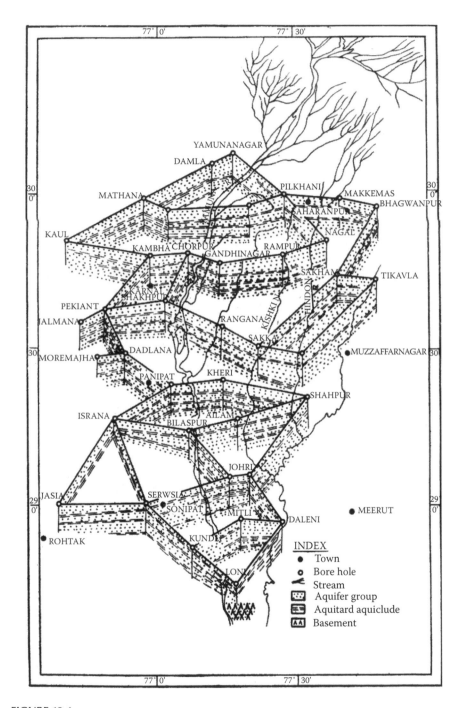

FIGURE 13.4
Groundwater management. (From Chaturvedi, M.C., *Water Resources Systems Planning and Management*, Tata-McGraw Hill, New Delhi, 1987. With permission.)

13.10 Societal–Environmental Systems Management

Societal–environmental systems management advances the conventional and the currently proposed Integrated Water Resources Management in several ways. It integrates the economic sectors in which the environmental vectors are used and emphasizes the development of the two vectors integrally. For example, given the current use of water for irrigation, the yields of the cereals in India are only about 2000 kg/ha, even less than the average productivity of 93 developing countries, excluding China, where the yield is almost twice that in India, as already emphasized earlier! It is emphasized that not only do water resources have to be developed for irrigation, but the use, over the cycle of production, application, beneficial economic return, and environmental sustainability, has to be made increasingly efficient as well. The agricultural system has to be modernized conjunctively. Thus, the life cycle of resource use has to be considered. Environmental impacts have to be considered conjunctively over the entire life cycle of activity, as leading to the new science of industrial ecology in the manufacturing area. Thus, development of the environment has to be viewed in the environmental systems perspective, as shown in Figure 13.3, and not in an ad hoc manner, as what is currently being done. All components of the study have to be backed by systems analysis and transparent participatory system dynamic modeling. These aspects have not been discussed in the chapter, as they are considered in detail in the text on the subject (Chaturvedi, 2011a).

13.11 Scientific Management of Water— Some Basic Perspectives

Water is available from a variety of sources and has to be managed for different objectives. It is also linked with other energy sources. For instance, groundwater supplies water conjunctively with surface supplies from storage projects. Its development, however, requires energy. On the other hand, the release of surface waters produces energy. The potential for generation of energy by surface water releases from storage projects, per unit of water, varies temporally. Demand for water also varies temporally. Surface waters are also available from nonstorage diversion works. The position is further complicated as hydropower supplies are integrated with the power system of thermal supplies and pumped storage and possibilty of water demand management in view of choice of cropping pattern. Furthermore, development potential and characteristics change on an annual and long-term temporal basis. Some complications of the scene will be presented by listing some issues, as in Table 13.3 which may be considered as scientific development

TABLE 13.3

Scientific Development Pointers

---

1. First focus is use and storage of water during monsoons in the plains through the Chaturvedi Ganga Water Machine. Groundwater recharge is managed at a local level, conjunctively with regional management of land–water configuration. Inflows from the Himalayas are spread over the region. Rivers are proposed to be managed as part of the local configuration.
2. Irrigation requirements are met from surface waters at local levels. Use of groundwater is to ensure that control is by the farmer.
3. During monsoons, nonirrigation demands will be met from river flows. Storage barrages will help.
4. River development may also be examined from navigation consideration, with storage barrages.
5. Note that reservoir capacity in the Himalayas is generally much less than monsoon inflows.
6. Operate it such that storage is maximized, flood impacts are also minimized, and energy generation is maximized, noting that floods are uncertain events.
7. Reservoir is full at the end of monsoons, hopefully.
8. It receives inflows during the nonmonsoon period, which is gradually decreasing.
9. River flows have to be managed such that requirements of environmental conservation, drinking water supplies, sanitation, and economy are met. If river flows are not adequate, releases are made from the reservoir to make up for it.
10. Return flows are treated up to the highest level and fed back into the river system.
11. While these in-stream and diversion demands at the local level are almost constant, river flows will be gradually decreasing with time.
12. Agriculture demands during nonmonsoon periods are met from groundwater and river surface diversions, if available beyond local level demands. As a last resort, these may be met by releases from reservoir.
13. During nonmonsoon periods, groundwater energy requirements are met from hydro from run-of-river and thermal and hydro from storage. If additional water is not required for irrigation, they can be pumped back.
14. Power generation is worked out conjunctively. Releases are made from reservoirs, and water is pumped back on a diurnal basis during the nonmonsoon period.
15. Hydrogenerations from run-of-river schemes will be substantial during the monsoon period. It is on an uncontrolled uniform basis round the clock. Hydrogeneration from storage schemes can be scheduled on a daily basis. Conjunctively the three, (1) hydro run-of-river generation, (2) thermal, and (3) hydrostorage generation, can be managed to meet the power demands. The groundwater demands will be variable on a daily basis and are uncertain while the domestic and industrial demands are variable over the day but constant on a day-to-day basis.
16. The hydroenergy can be stored by pumped storage in terms of the Chaturvedi India Water Power Machine on an intertemporal interspatial basis.
17. The system has to be managed on a yearlong basis from considerations of all these variable inputs and demands.
18. The water can be released into the Ganga system or for interbasin transfer, with some additional costs of energy generation, which will be small.
19. The Chaturvedi Water Power Machine can transform Ganga as well as Brahmaputra hydroenergy and thermal secondary energy into peaking energy.
20. In all these operations, impact on the delta has to be kept in mind on a priority basis, considering it as a very vulnerable dynamic system.

---

pointers. These are only indicative and have to be quantified through systems analysis, as will be discussed later (Chaturvedi, 1987).

13.12 GBM Systems Planning

The proposed developmental policies represent a complex of activities ranging from actions at the field level, watershed level, groundwater development and recharge, diversion canals, small and very large storage projects, developmental and management activities, management of rivers from a complex of considerations, and so on. They are all interrelated in space and time. Their creative development and estimation of their characteristics, capacities, and timing can be determined only through appropriate preliminary design backed by systems studies. Advances have been made that enable scientific transparent participative determination of optimal managememt in terms of these indicators (Chaturvedi, 1987, 2002, 2011a). An approach to water resources systems planning is briefly presented.

A tentative conceptual trajectory of developmental and management activities representing a preanalytic view is presented in Table 13.3. It may be emphasized that it is only a preanalytic vision focusing only on the conventional functions of water supply and power and must be enlarged in terms of perspectives discussed earlier.

Considerable development of major works has been undertaken in the GBM basin, as shown in Figure 13.5 (shown earlier as Figure 2.1). Reference to Figure 13.6 (shown earlier as Figure 5.3), which shows some development projects on the Yamuna–Ganga–Ramganga system, undertaken historically and continuing, will bring out the typical scheme of development and the challenge of engineering planning and design in the context of current policies of development. Large diversion canals like the Western Yamuna Canal (WYC) and Eastern Yamuna Canal (EYC) were constructed about one and a half centuries back to divert almost all the low season flows for irrigation as soon as River Yamuna debouched in the plains from the Himalayas. There was little urbanization or industrialization at that time. Even Delhi, which is situated on the banks of River Yamuna, had a population of about half a million until 1950 and did not have piped water supply or sanitation facilities in most of the regions. Now, with a population of about 15 million, which is growing at a rapid rate of about 4% annually, it faces one of the most severe problems of water supply and sanitation, as not only are there no low season water supplies, but the river has also been turned into a polluted drain with all the return flows and untreated sewerage from rapidly urbanizing and industrializing upstream areas and its own. Whatever low flows accrued was groundwater recharge from natural and irrigation return flows. They are being depleted and polluted on account of heavy unregulated groundwater development and polluted on

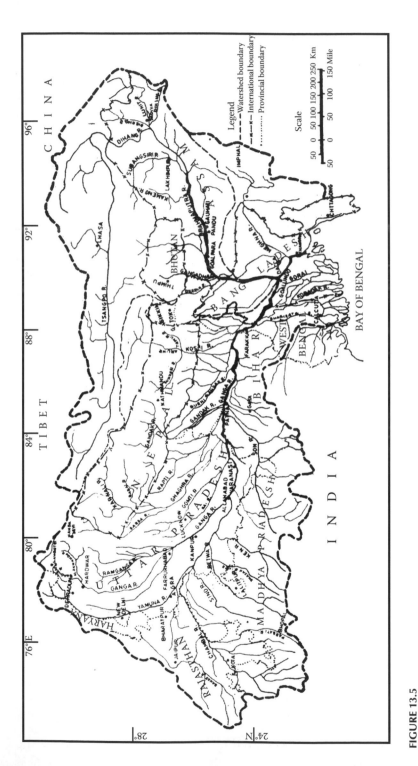

FIGURE 13.5
Ganga–Brahmaputra–Meghna river system. (From Chaturvedi, M.C., *Water Resources Systems Planning and Management*, Tata-McGraw Hill, New Delhi, 1987. With permission.)

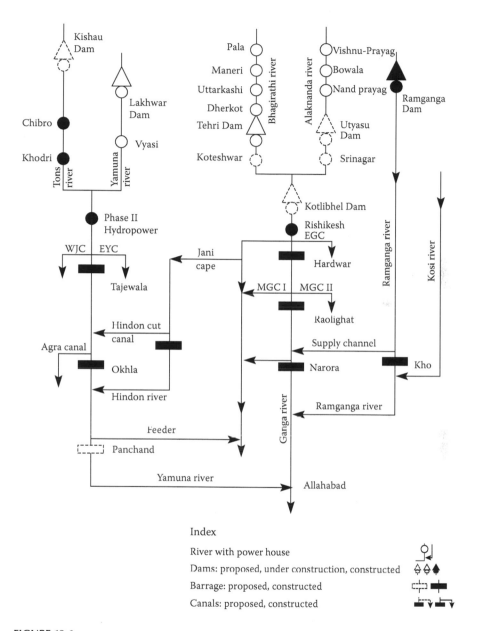

FIGURE 13.6
Yamuna–Ganga–Ramganga system. (From Chaturvedi, M.C., *Water Resources Systems Planning and Management*, Tata-McGraw Hill, New Delhi, 1987. With permission.)

account of unregulated fertilizer and pesticide use, besides industrial waste recharge. These return flows, along with the flows from the Ganga Canal through the Hindan cut, were again diverted downstream of Delhi for irrigation through the Agra Canal, constructed about one and a half centuries back. Rapid urbanization and industrialization are also taking place in all the cities located all along River Yamuna, upstream and downstream of Delhi. They face even worse problems of water supply and environmental degradation.

The story is repeated on River Ganges. All the low flows after monsoons were diverted for irrigation, through the Upper Ganges Canal (UGC) about one and a half centuries back, which was the pioneering irrigation work in India. The water supplies were inadequate for the large area brought under irrigation, and the Lower Ganga Canal (LGC) was constructed downstream to pick up the regenerated flows. The Ganga system and Yamuna system were interlinked to provide more balanced and enlarged water supplies to the region. The lower command of the UGC was served by the LGC, and some water from the UGC was diverted to River Yamuna for use through the Agra Canal. The diversion structures were initially temporary structures built to divert the postmonsoon flows, which were washed off every year. Permanent headworks, called "barrages," were built gradually only after 1920; the first one was constructed on the UGC at Hardwar.

As further irrigation development became necessary, storage on adjoining River Ramganga was developed, and waters so available were diverted to River Ganga, upstream of the LGC. Reallocation of surface withdrawals at various canal headworks was carried out, but conjunctive surface–groundwater has not been undertaken on account of the institutional limitations of the two technologies.[18]

The area is still short of water, and there is considerable potential to develop hydropower. Therefore, run-of-river and storage projects continue to be undertaken on River Yamuna and River Ganga, as shown in Figure 13.6. Some examples are the Tehri Dam, a 260-m-high dam completed recently, and three other equally high dams—Keshau, Lakhwar, Kotlibhel—which are in the process of being undertaken gradually. The run-of-river schemes are also major projects, with Maneri 1, Maneri 2, and Uttarkashi each having an installed capacity of 150, 300, and 400 MW, respectively. Canals for providing irrigation during the monsoon (Kharif canals), such as the Eastern Ganga Canal and Madhya Ganga Canal I, are in operation, and a Madhya Ganga Canal II has also been proposed.

Although the entire system is interlinked, spatially and temporally, in terms of supply and demand of water, water and power interactions, and economic and environmental considerations, projects are generally undertaken in an ad hoc manner, and the activities are classified. Some examples will illustrate the serious shortcomings on this account and bring out the need for the proposed advancements. Run-of-river hydroelectric schemes were currently designed on the principle of 90% reliability. This makes sense when considered in isolation and when financial profits are the objective. However, from economic considerations, the capacity has to be designed in terms of optimal marginal cost in terms of alternative energy modal choices. As discussed earlier, design on about 20% reliability turned out to be economical for Maneri Bhali 2. The problem thus becomes complicated, and systems planning becomes necessary, as discussed earlier. The scene gets further complicated as all hydroelectric development, in terms of the low flows, has been stopped from environmental considerations.

The Yamuna–Ganga–Ramganga system shown in Figure 13.6 is only one set of integrated projects on these tributaries. A similar set is proposed to be developed on almost all tributaries. Although all the projects will be undertaken over time and space individually, all are interrelated, with further interrelation with management activities. Systems planning is thus absolutely essential. Some issues of systems planning and continuing advancements in terms of sustainability were discussed in Chapter 12. Some work has been done and may be briefly presented to bring out the need for sophisticated planning in view of the colossal costs involved in developing water infrastructure and socioeconomic–environmental implications. It may be emphasized that it needs to be continually advanced and is given only as an example.

Following the overall outline presented in Chapter 12, a scheme of studies, as shown in Figure 13.7, was developed.[19] It consists of three sets of studies in the context of (1) socioeconomic systems, (2) physical systems, and (3) water resources development. These are related to project stage studies, shown on the extreme left, which give basic technoeconomic formulation. Focusing on the water resources development studies, shown in the middle, from an overall river basin perspective, the system is divided into regional terms for a more detailed study (Chaube, 1982). Next, each region is considered in detail for optimal development in terms of multiobjectives (Asthana, 1984). The subject of integrated and conjunctive land–surface and groundwater development needs to be studied in detail for a smaller region between two major tributaries, called *doab* in local terminology (Gupta, 1984). Some of the important problems which need to be studied in detail are as follows: surface and groundwater interaction (Prasad, 1981), microlevel land and water management (Khepar, 1980), and water demand studies from optimal cropping development in a dynamic context (Arya, 1980).

In the Ganges–Brahmaputra basin, with a number of very large dams to be undertaken, their integrated planning is extremely important to determine the optimum development (Bhatia, 1984). This has to be further advanced through intertemporal power sector studies of which hydroelectric development is a component (Thangraj, 1987). These studies have to be further integrated in the context of water resources development at the all-India level in the context of national agricultural development (Singh, 1980). Similarly, integrated development of energy at the national level has to be undertaken, as hydroelectric development is an integral part (Sarma, 1985).

Some pioneering work has been done as brought out in the studies referred to above, but the GBM basin is a tremendously large and challenging river basin. Continued advances have to be made, which, unfortunately have been lacking.

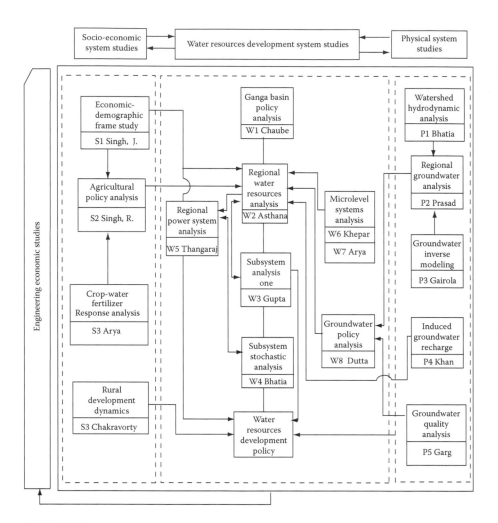

FIGURE 13.7
Scheme of river basin system studies (the name on the bottom is of the doctoral scholar). (From Chaturvedi, M.C., *Water Resources Systems Planning and Management*, Tata-McGraw Hill, New Delhi, 1987. With permission.)

13.13 Perspective—Societal–Environmental Systems Management

Intertemporal considerations were undertaken through taking time snapshots. However, integrated intertemporal and multipurpose demand can best be considered through system dynamic studies. These studies need to be advanced in the context of environmental systems management perspective and policies, as proposed and shown in Figure 13.8. A regional study based on the Pressure–State–Impact–Response approach, with system dynamic

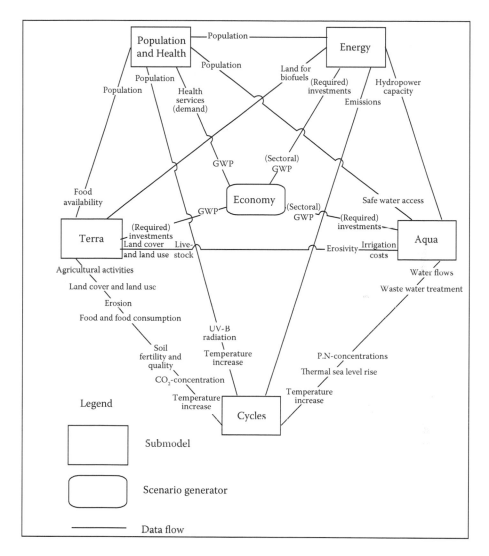

FIGURE 13.8
Societal–environmental systems development. (From Chaturvedi, M.C., *Environmental Systems Management for Sustainable Development*, 2011. With permission. Adapted from Rotmans and de Vries, 1997.)

modeling for one major tributary, Chambal, was accordingly undertaken (Kothari, 2000). National-level system dynamic modeling for a long-term perspective has also been undertaken (Hasan, 2004).

The development of the GBM basin presents one of the biggest challenges for sustainable environmental systems management. The foregoing discussion presented only some of the challenging issues. Flood management, challenge of urbanization and its impact on environment, groundwater quality problems as those of arsenic in the delta region, environmental problems at the water's edge, and the challenge of development and environmental sustainability of the Himalayan region are formidable concerns needing urgent attention. An attempt is being made to study the subject through collaborative professional and academic collaboration.[20]

13.14 Revolutionizing the Management

The institutional arrangement for developing and managing water resources, land, and the environmental system as they have evolved over time and as existing currently was discussed in Chapter 11. It was a typical activity of a colonial system. There was little engineering, except in designing the headworks for some canals. The essential task was management of distribution of water in the canal system. The design capability was poor, but a very efficient and honest official class with a core de spirit developed.

The tremendous challenge of developing water resources after Independence was undertaken in a commendable way.[21] There was a sense of excitement and commitment, inspired by the political leadership of Mahatma Gandhi, Jawaharlal Nehru, and other leaders, big and small. It infected all and sundry. However, it was dissipated fast, and currently, opportunism and corruption are rampant in the political system. They have equally infected all and sundry. In a way, infrastructural and environmental development, which comes under governmental action and management, has been worst affected.

Attempt has been made to study the subject of institutional modernization in a World Bank study in collaboration with the Government of India (WB, 1998a–f; Chaturvedi, 2011c). Recommendations have been made. It is doubtful if any action will follow.

Recommendations have also been made in a World Bank study regarding change in policies and institutions (Briscoe and Malik, 2006; Chaturvedi, 2011c). Their central message is "that India must build a 'new water state' at the Union and State levels. India must have a reinvigorated set of public water institutions." A transformation from a policy focused totally on development to one that increasingly incorporates management has been suggested. However, no details have been given. This is understandable and important to be emphasized. The Western world scientists do not have the understanding of the socioeconomic–environmental conditions of India, particularly the feeling that local people have.[22]

With our present suggestion that irrigation should be emphasized through groundwater development, which is already taking place on the farmer's initiative, a revolutionary change in management is proposed (Chaturvedi, 2011c). The role of Government should be in the realm of planning, management of the surface resources, and environmental systems management.

Development of groundwater may be privatized through agencies that may be interested in total agricultural and rural development. Government only allocates the surface waters to these private agencies through their canal network. Government does undertake planning in the sophisticated environmental systems perspective, undertakes development of major projects, and so on, but further interaction with society should be privatized. Indeed, the agencies should be encouraged to contribute to rural infrastructural development and modernization of agriculture, which would be the attractive activity. Experience of the currently developed countries could be profitably utilized.

13.15 Modernization of the Spectrum of Activities

While our emphasis is on policy aspects, first and foremost, modernization of engineering and the spectrum of activities is urgently required and needs to be emphasized at the outset. Water resources development involves complex scientific analysis. Construction of major engineering projects involves complex engineering activity and very large financial outlays. The engineering activity has a life cycle of data acquisition, scientific analysis, planning, design, and construction, followed by management. Advances in all these phases have taken place and are taking place. The developing countries have the advantage of utilizing the latest technology. It is a paradoxical characteristic of modern science that although it is freely available, it can be utilized only if the borrowers also make some contribution to this pool of knowledge. Unfortunately, this simple fact is generally not understood by the developing countries. The technological state over the spectrum of activities remains poor. Its modernization is the first priority. Indeed, the current low yields of Indian agriculture, one of the lowest among 93 developing countries, and the preoccupation with the interlinking of India's rivers are a testimony of the current inadequacy of the scientific state in the sector.

Engineering is only a part of the spectrum of activities required for the management of water as a vector of the environment closely linked with the management of the entire socioeconomic system. The entire system has to be modernized as discussed by Chaturvedi (2011c), where some revolutionary changes have been proposed.

The developmental personnel, which is currently totally engineering oriented and in a very limited bureaucratic context, has to be made broad based, consisting of persons of diverse specialization such as economics, management, and applied ecology. An essential requirement is that they should have a minimum understanding to be able to interact with each other on a common intellectual professional platform.

All these issues of management, as those of scientific development, are suggestions that need to be worked out in detail.

13.16 Culture Matters

Recent studies of the development of societies has brought out that over history, a culture of the society is developed, which shapes the future (Landes, 1998, 2000). One important issue is development of societal consciousness and institutionality (North, 2005). India is faced by monumental challenges. However, there are also great opportunities in view of the increasing capability of the societies. A paradoxical challenge faced by India, in terms of the ends and means, is achieving the highest imperative of development (Chaturvedi and Chaturvedi, 2011). Commitment to achieve our developmental objectives, by one and all, by the society, is the most important requirement.

13.17 Conclusion

Our thesis is that in contrast to the current ad hoc project-oriented and technologically biased development with a sense of technological hubris, development of the environment in the environmental systems perspective has to be undertaken. Management of water has an important role. It has to be undertaken considering the entire hydrologic cycle and sector activities integrally with increasing sensitivity and creativity and backed by appropriate systems studies considering development and conservation, and technology and management integrally embedded in the integrated management of the socioeconomic–environmental system. Our focus has been on planning, but we hasten to emphasize that the totality of the spectrum of activities, including management, has to be revolutionized.

Technology is a crucial activity for development and conservation of the environment, but it has to be practiced with ever-increasing sensitivity and creativity according to the modern concepts of environmental systems management. GBM poses difficult challenges but also offers promising developmental potential. Our conclusion is that sustainable development of the region can be undertaken to achieve the highest level of socioeconomic development. However, this requires much committed, sensitive, and scientific planning, development, and management.

Novel technologies and management have been suggested which will revolutionize the management of the environment and allow the challenge of development to be met with sustainability, but these have to be studied in detail. Reference may also be made to the serious problem of arsenic, as discussed in Chapter 7. It requires urgent attention.[23]

Above all, however, there has to be a commitment to achieve the developmental goals. Culture matters.

Notes

1. The population of India is estimated to stabilize at about 1.640 or 1.581 billion by 2050, according to the current high or low development scenarios. Assuming about half the population in the GBM basin and increase in other riparian countries of the GBM basin in about the same proportion, yield the figure of about 1 billion. It is just an indicative estimate.

2. The author designed the Ramganga Dam, the first storage project in the Eastern Himalayas and has been on the Board of Consultants of the major storage projects and hydroelectric projects in the Eastern Himalayas.

3. The author designed the Rihand Dam, the first major storage project in the Central Highlands and was on the Board of Consultants of Tehri Dam.

4. The author studied the Swedish and the German system while in these countries during his collaborative research activities. He advised them on modernization of the system on the basis of the modern advances in systems engineering, which was the subject of collaborative research.

5. Hall (1980). Personal communication. The California system and pumped storage was discussed while working with him in California.

6. The author has been associated with the design of several Himalayan projects as designer or as Member, Board of Consultants. Pumped storage could not be introduced on Ramganga Project and Beas Project because the machines had already been ordered when the author took charge. The initial design of the Tehri Project did not provide for pumped storage and was changed by the author to a pumped storage system, while working on its Board of Consultants.

7. The author was working on the Board of Consultants of the hydro-electric projects when the design of Maneri Bhali II was undertaken, Maneri Bhali I, upstream, having already been completed. Both have the same head and same discharge and, according to the basic principles, should have the same installed capacity. However, consideration of all these factors led to the decision of doubling the installed capacity of Maneri Bhali II.

8. Rihand Dam, in the Central Highland area, had one of the world's best height/storage ratios until Kariba, on Zambezi, came up. Both are for production of aluminum, where the raw material is basically energy!

9. The scheme has been proposed for a long time. The scheme was first proposed in 1973; the context of the National Water Grid proposed by then Minister of Irrigation, Dr. K.L. Rao, with whom the author was working closely and was presented in the First World Congress of the International Association of Water Resources in 1973 (Chaturvedi, 1973). However, as Dr. K.L. Rao quit the Ministry, his proposal was dropped. The subject was presented in the present form in an International Workshop on Energy for Growth and Sustainability, organized by the Indian National Academy of Engineering (Chaturvedi, 1998).

10. The scheme was also discussed in the Working Group of the NCIWRDP (1999) dealing with the interlinking of rivers (ILR), of which the author was a member. The Working Group did not approve of the IIR proposal except for some specific schemes in the peninsular region. The author's Water Power Machine proposal was found to be attractive and was recommended for undertaking feasibility studies (NCIWRDP, 1999, Working Group ILR Report).

11. The idea of groundwater recharge was discussed in a seminar of the Indian National Academy of Science and the U.S. Academy of Science, held in New Delhi, attended by Prof. Roger Revelle and the author. Studies were undertaken by both independently and later collaboratively (Revelle and Herman, 1972; Lakshminaryana and Revelle, 1975; Chaturvedi et al., 1975; Srivastava, 1976; Chaturvedi and Srivastava, 1979), which confirmed the feasibility of groundwater recharge. Revelle gave the title Ganga Water Machine to the idea. The scheme was discussed by the author with then Minister of Irrigation, Dr. K.L. Rao, who got a field study undertaken that confirmed the feasibility of the proposal.

12. It was proposed by Roger Revelle and the author that the subject be discussed by scientists and that field level studies be undertaken. It was accordingly discussed in an Indo-U.S. seminar held in New Delhi in 1981 and considered to be scientifically and economically attractive (Chaturvedi, 1981, 1987).

13. Plans were made in the seminar for a further detailed study beyond the one undertaken earlier. According to the rules for international collaborative research, approval was required from the concerned Ministry. A meeting was held in the Planning Commission under the Chairmanship of then Member, Planning Commission, Dr. M.S. Swaminathan, attended by Prof. Roger Revelle, Prof. Peter Rogers, Mr. C.C. Patel, who was then Secretary of Water Resources, GOI, and the author. The study was approved by the Planning Commission but was scuttled by Patel who was obsessed with the idea of interlinking the rivers of India, on the pleas that the field study will be conducted by the GOI, which was never undertaken.

14. World Bank held a Conference on the subject of scientists in 1986 in Washington, DC to which the author was also invited (WB, 1986). Reference is also contained in GOI (2002, p. 121).

15. As Hagerstrand and Lohm (1990, p. 608) observe in the context of environmental change in Sweden, "Crown seems to have viewed the peasantry predominantly as an object of taxation and a producer of soldiers for army and navy." It is interesting to note that the papal guards by tradition are still from Switzerland, as they constituted the most valiant fighters. Similarly, Gurkhas of Nepal are famous for their fighting qualities. The occupation by women has equally been noted and continues with rising population and lack of economic development in the Himalayas. Some personal observations from the author's visits to Switzerland and Sweden may illustrate the analysis. In one of the lecture visits to the Swiss Institute of Technology, Zurich, as the author genuinely conveyed his appreciation of the beautiful view from host's window of the rich nature and society, the host told the author that the Swiss people had petitioned Ferdinand the Great in the tenth century that they be allowed to move over to Sardinia, where he had established the capital of his empire! Second, he was surprised to see a large enrollment in the water resources area. He was told that they obtained good employment in the consulting firms working for the developing countries in the design and construction. The author visited the offices of these firms. The same situation was encountered in Sweden, where the author spent several months at a research institute for collaborative research. It was clear that hydrodevelopment had contributed signif icantly to the economic development and attendant environmental enhancement. In contrast, as the author traveled on foot along the length and breadth of the Himalayas in India, it was painful to see how poverty has ruined the environment and life of the people.

16. In a study in the context of the design of the Ramganga Dam, the author found that the investment on soil conservation was much more remunerative than the investment on the storage project! The interesting result follows from the simple logic that the investment on soil conservation enhances the life and therefore return from a highly expensive capital-intensive investment.

17. The author, along with Prof. M.S. Swaminathan and Dr. N.S. Saxena, then Secretary Planning Commission, was invited by the Centre for Science and Environment to intervene and save a small dam or *bunda* as it is called in local parlance, called Lakha ka Vas, in the Aravali region. That gave an idea how totally unscientific unguided local level work can be. The bunda was not demolished on account of the intervention of the author, but was washed away in the first monsoon!

18. The Ramganga Project was designed by the author.

19. Chaturvedi, 1987, 1997, 2002. The studies were undertaken with the twin objective of contributing to the science and developing indigenous capability in the context of the Indus basin and Ganga basin. The former was sponsored by the Planning Commission. The experience may be stated as it brings out difficulties in developing capability in the profession. Two research scholars were hired to undertake some studies for the Indus basin. One got married to the daughter of the Chief Engineer as he finished his study. He did not submit his completed doctoral dissertation as he was advised that if he got a Ph.D., he will be posted in Design and Planning, which had no scope for moneymaking. The second student was taken with the author to Harvard toward his final stage of research. He gave up the work to go back and stay in the USA, and to get over the visa problem, he married the secretary of the Department. In the second study, 12 officers were sponsored from the State of UP on a Ford Foundation sponsored project. Nine refused to go beyond Masters Level because if they earned a Ph.D., they will not have any scope of making money! Most of the studies were undertaken by faculty from engineering colleges, working under the Government of India scheme of Quality Improvement of Faculty. Each study is a doctoral research study. Details are given in Chaturvedi (1987).

20. Attempt is being made to develop a collaborative applied research program by government agencies and the five Indian Institutes of Technology in the GBM basin, focusing on comprehensive development of GBM as proposed in the chapter.

21. The author had the privilege of starting his professional career under one of the most distinguished engineers of the time, Dr. A.C. Mitra, and working for long under his guidance on many major projects such as the Rihand Dam and Ramganga Dam and then joining him and the other leading engineers of the country such as Dr. A.N. Khosla, Dr. M.R. Chopra, and Dr. Kanwar Sain as Member of the Board of Consultants.

22. John Briscoe is an old associate and a good friend. He was a research scholar when the author first visited Harvard. He was paired up with the author to look after him. They have been interacting until the present, particularly as he is at Harvard and the author spends his summers at Harvard.

23. Arsenic has been detected in Bangladesh and is under study. The problem is further compounded by the fact that it appears that the arsenic problem permeates the entire GBM basin as high pollution has been identified in Bihar and Harayana. A new dimension of water management opens up. The arsenic problem is not related

to drinking water only. According to a recent report, research conducted at the University of Manchester has revealed that rice grown in West Bengal, which will be applicable to Bangladesh also, is likely to cause cancer on account of the presence of arsenic in the water. It was also reported that modifying the way rice is cooked could be a way to reduce arsenic exposure. Presence of naturally occurring petroleum in West Bengal, as in Cambodia, which also suffers from an arsenic problem, could be a reason for the presence of arsenic.

14

Total Integrated Development of GBM—Policy and Implementation

14.1 Introduction

We argued for a revolutionary concept of societal–environmental systems management. Second, we argued for an integrated perspective of development of the GBM basin over time and space, irrespective of the political boundaries, in terms of which implementation plan has to be worked out. We spell out these proposals briefly.

We propose that (1) integrated development of the GBM basin over space, irrespective of political international or national political boundaries, be undertaken; (2) integrated development of the GBM over time be undertaken, considering it at appropriate operational space; and (3) implementation be undertaken at the smallest spatial unit so that people are involved. We will demonstrate that, besides the scientific rationale, this is essential for the sustainable development of the region, is in the best interest of each riparian country, and has to be undertaken urgently in view of the enormous challenge, which will not be met otherwise, with disastrous socioenvironmental implications. We consider these ideas a bit in detail to work out how to foster their implementation.

Before formulating the scientific approach of integrated systems study, some specific issues pertaining to the two riparian countries, Nepal and Bangladesh, which represent two different environmental characteristics of the basin, are examined. These have been discussed in Chapters 6 and 7, respectively, but may be overviewed in the context of the proposed approach to develop integrated basin development. Nepal is considered in Section 14.3 and Bangladesh in Section 14.4. India contains regions featuring each of these two characteristic regions, besides representing the third characteristic region of alluvial plains and southern highlands. This has been considered in detail in Chapter 13 in the context of environmental systems management, which is extended to cover the entire GBM basin.

The current interaction between India and Bangladesh is essentially related to making up for the loss of water to Bangladesh on account of the transfer of Ganga waters to the western section of the delta by India, as discussed in Chapter 7. A brief assessment of the proposals of the two countries in this context is undertaken in Section 14.5.

The issue, however, is not merely of transferring water. A delta is a dynamic configuration, and its management should be seen in a wider perspective of dealing with this challenge. It is severely threatened by many problems, with climate change likely affecting it further in the most serious manner. Management from these perspectives is briefly discussed in Section 14.6, which has to be considered in terms of the integrated development of the GBM basin.

Based on the proposition that integrated development of the GBM basin is the best approach from a scientific perspective and based on political, humanistic, sustainability, and technoeconomic considerations, a policy of environmental systems development and management of the GBM basin integrally or the Greater Ganga, as it can be called, is developed in Section 14.7.

Corresponding to development, revolutionary changes have also to be brought out in management. These have been considered in detail in the context of the development of the waters of India (Chaturvedi, 2011c). These have been briefly presented in the context of management of the GBM basin, focusing on India, in Sections 13.11–13.13. These will also apply to the two other riparian countries.

It is understood that it will not be easy for this integrated development of the Greater Ganga to be undertaken. An approach that would contribute to its implementation is therefore proposed in Section 14.8.

14.2 Rationale for Considering Total Integrated GBM Development

Consideration of the total integrated GBM development as one socioeconomic–environmental system follows from several perspectives.

The environment does not understand and recognize man-made political divisions of space. The disturbing perspective of climate change brings out the central fact of spatial and temporal unity as far as the environment is considered. This is also brought out vividly in the case of water. Precipitation in one part leads to flowing water, which does not recognize man-made artificial boundaries. Any river basin is one integrated environmental entity. Consideration of river basins integrally is thus imperative based on scientific considerations.

The technoeconomic perspective echoes the same tune. The benefits of harnessing water for flood mitigation, generation of energy, provision of water, and mitigation of salinity intrusion can be best obtained through collaborative development. The novel technologies of the Ganga Water Machine, arising out of the unique geophysical hydrologic characteristics of the GBM basin, bring it out even more forcefully. It will be seen that the proposed revolutionary technologies, open a new perspective of development, not only for India but also for the riparian countries Nepal and Bangladesh. The novel technologies show that, paradoxically, the optimal sustainable development of the basin from the perspective of each of the riparian countries is synonymous with the optimal development of the basin considered as one integrated whole.

There is no question of sharing of water because there is abundant water and, for several months, too much water in the form of floods. The challenge is of sharing the effort for its most efficient development and management to the best of each person's capability in the best interest of all, equitably.

The third reason for integrated development is that it is in the interest of each country that each of the three countries gets out of the current world's most abysmal poverty and sets on a rapid path of socioeconomic development. Development of water is essential for it. It can be undertaken best, particularly in the context of the novel technologies in view of the integrated characteristics of the basin, if integrated development of the basin is undertaken, above the consideration of the political boundaries. Therefore, integrated development of the GBM basin is in the enlightened self-interest of each country.

It is difficult to convince policy makers and politicians of this simple truth. Advances have been made in the science of environmental systems management that enable scientists to present these facts in terms of participatory transparent models. The present analysis is a prelude to undertaking this scientific analysis.

14.3 Nepal—Characteristics and Challenges

Nepal is entirely in the Ganga basin in the Himalayas. The Himalayan region of the Ganga basin west of Nepal is in India and has been considered with the rest of it in Chapters 5 and 13, but the Ganga basin Himalayas are dominantly in Nepal.

As discussed earlier, a historical perspective of the development of Switzerland provides guidance and inspiration to the development of Nepal. Switzerland was once the poorest country in Europe but is now the richest country in the world. Scientific management of the environment has contributed significantly in this transformation. The difficult terrain of the mountain region was a big constraint but also provided a valuable resources. Hydroelectric development contributed much to it. The science of hydroelectric development was established in Switzerland.

A perspective of rapid transformation from abysmal poverty to the wealth of nations corresponding to the environmental characteristics, as for all characteristic environmental regions, has to be worked out. There are the immediate basic problems, common all over the GBM, of providing drinking water to the people, developing the infrastructure, providing water for agriculture, and conserving the land. They are most challenging in the hill areas.

Hydroelectric development has to be undertaken urgently. In view of the phenomenal benefits of the Chaturvedi India Water Power Machine, close collaboration of Nepal with India is in Nepal's interest. The development can be undertaken only through integrated spatial environmental development, totally losing sight of the political boundaries. The sharing of costs and benefits is a matter of wisdom and not of petty gains.

The development of the storage projects will contribute significantly to the mitigation of floods, which is beneficial to the downstream riparian countries, particularly Bangladesh, as it is the one suffering the most. An interesting characteristic of the Chaturvedi India Water Power Machine is that it will contribute additionally to the mitigation of floods, as the energy of these waters can be profitably transformed.

Management of land erosion is often overlooked, as it is perceived to be something concerning others in its impact. This is patently wrong. The wealth of Nepal depends on conserving the land resources and minimizing sedimentation in the reservoirs. Studying the subject in the context of a high dam that the author designed in the Himalayas, a paradoxical finding emerged, as discussed in Chapter 13. The return of investment on land management in the reservoir catchment was an order of magnitude higher than that of the construction of the high dam!

14.4 Bangladesh—Characteristics and Challenges

The GBM basin has one of the world's largest deltas. Paradoxically it was the most prosperous region of the world in a not too distant past. Currently, it faces the most serious environmental challenge that concerns the entire GBM basin. The development of the delta has been considered in detail in Chapter 7. Some essential aspects are presented briefly for easy reference, to highlight the issue of integrated GBM development in the interest of Bangladesh.

14.4.1 Characteristics

The region is intensely dominated by the impact of water. About 80% of the land is flood plains. Of the balance, 8% are terraces and 125 are hills. Two-thirds of the land is cultivated.

The water ecosystem comprises the tributaries and distributaries of the three major river systems—the Ganges, the Brahmaputra, and the Meghna—and the numerous perennial and seasonal wetlands like *haors*, *baors*, and *beels*. The quaternary alluvium of Bangladesh constitutes a huge aquifer with reasonably good transmission and storage characteristics.

However, a very serious problem has been noted of late. The groundwater is severely polluted by arsenic. This problem is gradually being noted all over the basin, but it is most severe in the delta region.

The hydrology is characterized by monsoonal features. The average annual rainfall varies from 1200 mm in the extreme west to over 5000 mm in the northeast, with 80% of the rainfall concentrated in the monsoon months, June–October.

Water has been considered to be significant in the culture and life of the region. Life had been concentrated in villages with few urban centers. The people had adjusted to the environmental conditions in terms of the agricultural activities. These are in terms of three cropping seasons, which coincide approximately with the three meteorological seasons: *Kharif I* (premonsoon), *Kharif II* (monsoon), and *Rabi* (winter). *Aus, aman,* and *boro* are the three rice varieties grown in these three cropping seasons, respectively. The yields are poor on account of poor water management.

The following particular hydrologic features result from the unique geographical situation of the Delta of which Bangladesh is the overwhelmingly dominant constituent. The figures relate to Bangladesh but are expressed in terms of the Delta because we are emphasizing that it is the problem of the GBM basin and not one country.

- All of the three major river systems originate outside the Delta, and the major catchments of these rivers (93%) are outside the Delta. It means that the entire GBM has to particularly take care of the impact of the activities therein on the Delta.

- Of the total annual stream flows in the Delta—85% occurring between June and October—about 67% is contributed by the Brahmaputra, 18% by the Ganges, and about 15% by the Meghna and other minor rivers. About 93% of the annual flows originate outside the Delta.

- The hydrographs of the main rivers have dominantly the monsoonal features. The peak discharges are reached in July/August, and the lowest are in March/April. The range between the average high and low flow is very high: about 20 in Brahmaputra and 30 in Ganges.

- The annual volume of surface water in the Delta would form a lake the size of the country and having a depth of 10.2 m.

- The Delta has to drain water from an area that is 12 times its size.

- One-third of the area in the Delta is influenced by the tides in the Bay of Bengal.

14.4.2 Problems of Water Resource Development

The critical physical/natural problems that the delta region encounters in developing a water management strategy are flooding, arsenic management, water scarcity, riverbank erosion, sediment/salutation, salinity management, and threats from the climate change. They are interrelated; they are considered individually but will then be considered in totality as their management is integrated.

14.4.2.1 Flood Management

Floods are a recurring phenomenon in Bangladesh. Even in a normal year, 20%–30% of the country is flooded. About 60% of the country is submerged in flood, with an approximately 100-year return period. Up to 80% of the country is considered flood prone.

Flooding in Bangladesh is the result of a complex series of factors. These include a huge inflow of water from upstream catchment areas coinciding with heavy monsoon rainfall in the country, a low flood plain gradient, and congested drainage channels; the major rivers converging inside Bangladesh; tides and storm surges in coastal areas; and polders that increase the intensity of floodwaters outside protected areas. Different combinations of these factors give rise to different types of flooding.

14.4.2.2 Arsenic Problem

The arsenic problem is encountered globally, but it is extremely variable depending upon the dependence on groundwater. For example, there is more arsenic in the groundwater of Massachusetts, USA, than in Bangladesh, but they are not dependent on groundwater for domestic use and for agriculture (Chaturvedi, 2009). It has very high importance in Bangladesh because of the critical dependence of modernization of agriculture on groundwater.

14.4.2.3 Water Availability

Estimates of water supply and demand have been made in Bangladesh. It is considered that the minimum dry season availability is 3.7 billion cm^3 and the maximum availability in August is 111.25 billion cm^3. The estimates of dependable groundwater availability are tentative but are in the order of 21 billion cm^3. There are hardly any storage opportunities.

As in the Ganga basin, it was estimated by the World Bank that extensive artesian groundwater is available. However, they have become silent on the subject of late.

14.4.2.4 Riverbank Erosion

Most of the rivers of Bangladesh flow through unconsolidated sediments of the GBM flood plain and delta. The riverbanks are susceptible to erosion by river current and wave action. River erosion includes channel shifting, the creation of new channels during floods, banks slumping due to undercutting, and local scour from turbulence caused by obstruction. The rivers of the GBM basin flow within well-defined meander belts on extensive flood plains where erosion is heavy. Sudden changes are common during floods that cause rapid bank erosion. In the lower deltaic area, riverbank erosion is caused by tidal currents and storm surges from the sea.

Studies have shown that riverbank erosion is a very serious problem. For example, the Brahmaputra–Jamuna has changed course completely after 1762. The Bangladesh Water Development Board (BWDB) has estimated that about 1200 km of the riverbank is actively eroding and that more than 500 km faces severe problems related to erosion. GBM rivers show that a net area of 87,000 ha was lost within a decade, 1982–1992, most of it being agricultural land, equivalent to an annual erosion rate of 8700 ha.

14.4.2.5 Sedimentation

Any delta is the outlet of all the major upstream rivers, and the very creation of the delta is a consequence of sedimentation. A quasi-stable state is generally reached, but due to floods and upstream human activities, changes keep on taking place.

It has been estimated that the annual sediment load that passes through the country to the Bay of Bengal ranges between 0.5 billion to 18 billion. The problem arises because the rivers have one of the world's highest and youngest mountains, which experience very heavy monsoon precipitation, leading to the world's biggest alluvial lands, where the naturally occurring forests have been cleared and the land has been put under agriculture, leading to the world's biggest delta. The processes also bring out that management has to be considered based on the totality and integrity of geophysical management and agricultural activities.

The impacts of sedimentation are varied. The problem led to the construction of the Farakka Barrage. Its implications may not be as severe in other parts of the region as in the case of activities of a port, but they have to be duly managed.

14.4.2.6 Salinity Problem

Salinity is a major problem in deltas and is particularly severe in the large alluvial delta of the GBM basin. The coastal zone directly affected by salinity is extensive and is inhabited by a large population.

The rivers of the delta combine to form a single, broad, and complex estuary. The greatly diminished dry flow in the dry season allows salinity to penetrate inland through this estuarine river system. It adversely affects the agricultural activities and water availability. It creates diverse environmental problems.

14.4.2.7 Climate Change

Climate change is a serious threat to the entire humanity. It will affect the world in many ways such as rise in temperature, change in precipitation, melting of glaciers, and rise of the sea levels. The entire GBM basin will be very severely affected, but the Delta and Bangladesh face the most severe problem of sea rise acutely. Climate change implications are extremely serious, but, unfortunately, these have not been given due attention so far.

14.4.3 Management of Delta

A perspective of rapid transformation from abysmal poverty to the wealth of nations corresponding to the environmental characteristics, as for all characteristic environmental regions, has to be worked out. There are first the immediate basic problems, common all over the GBM, of providing drinking water to the people, developing the infrastructure, providing water for agriculture, and conserving the land. In addition, there are the following specific problems (Ahmad et al., 2001; WB, 1989, 1998).

14.4.3.1 Flood Management

Flood management received considerable international and national attention in the aftermath of the 1987 and 1988 floods. Varied and novel approaches of developing embankments have been suggested. However, a final agreed upon and scientifically well-tested approach has not been developed.

The management plan has to be embedded in flood management for the entire river basin. For example, management of Brahmaputra and Meghna floods in the upstream reaches are entirely within the jurisdiction of India. It will be in India's interest to give highest emphasis to flood management. However, it has been neglected so far. Further collaboration with the deltaic region is in terms of not only data, as generally focused, but also integrated flood management. There should be no difficulty in this respect.

14.4.3.2 Arsenic Problem

The arsenic problem is extremely serious. The problem is not merely of drinking water but of the implication of using arsenic waters in irrigation and its corresponding impact on the produce. Some alarm has already been raised. As has been said, the delta region is dealing with poison. Unfortunately, the subject has not been given the attention it deserves.

14.4.3.3 Water Supply–Demand Management

From the preliminary studies made by the author, it appears that the Bangladesh authorities feel reasonably comfortable with the subject. Comments cannot be made in the absence of study of the subject, but the author feels very worried on account of the arsenic problem. Secondly, the subject requires a much higher level of analysis than is currently bestowed.

14.4.3.4 Riverbank Erosion, Sedimentation, and Salinity

These are classic hydraulic problems that can be analyzed with an increasing level of sophistication with advances in the field.

14.4.3.5 Climate Change

Management of the problem is an international issue, but measures to deal with the impact should be undertaken immediately.

14.5 Assessment of the Current Official Proposals

Distribution of water in the Delta is imbalanced. This follows from the configuration and characteristics of the two major rivers, Ganga and Brahmaputra. The configuration is not static but is dynamic, which has been changing over time and will keep on changing. This applies to all parts of the GBM basin, being more pronounced in some places than others. The meandering of River Kosi over a hundred miles is perhaps the most severe.

This imbalance started affecting the port of Calcutta and raised particular concern. The solution adopted was the construction of the Farakka Barrage. There have been controversies about it, as is usual with all major engineering activities. For example, the development of irrigation in India through the pioneering construction of the Ganga Canal by Cautley in 1854 led to bitter criticism from the leading water scientist of the time, Sir Cotton, leading to the impeachment of Cautley and his later redressal and honoring by the Royal Scientific Society.

The differences in opinion about the proposals by the Indian officials and the Bangladeshi officials have been intensified by the political circumstances and have led to the current impasse in the scientific development of the GBM basin. Unfortunately, the proposals have not been scientifically discussed and have been vested with national honor. This violates the basic principles of science and leads to all the trouble. Therefore, we will briefly examine the proposal. The analysis is undertaken because there is a fundamental error in formulating the proposals and a scientific approach leads to a very simple solution in the best interest of scientific development of the GBM basin and in the best interest of both countries.

The proposals and response have been discussed in Chapter 9. They are again briefly presented.

14.5.1 Indian Proposal

The Indian proposal, in principle, consists of transferring water from Brahmaputra to Ganga upstream of the Farakka Barrage.

Anybody who has done real-life engineering in this area will understand that it is sheer folly. Construction of such a major canal in the foothills of the Himalayas, transversing numerous meandering hill channels with mighty floods, is contrary to all engineering sense and will be prohibitively expensive. Study of the 5000 cusec canal transferring Ramganga waters to Ganga, designed by the author, taught the author about the difficulty and the technoeconomic absurdity of the proposed Brahmaputra–Ganga link. Reference may be made to Dr. K.L. Rao's decision to transfer water from Ganga after it had been charged by the northern tributaries Ghagra and Gandak, rather than transferring water from Brahmaputra as currently proposed in the Interlinking of Rivers proposal. This is another testimony to the impracticality of the proposal (Rao, 1975). An interesting experience gained by the author was that the cost reduced rapidly as the alignment was shifted further south of the Himalayas, a characteristic that we will use later to devise another scheme for water transfer.

14.5.2 Bangladesh Proposal

The Bangladesh proposal is another example of technical naiveté. Any undergraduate student of engineering knows that the height of a dam is set based on certain cost–benefit considerations and that any further increase is exponentially prohibitive economically, introduces serious safety considerations, and is increasingly unacceptable environmentally. Secondly, it introduces a third party that has no interest in the matter and does not stand to gain anything from it. Thirdly, granted that it is agreed upon, its implementation will take a millennium, besides being prohibitively expensive and risky in the seismic active Himalayas.

14.5.3 Assessment of Indian and Bangladesh Proposals and Out Perspective

The only rationale of the proposals is to score a political point against the other party. In contrast, we are emphasizing that the matter should be considered scientifically as a challenge to a most distressed region of the world, socioeconomically and environmentally, which is one cultural and environmental entity.

14.6 Dealing with Dynamics of Delta Configuration

Ganga and Brahmaputra are two great hydraulic forces. The Delta is a byproduct of matter—solids and water—in this interaction. It is not a static permanent configuration. Increasingly, man-made infrastructure for habitat, transportation, and flood control will be built on it. It is important that while developing the infrastructure, the dynamic nature of the delta should be kept in view. Further, the issue is on how we can deal with this basic characteristic.

A connecting Brahmaputra–Ganga link, as downstream as appropriate, because it will be increasingly economical the more downstream it is, from considerations of interregional dynamics management, appears to confer some hydraulic control on the system. It will have another great benefit. It will provide a line of high potential, which may contribute to dealing with the arsenic problem.

The internal Bangladesh proposal of transferring Brahmaputra waters to Ganga was found to be technically and economically feasible (Crow, 1995). It may be developed in the perspective of not merely transferring water but of managing the dynamic complex, as proposed. With the modern scientific advances, it is possible to study the problem and come to a reasonable solution. Until then, it is just a suggestion. But the conclusion is that this problem should be considered in a proper perspective and not as an issue of just making India compensate for the waters transferred at the Farakka Barrage.

14.7 Scientific Management of GBM— Environmental Systems Perspective

Thus, we look at the development of the GBM basin integrally, or the Greater Ganga basin as it may be more scientifically called, in terms of the scientific development in accordance with the basic principles of environmental systems management.

Several encouraging features may be brought out. One, as the physical features are optimal, scientific management of the water is beneficial to all, irrespective of the political boundaries. For instance, flood mitigation and land erosion are beneficial at the local as well as the downstream level. Development of water resources during the lean season for irrigation leads to increased return flows for the downstream needs. Hydroelectric development can be undertaken best by integrating it with the thermal components, which are further enhanced through pumped storage, even as conventionally undertaken, and even more so in terms of the Chaturvedi India Water Power Machine.

Two, the proposed novel technologies encourage increased water availability and regional interlinkage. For example, the Chaturvedi Water Machine minimizes high monsoon flows and enhances the total water availability, leading to increased low flows in the nonmonsoon period, to the benefit of the entire basin. With the policy of maintaining low flows in the river from environmental considerations, low period flows in the delta region are bound to be enhanced. The Chaturvedi Water Power Machine also contributes to mitigation of floods and enhancement of low flows, besides generation of valuable peaking power for the basin.

Third, the GBM configuration is such that development of each of the three rivers, Ganga, Brahmaputra, and Meghna, can be carried out in the context of national development and integrated in terms of the Delta. The development of Nepal and India is synonymous and has to be carried out integrally. It is even more so in the context of the Chaturvedi Water Power Machine. The development of the upstream regions of Brahmaputra and Meghna to manage the beneficial and harmful impacts of the abundant waters is particularly beneficial to the lower regions. There is considerable surplus water for the needs of the entire basin.

We adopt the scientific indicators developed in Table 13.2, which lead to the development of the scientific policy for the GBM basin considered from the upland regions, and modify it by taking into consideration the challenges presented by the Delta, to lead to Table 14.1, the scientific policy for the Greater Ganga.

TABLE 14.1

Scientific Development Policy of the Greater Ganga Basin

1. Undertake storage developments all over the basin phased in terms of flood mitigation in the Delta basin, on the basis of comparative priority of possible completion and comparative cost–benefit analysis. Storages on Brahmaputra, as proposed in the Indian proposal, will have higher benefits in terms of flood mitigation and hydropower generation because of higher discharges, but benefits on irrigation may be comparatively lower. The possibility of their earlier completion also appears more feasible. All storages are to be developed in terms of the India Water Power Machine to obtain their maximum technoeconomic–environmental benefits.
2. For enhancement of flows in Bangladesh, transfer from Brahmaputra appears to be more reasonable, economical, and having the possibility of earlier implementation than transfer from Ganges. The Indian proposal from Jogipatta appears to be techno-economically infeasible in view of the cross drainages to be negotiated. The lower the interbasin transfer, the better its technoeconomic feasibility. Feasibility of the Bangladesh alternative proposal is much better. It has also the advantage that it integrates the delta meaningfully and provides high potential for the groundwater flow, which may have positive impact on arsenic characteristics.
3. Development of Ganga, Brahmaputra, and Meghna to be undertaken as brought out in Table 13.2 from consideration of planning, developing each river basin individually with due consideration of integrated management of the delta and further integrated in terms of the Delta. Development of the delta region to be undertaken as currently proposed by the scientists and authorities concerned.
4. The biogeochemical cycles aspect has not been given due attention so far. The arsenic problem in the delta region is extremely serious and needs urgent attention. But the cycle problem has to be investigated all over the region, as there have been reports of arsenic problem in the mid Ganga basin and western parts of the Yamuna basin. The biogeochemical cycle is much more than merely its arsenic manifestation.
5. The issues at the water's edge have generally not been given due attention.
6. Climate change is a certainty. Its contours are uncertain. Its implications must be studied urgently and adaptive measures started to be contemplated and implemented. While the impacts will be all over the basin, the Delta region will be worst affected, and adaptive and mitigating measures should start to be undertaken at the earliest.

As stated at the outset, the suggestions are tentative and should be studied by scientists from the three countries collaboratively in terms of the perceptions of one people and one region as proposed. With the advancement of modern technology, transparent participatory models can be developed. The challenge is to contribute to science as well at the frontiers. Very specifically, no foreign so-called expert or agency should be involved.

14.8 New International Perspective of GBM Basin

In developing the GBM basin, it has to be considered as one environmental–socioeconomic entity, irrespective of the political divisions, with the efforts perceived as one humanitarian and environmental activity. It is in the overall socioeconomic–environmental interest of each country, as it will contribute to peace and stability, besides the direct economic benefits. In view of the abysmal poverty and backwardness, the serious environmental threats, and the context of the overwhelmingly large population, it has to be undertaken most urgently, with the highest commitment.

Thus, the following conclusion emerges. The development of the GBM may be considered in terms of three physical configurations—Ganga, Brahmaputra, and Meghna, considering in each the totality of the configuration, which is in terms of the Himalayan plains, central highlands, and delta in differing degrees. The three physical configurations may, in addition, be considered independently to further enlarge the totality of the development and the Delta. It may then be integrated to give the optimal development of the totality. Scientific analysis of the development of the GBM basin should be carried out in terms of environmental systems management backed by transparent participatory system dynamic models. The analysis may be carried out by each country working independently and collaboratively, or as one team. The results will be the same. The science and tools are available (Chaturvedi, 2011a, 2011c).

14.9 Implementation of Integrated Development of GBM Basin

The implementation of the study of integrated development of the GBM basin will not be easy. Some earlier experiences may therefore be taken note of, and an appropriate policy may be developed.

The Union Minister of the Government of India decided as early as 1969 that systems studies of the GBM be undertaken by the scientists of IIT Delhi and IIT Kanpur, led by the author. A three-month short course was organized to which officers from the Government of India were deputed. But despite the Minister's personal involvement, it could not be undertaken on the plea that the data is classified.

Scientists had been sponsored to work at Harvard University on the problems of India, but they were just scientific studies (Chaturvedi and Rogers, 1985). The Ford Foundation supported the development of capability in the profession in India, and a comprehensive applied research river basin level developmental planning study was carried out, as has been briefly described in Chapter 13 (Chaturvedi, 1987).

Efforts have been made to contribute to the formulation of a scientific policy of developments by knowledgeable persons from the three countries working together collaboratively on problems of the Ganga basin by Jagat S. Mehta, former Foreign Secretary, Government of India. Mehta and the author had been Visiting Professors at the University of Texas at Austin, USA. As a preliminary activity, a workshop involving knowledgeable persons from the three countries and some other international experts was organized by David Eaton in 1986 in Australia as part of the XXVII Conference of the Institute of Management Sciences (Panandikar, 1992). However, the studies could not be undertaken, as there was some objection from the officials of the countries directed to the Ford Foundation (Eaton, 1992).

George Verghese, an eminent journalist, was also invited to the workshop, as he had been interacting with the author regarding the development of the GBM basin for several years in India and had also been covering the Indo–Bangladesh disputes. Another workshop was organized by Verghese and Iyer from the Centre for Policy Studies and representatives from Bangladesh Unnayan Parishad (BUP) Dhaka and the Institute of Integrated Development Studies (IIDS) Kathmandu in New Delhi. Verghese and Iyer obtained inputs from a number of persons in India (including Jagat Mehta and the author) and brought out a contribution. Similar studies were brought out by BUP and IIDS. An integrated study was also compiled. The exercise was repeated in 2001 by the same institutions. These are valuable scientific studies but are also informative documents for the general public, as the authors themselves have emphasized.[1]

The implementation of the integrated development of the GBM basin will not be easy. Initiative has to be taken by the scientific community. Attempts are being made for the Indian scientific community and governmental agencies to work together in contributing to the scientific development of the GBM basin in terms of modern advancements. Hopefully, such efforts could be made to the extent of having some scientists in India undertake a scientific study of the integrated GBM basin and invite scientists from the two other riparian countries for collaboration. Initiative has to be taken by the Indian scientists because India is the dominant riparian.

Some efforts have been made by the author. He met the Ambassador of Nepal, Dr. Durgesh Mansingh. He was extremely knowledgeable and enthusiastic about the matter. The author also met the Ambassador of Bangladesh. The subject was the theme of the author's talk at his invitational lecture during World Water Day on March 22, 2009, spearheaded by the Ministry of

Water Resources, Government of India. Collaborative development is scientific development in the interest of all political entities from socioeconomic and sustainability considerations.

14.10 Conclusion

From scientific, technological, and sociopolitical considerations, it is essential and in the best enlightened interest of each country that collaborative development of the environment, particularly the waters, is undertaken. As the challenge of meeting the threat of climate change has been brought out, the entire humanity will have to face the challenge collectively. Similarly, management of the waters of the GBM has to be undertaken collectively, as the study has brought out. It has to be undertaken urgently in view of the most serious socioeconomic–environmental implications. Sadly, it has been totally neglected. It is difficult to implement it because of the usual political prejudices. Hopefully, the Indian academic community takes the initiative and pursues the matter urgently with commitment.

Notes

1. A collaborative scientific study of the GBM basin that was proposed to be undertaken by scientists from the three countries was done in 1989 under the guidance of the author; Prof. Jagat Mehta, one time Foreign Secretary, Government of India, who was working at the University of Texas at Austin; and Prof. David Eaton, who was a University of Texas professor in this area. The funding for the study was authorized by the Ford Foundation. As a startup, a conference of academic scholars and practitioners from Bangladesh, India, and Nepal was organized to examine the substantive issues as part of the XXVII Conference of the Institute of Management Sciences, on July 21–23, 1986. The papers have been brought out in Eaton (1992). Unfortunately, it did not materialize.

2. Representatives from the three countries have met and exchanged ideas as reported in Verghese and Iyer (1992), Ahmad et al. (1994), Thapa and Pradhan (1995), and Ahamad et al. (2001). However, as the organizer of one set of these studies clearly stipulated, "these studies are addressed to the lay reader and are not intended to be technical treatises" (Panandikar, 1992).

15

Conclusions

Development of water is crucial in the Indian subcontinent on account of the hydrologic–climatic considerations. It has been undertaken throughout history. Very large-scale development of the water resources development has taken place after Independence, particularly in India. The GBM basin is the centerpiece of India.

Besides the hydrologic–climatic factors, India faces a very serious challenge of development of water on account of two additional factors. One is the large increase in population and economic development and the corresponding impact on the environment in terms of the cycle of the demands and return flows. Second is climate change.

There has been consciousness of the importance and urgency of development. A National Commission on Integrated Development Policy (NCIWRDP, 1999) was established by the Government of India. It has given detailed directions. One of the recommendations is the interlinking the rivers of India. It has two components—the Himalayan Component and the Peninsular Component. The GBM basin is the kingpin of development of the Himalayan Component. The Peninsular Component is also considered as being supplied water from the GBM basin. The development of India's water, according to the NCIWRDP (1999), is considered in isolation with that of Nepal. In view of disputes about sharing of GBM waters between Indian and Bangladesh, each country has developed proposals for its development. There is continued conflict about it.

A new activity of management of water has been in terms of the National Water Mission as part of the Prime Ministers' Committee on Climate Change.

Study of the development of India's waters by the established departments at the State or the Center or these new missions brings out that the development policy has been an extension of the historic policy, which was developed to minimize the vagaries of the weather, mitigate famines, and not, essentially, to improve the productivity. An important fact, which is totally ignored, is that the development of the waters of India is very poor in socioeconomic–environmental terms. The agricultural yields are one of the poorest in 93 developing countries, excluding China, the yields in China being almost twice and in industrial countries thrice of those in India (WR, 1986–1987). Further, the water per unit of land irrigated is one of the highest even in developing countries and more than twice that of the developed countries. Development has been even worse in the two countries of the basin—Nepal and Bangladesh.

We have studied the development of the GBM. Going beyond the historic and current development, we have demonstrated that, on account of the unique physiographic and hydrologic characteristics, several novel technologies are possible, which almost double the water availability and put the water in the hands of the farmer. They increase the hydroelectric potential several fold and promote collaborative development between the countries of the region, as they immensely increase the returns and environmental sustainability of each country. Collaborative development is inherent on account of the integrity of the environmental system, which does not know political divisions. Collaborative international development is essential for implementing the revolutionary technological changes proposed, which will vastly increase the water and energy potential for each country. Thus, considering the development of the GBM basin in an international perspective is very important for each constituent country. They also vastly contribute to meeting the dangers of the climate change.

Development of the Himalayan region is crucial for the development of the GBM waters. This will depend crucially on Nepal. Unfortunately, this has been badly lacking. Two factors are particularly responsible for this sad state of affairs: one, Nepal's poor socioeconomic state, and two, Nepal's psychological distrust of India. India will have to take the initiative, very carefully, to bring out the tremendous opportunities that Nepal has, in her interest, in undertaking her developmental potential, because it can be undertaken only through collaborative development with India.

It must be emphasized that the development and management of the water resources require the socioeconomic–cultural advancement of the society, particularly the farmers. We have concentrated only on the macrolevel aspect because of the issue of technology. There is immense possibility of improving potential, economic returns, and sustainability of the water sector by advancing the use aspect. But that will require even greater emphasis on the development of the capability of the user.

Another aspect that must be emphasized is that the countries of the GBM have to develop and manage their water resources themselves. The so-called experts from foreign countries should not be even allowed to meddle in our affairs because they hardly know our conditions of development, and it is against our indigenous development interests.

As modern research has demonstrated, above technology, the central issue is "culture" (Lands, 1998). The "culture" of the society has to be changed completely. Presently, while on one hand the poor farmer is considerably handicapped in efficient use of water on account of the socioeconomic conditions, the politicians and the bureaucracy are more concerned about making money than in the development of water or of the country. A revolution is urgently required in all aspects of water resources development and management. This is the central issue. The study is only an introduction to foster these activities and culture, as part of the author's long involvement in the development of India's waters, particularly of the GBM basin.[1] It appears that some fruits have started growing.[2]

Notes

1. The author's background is given as an annex so that the perspective of the observations is brought out. The author has designed all the major dams of the GBM basin and later became a Member of the Board of Consultants for the development of the GBM basin, the highest body of the Government of India in this context. Foreign scientists have been interested in studying and contributing to the development of the Ganga basin for a long time. The Harvard University water group, led by Prof. Roger Revelle, a former member of President Kennedy's Cabinet, has been particularly interested in view of its involvement. The author had a long collaborative association with the Harvard group, which continues today, emphasizing all the time that the leadership rests with us.

2. On the recommendation of the author, the Ministry of Water Resources and the Ministry of Environment and Forests, Government of India, each have sanctioned projects under which scientists from the nine Indian Institutes of Technology (IIT) and sponsored engineers from the GBM basin official departments are working together on topics of interest, as applied research activity at the highest level.

   The University of Iowa, Iowa City, Iowa, USA, the author's alma mater, is undertaking a detailed study of the Ganga Basin through a team of research scholars, which is proposed to be followed by detailed collaborative research to contribute to the art and science of the subject.

   The author has devoted whatever it is that is needed to the development of the country, establishing the Vipula and Mahesh Chaturvedi Foundation. A Professorial Chair is established at the IITs, in accordance with the recent policies and terms of the IITs, every year. We have established one in the area of Policy Studies at IIT Delhi; trying to establish one at IIT Roorkee, which is the oldest engineering institute in India and the author's alma mater, to contribute to the development of GBM; and one at the IIT Kanpur, where the author was the Founding Head of the Department of Civil Engineering.

References

Chapter 1

Agenda 21. 1992. United Nations Conference on Environment and Development, Rio de Janeiro, Brazil, June 3–14, 1992.

Ahmad, Q.K., A.K. Biswas, R. Rangachari, and M.M. Sainju (Eds.). 2001. Ganges–Brahmaputra–Meghna Region—A Framework for Sustainable Development. The University Press Limited, Dhaka.

Boserup, E. 1965. *The Conditions of Agricultural Growth*. Aldine Press, Chicago.

Boserup, E. 1981. *Population and Technological Change. A Study of Long Term Trends*. Chicago University Press, Chicago.

Chaturvedi, M.C. 1976. *Second India Studies—Water*. Macmillan, New Delhi.

Chaturvedi, M.C. 1987. *Water Resources Systems Planning and Management*. Tata McGraw Hill, New Delhi.

Chaturvedi, M.C. 2001. Sustainable development of India's waters—some policy issues. *Water Policy*, 3, 297–329.

Chaturvedi, M.C. 2011a. *Societal Environmental Systems Management* (in process of publication).

Chaturvedi, M.C. 2011b. *India's Waters: Environment, Economy and Development*. Taylor & Francis, Boca Raton, FL.

Chaturvedi, M.C. 2011c. *India's Waters: Advances in Development and Management*. Taylor & Francis, Boca Raton, FL.

Daly, H. 1993. The perils of free trade. *Scientific American*, Nov., 50–57.

Damodaran, A. 2010. *Encircling the Seamless*. Oxford University Press, New Delhi.

Galbraith, J.E. 1989. *The Nature of Mass Poverty*. Pelican, London.

Kumar, M.D. 2010. *Managing Water in River Basins*. Oxford University Press, New Delhi.

Landes, D.S. 1998. *The Wealth and Poverty of Nations—Why Some Are So Rich and Some So Poor*. W.W. Norton, New York.

Pasucal, U., A. Shah, and J. Bandopadhyay (Eds.) 2009. *Water, Agriculture and Sustainable Well-Being*. Oxford University Press, New Delhi.

Pai Panandikar, V.A. 1992. Foreword. In Verghese, B.G. and R.R. Iyer (Eds.). *Harnessing the Eastern Himalayan Rivers*. Konarak Publishers, New Delhi.

Ruttan, V.M. 1987. *Lectures on Technological and Institutional Change in Agricultural Development*. Pakistan Institute of Development Economics, Islamabad.

Spear, P. 1978. *A History of India*, Vol. 2. Penguin, London.

Thapar, R. 1977. *A History of India*, Vol. 1. Penguin, London.

UNDP. 2006. *Human Development Report*. Oxford University Press.

Vaidyanthan, A. 1999. *Water Resources Management—Institutions and Irrigation Development in India*. Oxford University Press, New Delhi.

Vaidyanathan, A. 2010. *India's Water Resources*. Oxford University Press, New Delhi.
Verghese, B.G. 1999. *Waters of Hope*, Second Edition. Oxford and IBH Publishing Co. Pvt. Ltd, New Delhi.
World Bank. 2010. *Annual Report*. Washington, D.C.

Chapter 2

Bhatia, B.M. 1967. *Famines in India*. Asia Publishing House, Bombay.
Bruijzneel, L.A. and C.N. Bremmer. 1989. Highland–lowland interactions in the Ganges–Brahmaputra river basin: A review of published literature. ICIMOD Occasional Paper No. II, Kathmandu.
Carson, B. 1985. Erosion and sedimentation processes in the Nepalese Himalaya. ICIMOD Occasional Paper No. 1, 15. International Centre for Integrated Mountain Development, Kathmandu.
Chaturvedi, M.C. 1987. *Water Resources Systems Planning*. Tata McGraw-Hill, New Delhi.
Coleman, J.M. 1969. Brahmaputra River: Channel processes and sedimentation. *Sedimentary Geology*, 3, 129–239.
Curray, J.R. and D.G. Moore. 1971. Growth of the Bengal deep-sea fan and denudation in the Himalayas. *Geological Society of America Bulletin*, 82, 563–572.
Das, H.P., D.K. Singh, and H.N. Sharma. 1987. Meghalaya–Mikir Region. In R.L. Singh (Ed.), *India, a Regional Geography*. National Geographical Society of India, Varanasi.
Dhar, O.N., B.N. Mandal, A.K. Kulkarni, O.N. Dhar, M.K. Soman, and S. Mulye. 1982. Distribution of rainfall in the Himalayan and Sub-Himalayan regions during "breaks" in the monsoon. In *Proceedings of the International Symposium on Hydrological Aspects of Mountainous Watersheds*, Roorkee, Nov. 4–6, 1982, Manglik Prakashan, Saharanpur, pp. I-22–26.
Dutt, D.K. 1992. *Groundwater Development: Potential and Possibilities*. Centre for Population Studies, New Delhi. (mimeo)
Encyclopaedia Britannica. 1984. Himalayan Mountain Ranges. Encyclopaedia Britannica, Chicago.
Galay, V. 1987. Erosion and sedimentation in the Nepal Himalaya. An assessment of river processes. Ministry of Water Resources, HMG, Nepal, Report No. 4/3/010587/1/1 Seq. 259.
Gansser, A. 1961. *Geology of the Himalaya*. John Wiley, New York.
Gole, C.V. and S.V. Chitale. 1966. Inland delta building of Kosi River. *Journal of the Hydraulics Division, ASCE*, 92 HY2.III-126.
Habib, J. 1992. The geographical background. In T. Rayachaudhri and J. Habib (Eds.), *The Cambridge Economic History of India*, Vol. 1. Orient Longman, New Delhi.
Haroun-er-Rashid. 1977. *Geography of Bangladesh*. University Press Ltd, Dacca.
Holeman, J.N. 1968. The sediment yield of major rivers of the world. *Water Resources Research*, 4, 737–747.

Jangpungi, B.S. 1978. Stratigraphy and structure of Bhutan Himalaya. In P.S. Saklani (Ed.), *Tectonic Geology of the Himalaya*. Today and Tomorrow's Printers and Publishers, New Delhi.

Hingham, A.G., V.C. Thakur, and S.K. Tandon. 1982. Structure and tectonics of the Himalaya. In *Himalayan Geology Seminar*, G.S.I., Miscellaneous Publications, No. 41, Pt. III.

Kalayanaraman, S. 2000. Status of research and revival of River Saraswati. In *Proc. International Conference on Sustainable Development of Water Resources*, Institute for Resource Management and Economic Development, New Delhi, Nov. 27–30, 2000.

Kaul, O.N. 1992. *Forests and Foresting in the GBM Basin*. Centre for Policy Research, New Delhi. (mimeo)

Klaassen, G.J. and K. Vermeer. 1988. Channel characteristics of the braiding of Jamuna River, Bangladesh. In W.R. White (Ed.), *International Conference on River Regime*, pp. 173–189. Wiley and Sons, New York.

Millikan, J.D. and R.H. Meade. 1983. World-wide delivery of river sediment to the oceans. *The Journal of Geology*, 91, 1–21, 1983.

Molnar, P. 1986. The structure of mountain ranges. *Scientific American*, 225, 70–79.

Mooley, D.A. and B. Parasarthey. 1983. Droughts and floods over India in summer monsoon seasons 1871–1980. In A. Street-Perrot et al. (Eds.), *Variations in the Global Water Budget*, pp. 239–252. D. Reidel, Dordrecht.

Morgan, J.P. and W.G. McIntire. 1959. Quarternary geology of the Bengal Basin, East Pakistan and India. *Bulletin of the Geological Society of America*, 70, 319–342.

Pal, S.K. and K. Bagchi. 1975. Recurrence of floods in Brahmaputra and Kosi Basins: A study in climatic geomorphology. *Geographical Review of India*, 37, 242–248.

Parpola, A. 1994. *Deciphering the Indus Script*. Cambridge University Press, Cambridge.

Ramaswamy, C. 1962. "Breaks," in the Indian summer monsoon as a phenomenon of interaction between the easterly and subtropical westerly jet streams. *Tellus*, 14, 337–349.

Rao, Y.P. 1981. The climate of the India subcontinent. In K. Takahashi and H. Arakawa (Eds.), Climates of Southern and Western Asia. *World Survey of Climatology*, Vol. 9, pp. 67–182. Elsevier, Amsterdam.

Rogers, P., P. Lyden, and D. Seckler. 1989 (April). *Eastern Waters Study*. Office of Technical Resources, Agriculture and Rural Development Division, Bureau for Asia and Near East, U.S. Agency for International Development, Washington, D.C.

Sharma, C.K. 1977. *River Systems of Nepal*. Mrs. Sangeeta Sharma, Kathmandu.

Sharma, C.K. 1985. *Water and Energy Sources of the Himalayan Block*. Sangeeta Sharma, Kathmandu.

Shrestha, T.B. 1988. *Development of Ecology of the Arun River Basin in Nepal*. International Centre for Integrated Mountain Development, Kathmandu.

Singh, R.L. (Ed.). 1987. *India, a Regional Geography*. National Geographical Society of India, Varanasi.

Singh, V. and R.V. Verma. 1987. Upper Ganga Plain. In R.L. Singh (Ed.), *India, a Regional Geography*. National Geographical Society of India, Varanasi.

Valdiya, K.S. 1996. River piracy, Saraswati that disappeared. *Resonance*, 1, 19–28.

Wadia, D.N. 1975. *Geology of India*. Tata McGraw Hill, New Delhi.

Chapter 3

Chaturvedi, V. and M.C. Chaturvedi. 1985. Development, technology and education. *Proceedings of the 12th Comparative Education Society of Europe*, Antwerp, July 1985.

Chaturvedi, V. and M.C. Chaturvedi. 1994. Educational and developmental policies of India. *International Perspectives on Education and Society*, Vol. 4, pp. 247–267. JAI Press Inc., USA.

Encyclopaedia Britannica. 1984. Himalayan Mountain Ranges. *Encyclopaedia Britannica*, Chicago.

Myrdal, G. 1971. *Asian Drama* (Abridgement by Seth S. King). Allen Lane, The Penguin Press, London.

Nehru, J. 1947. *Discovery of India*. The Signet Press, Calcutta.

Rothermund, D. 1988. *An Economic History of India*. Cromm Helm, London.

Sagisti, F.R. 1979. *Technology, Planning and Self-Reliant Development*. Prager, New York.

Spear, P. 1978. *A History of India*, Vol. 2. Penguin, London.

Thapar, R. 1977. *A History of India*, Vol. 1. Penguin, London.

Toynbee, A.J. 1947. *A Study of History* (abridgement of Vol. I–VI by D.C. Somerville). Oxford University Press, New York.

Vohra, R. 1997. *Making of India*. M.E. Sharp, New York.

World Bank. 1991. *India—Irrigation Sector Review*, Vol. 1 and 2. Agriculture Operations Department, India Department, Asia Region.

Chapter 4

Bhatia, B.M. 1967. *Famines in India*. Asia Publishing House, Bombay.

Brown, J. 1978. Sir Proby Cautley (1802–71). A pioneer of Indian irrigation. *History of Technology*, 3, 35–89.

Buckley, R.B. 1880. *The Irrigation Works of India and Their Financial Results*. W.H. Allen and Co., London.

Cautley, P.T. 1860. *Report on the Ganges Canal Works*. Smith Elder, London.

CBIP (Central Board of Irrigation and Power). 1954. *Irrigation in India through Ages*. Government of India Press, New Delhi.

Chaturvedi, M.C. 1967a. *Design of Ramganga Canal, Memo to Board of Consultants*. Government of Uttar Pradesh, Lucknow.

Chaturvedi, M.C. 1976b. Hydrologic and hydraulic considerations in water resources projects. Ninth Congress of Large Dams, Istanbul.

Darian Steven, G. 1978. *The Ganges in Myth and History*. The University Press of Hawaii, Honolulu.

Deakin, A. 1893. *Irrigated India: An Australian View of India and Ceylon; and their Irrigation and Agriculture*. W. Thacker and Co., London.

Gadgil, M. 1985 (November). Towards an ecological history of India. *Economic and Political Weekly*, Vol. XX, Nos. 45, 46, and 47, 1909–1918.

Ganguli, B. 1938. *Trends of Agriculture and Population in the Ganges Valley*. Methuen, London.

Government of India (GOI). 1922. *Triennial Review of Irrigation in India, 1918–21*. Public Works Departments, Calcutta.

Guha, R. 1994. Colonialism and conflict in the Himalayan forest. In R. Guha (Ed.). *Social Ecology*. Oxford University Press, New Delhi.

Gustafson, W.E. and R.B. Reidinger. 1971 (December). Delivery of canal water in North India and West Pakistan. *Economic and Political Weekly*, Vol. XV, No. 52, p.a. 157.

IC (Report of the Irrigation Commission). 1972. Ministry of Irrigation and Power. Government of India, New Delhi.

ICAR (Indian Council for Agricultural Research). 1964. *Agriculture in Ancient India*. New Delhi.

Kangle, R. 1969. *Kautilya's Arthshastra, An English Translation* (with critical notes), Vol. 1–3. University of Bombay, Bombay.

Klein, I. 1993. *Irrigation, Environmental Decay and Diseases in British India*. Department of History, The American University, Washington, D.C. (mimeo)

Kosambi, D.D. 1970. *The Culture and Civilization of India—A Historical Outline*. Vikas, Delhi.

McPherson, W.J. 1972. Economic development in India under the British Crown, 1858–1947. In A.J. Youngson (Ed.), *Economic Development in the Long Run*. George Allen and Unwin, London.

NCA (National Commission on Agriculture). 1976. *Part V, Resource Development*. Government of India, Ministry of Agriculture and Irrigation, New Delhi.

Nicolls, L.R. 1988. Agricultural engineering in India (I–XII). *Engineering*, XLV (May) and XLVI (July).

NIH (National Institute of Hydrology). 1990. *Hydrology in Ancient India*. NIH, Roorkee.

Norton, C.E. 1853. Canals of irrigation in India. *North American Review*, Oct.

Paranjpaye, V. 1988. *Evaluating the Tehri Dam*. Indian National Trust for Art and Cultural Heritage, New Delhi.

Rogers, P. 1983. Irrigation and economic development: Some lessons from India. In K.C. Noble and R.K. Sampath (Eds.), *Issues in Third World Development*, Chapter 16, pp. 347–390. West View Press, Boulder, CO.

Rogers, P. 1989. History of Ganges water resources with an eye to the future. *International Workshop on Economic Resurgence of Bihar through Water Resources Development: Issues and Strategies*, Centre for Water Resources Studies, Bihar College of Engineering, Patna University, 3–4 March 1989.

Sengupta, N. 1985. Irrigation: Traditional versus modern. *Economic and Political Weekly*, Special Number, August.

Spear, P. 1978. *A History of India*, Vol. II. Penguin, London.

Stone, I. 1984. *Canal Irrigation in British India*. Cambridge University Press, Cambridge.

Thapar, R. 1977. *A History of India*, Vol. I. Penguin, London.

Ucko, P.J. and G.W. Dimbleby. 1969. *The Domestication and Exploitation of Plants and Animals*. Gerald Duckworth and Co., London.

Whitcombe, E. 1972. *Agrarian Conditions in Northern India, Vol. I: The United Provinces under British Rule 1860–1900*. University of California Press, Berkeley.

Whatcom, E. 1980. Irrigation. In D. Kumar (Ed.), *The Cambridge Economic History of India*, Vol. II. Cambridge Press, Cambridge.

Wilson, H.M. 1892. *Irrigation in India*. U.S. Geological Survey, Washington, D.C.

Wittfogel, K.A. 1956. The hydraulic civilizations. In W.L. Thomas Jr. (Ed.), *Man's Role in Changing the Face of the Earth*. University of Chicago Press, Chicago.

Wittfogel, K.A. 1957. *Oriental Despotism: A Comparative Study of Total Power*. Yale University Press, New Haven.

Chapter 5

Agrawal, A. and S. Narain. 1989. *Towards Green Villages: A Study for Environmentally Sound and Participatory Rural Development*. Centre for Science and Environment, New Delhi.

Bali, J.S. 1988. *A Critical Appraisal of Past and Present Policies and Strategies of Watershed Development and Management in India and Role of Governmental and Non-Governmental Organizations in Small Scale Watershed Development*. Ministry of Environment, New Delhi. (mimeo)

Centre for Science and Environment (CSE). 1991. *State of India's Environment: A Citizens Report 3, Floods, Flood Plains and Environmental Myths*. CSE, New Delhi.

Central Water Commission (CWC). 1989. *Water and Related Statistics*. Government of India, New Delhi.

Chaturvedi, M.C. 1985. Water in India's development, issues, developmental policy and programmers, and planning approach. In M.C. Chaturvedi and P. Rogers (Eds.), *Water Resources System Planning—Some Case Studies for India*. Indian Academy of Sciences, Bangalore.

Chaturvedi, M.C. 1986. Modernization of the development of the water resources of the Ganga–Yamuna Valley. Report XII Meeting, Board of Consultants, Ganga–Yamuna Valley Multipurpose and Hydroelectric Projects, Dehradun, U.P., 15–16 April 1986. (mimeo)

Chaturvedi, M.C. 1988. System and environmental impact studies of Ganga–Yamuna Valley projects. D.D. Chief Engineer Planning, Irrigation Department, Government of Uttar Pradesh, Lucknow, 8 November 1988. (mimeo)

Chaturvedi, M.C. 1990. *Policy Study of Some Aspects of Water Resources Development in India*. Planning Commission, Government of India. (mimeo)

Chaturvedi, M.C. 1990. *Eastern Water Study: Water Management*, Centre for Policy Research, New Delhi. (mimeo)

Chaturvedi, M.C. 1992. Irrigation and drainage system policy analysis and India case study. *ASCE Journal of Water Resources Planning and Management*, 118, (4) 445–464.

Chaturvedi, M.C. 2001. Sustainable development of India's waters—some policy issues. *Water Policy*, 3, 297–320.

Chaturvedi, M.C. 2006. India—The challenge of development and management of water, hard or soft landing in closing basins. 2006 SIWI Seminar, World Water Week 2006, August 20–26, 2006, Stockholm.

Chaturvedi, M.C. 2011a. *Societal Environmental Systems Management* (in process of publication).

Chaturvedi, M.C. 2011b. *India's Waters: Environment, Economy and Development*. Taylor & Francis, Boca Raton, FL.

Chaturvedi, M.C. 2011c. *India's Waters: Advances in Development and Management*. Taylor & Francis, Boca Raton, FL.

Chaturvedi, M.C. 1985. Ganga–Brahmaputra–Barak basin. In M.C. Chaturvedi and P. Rogers (Eds.), *Water Resources Systems Planning—Some Case Studies for India*. Indian Academy of Sciences, Bangalore.

Datya, K.R., V.N. Gore, S. Paranjape, and R.K. Patil. 1988. *Techno-economics of Small-scale Watershed Development in Major Agriculture Regions of India*. Institute of Economic Growth, New Delhi.

Dhruvanarayan, V.V., G. Shastry, and V.S. Patnaik. 1990. *Watershed Management.* Indian Council of Agricultural Research, New Delhi.

Government of India (GOI). 1972. *Report of the Irrigation Commission,* Vol. I–III. Ministry of Irrigation and Power.

Government of India (GOI). 1976. *Report of the National Commission on Agriculture,* Vol. V. Ministry of Agriculture.

Government of India (GOI). 1980. *Report, National Commission on Floods.* Ministry of Energy and Irrigation, New Delhi.

Government of India (GOI). 1987. *National Water Policy.* Ministry of Water Resources, New Delhi.

Government of India (GOI). 1988. *Report of the Committee of Experts on Draft Outline of National Land Use Policy.* National Land Use and Conservation Board, Ministry of Agriculture, New Delhi.

Government of India (GOI). 1989. *Report of the Working Group on Wastelands Development Sector, Eighth Five Year Plan.* Ministry of Environment and Forests, New Delhi.

Government of India (GOI). 1990. *National Strategy for Conservation and Sustainable Development.* Ministry of Environment and Forests, New Delhi.

Government of India (GOI). *Eighth Five Year Plan, 1992–97.* Planning Commission, New Delhi.

Gyawali, D. 2001. *Water in Nepal.* Kathmandu. Himal Books.

Hagman, G. 1984. *Prevention Better than Cure—Report on Human Environmental Disaster in the Third World.* Swedish Red Cross, Stockholm.

Narsimhan, S.R. and R. Singh. 1994. Hydropower development—main issues. Fifth National Water Convention, National Water Development Agency, Ministry of Water Resources, Government of India, New Delhi, 25–27 February 1994.

National Commission on Integrated Water Resources Development Policy. 1999. Ministry of Water Resources, Government of India, New Delhi.

Puri, P. 1992. *Dimensions of a Land Policy in India.* CBS Publishers and Distributors, Delhi.

Rangachari, R. 2006. *Bhakra–Nangal Project—Socio-economic and Environmental Impacts.* Oxford University Press, New Delhi.

Rao, K.L. 1975. *India's Water Wealth, Its Assessment, Uses and Projections.* Orient Longman, New Delhi.

Singh, P. (Ed.). 1989. *Problem of Wasteland and Forest Ecology in India.* Ashish, New Delhi.

Singh, P. 1990. Flood management—possibilities and limitations. Seminar on New Perspectives in Water Management, Indian National Academy of Engineering, New Delhi.

Society for Promotion of Wasteland Development (SPWD). 1988. National Workshop on Small-Scale Watershed Development, 30 October–1 November, New Delhi.

United Nations Development Program. 1994. *Human Development Report, 1994.* Oxford University Press, Delhi.

Vaidyanathan, A. 1994. Integrated watershed development; some major issues, society for promotion of wastelands development. Foundation Day Lecture, 1 May 1994. Center for Development Studies, Trivandrum.

World Bank, India. 1991. *Irrigation Sector Review,* Vol. I—Main Report; Vol. II—Supplementary Analysis, New Delhi.

World Resources 1996–97. Washington, D.C.

Yadav, H.R. (Ed.). 1988. *Dimensions of Wasteland Development.* Concept, New Delhi.

Chapter 6

Achnet, S.H. 1991. *Environment Sectoral Review—Nepal Country Study*. Institute for Integrated Development Studies, Kathmandu.

Agarwal, N.K. and Associates. 1991. *Energy Sectoral Review*. Institute for Integrated Development Studies, Kathmandu.

Agarwal, N.K. 1991. *Agriculture and Irrigation—Nepal Country Study*. Institute for Integrated Development Studies, Kathmandu.

Bruijzneel, L.A., with C.N. Bremmer. 1989. Highland–lowland interactions in the Ganges–Brahmputra river basin: A review of published literature. ICIMOD Occasional Paper No. II, Kathmandu.

Carson, B. 1985. Erosion and sedimentation processes in the Nepalese Himalayan region. ICIMOD Occasional Paper. No. I, Kathmandu.

Hamilton, L.S. 1987. What are the impacts of Himalayan deforestation on the Ganges–Brahmaputra lowlands and delta: Assumptions and facts. *Mountain Research and Development*, 7, (3), 258–263.

Ives, J.D. and Messerli, B. 1988. *The Himalayan Dilemma—Reconciling Development and Conservation*. United Nations University, London and New York.

Ives, J.D. 1997. The study of Himalayan degradation, its validity and application challenged by recent research. *Proceedings Mohawk Mountain Conference on the Himalayan–Ganges Problem*.

ICIMOD (International Centre for Integrated Mountain Development). 1985. Managing the watersheds. *Report of the International Workshop on Watershed Management in Hindukush–Himalayan Region*, Kathmandu, Nepal.

Malla, S.K. 1991. *Hydropower Development—Nepal Country Study*. Institute for Integrated Development Studies, Kathmandu.

Malla, S.K., S.K. Srestha, and M.M Sanju. 2001. Nepal's water vision and the GBM basin framework. In Q.K. Ahmad, A.K. Biswas, R. Rangachari, and M.M. Sanju (Eds.), *Ganges–Brahmaputra–Meghna Region—A Framework for Sustainable Development*. The University Press Ltd., Dhaka.

MPFS (Master Plan for the Forestry Sector Nepal). 1988. Ministry of Forests and Soil Conservation, Kathmandu.

NCS, National Conservation Strategy of Nepal. 1989. Building on Success, IUCN/HMG, Kathmandu, Nepal.

Nelson, A. 1980. Reconnaissance Inventory of Major Ecological Land Units and their Watershed Condition in Nepal, FAO/HMGN/UNDP, Report, IWM/WP/17.

Pandey, K.K. 1982. *The Livestock Fodder Situation and the Case for Additional Fodder Resources in Mountain Environment and Development*. Swiss Association for Technical Assistance, Kathmandu.

Rao, A.R. and T. Prasad. 1980. *Water Resources Development of the Indo–Nepal Region*. (mimeo)

Sharma, C.K. 1983. *Water and Energy Resources of the Himalayan Block*. Sangeeta Sharma, Kathmandu.

Thapa, B.B. and B.B. Pradhan. 1996. *Water Resources Development—Nepalese Perspective*. Konark Publisher Pvt. Ltd., Delhi.

World Bank. 1995. Trends in Developing Economics. Washigton, D.C.

World Bank. 1997. Nepal-Poverty and Incomes. Washigton, D.C.

Wyatt, S.J., 1982. *The Agricultural System in the Hills of Nepal, the Ratio of Agriculture to Forest Land and the Problem of Animal Fodder*. Agricultural Projects Service Centre, Kathmandu.

Chapter 7

Ahmad, Q.K., A.U. Ahamad, U. Ahmad, H.R. Khan, and K.B.S. Rasheed. 2001. 2001 GBM regional water vision. In Q.K. Ahmad, A.K. Biswas, R. Rangachari, and M.M. Sanju (Eds.), *Ganges–Brahmaputra–Meghna Region—A Framework for Sustainable Development*. The University Press Ltd., Dhaka.

Ahmad, N. 1968. *An Economic Geography of Bangladesh*. Oxford University Press, London.

BCAS and DOE. 2001. *Bangladesh Climate Change Country Assessment of Vulnerability and Adaptation to Climate Change.* Bangladesh Centre for Advanced Studies and Department of Environment, Dhaka.

Encyclopaedia Britannica. 1984. Bangladesh. *Encyclopaedia Britannica*, Chicago.

Huque, C.E. and M. Rahman. 1994. *Water Resources Management*. In H. Zafaruqab, M.A. Tashim, and A. Chowdhury (Eds.), *Policy Issues in Bangladesh*. South Asian Publishers, New Delhi.

Hindustan Times. 2007. Poisonous Paddy Fields. 31 October 2007, New Delhi.

IBRD/IDA. 1972. *Land and Water Resources Sector Study*, Vol. 1–9. The World Bank, Washington, D.C.

Khan, T.A. 1987. *Water Resources Situation in Bangladesh*. In M. Ali, G.E. Radoserich, and A.A. Khan (Eds.), *Water Resources Policy in Asia*, pp. 139–164. Balkema, Rotterdam.

Majumdar, S.C. 1940. *River of the Bengal Delta*. Calcutta University Press.

MPO. 1986. National Water Plan Summary Report, 1986, Master Plan Organization, Ministry of Irrigation, Water Development and Flood Control, Government of Bangladesh, Dacca.

Nishat, A. and S.K. Chowdhury. 1987. *Water Quality*. In M. Ali, G. E. Radoserich, and A.A. Khan (Eds.), *Water Resources Policy in Asia*, pp. 349-362. Balkema, Rotterdam.

Rashid, H.E. 1977. *Geography of Bangladesh*. Dhaka University Press, Dhaka.

Rogers, P.P.L. and D. Seckler. 1989. *Eastern Water Study; Strategies to Manage Flood and Drought in the Ganges, Brahmaputra Basin*, Office of Technical Resources, Agriculture and Rural Development, Discussion, Bureau for Asia and Near East, USAID, April 1989.

Siddiqui, M.F.A. 1981. Management of River Systems in the Ganges and Brahmaputra Basins for Development of Water Resources. In Z. Munir (Ed.), *River Basin Development*, Tycooly International Publishing Ltd., Dublin.

Spear, P. 1978. *A History of India*, Vol. 2. Penguin, London.

World Bank. 1990. *Bangladesh Action Plan for Flood Control*. Unpublished.

World Bank. 1998. *Water Resources Management in Bangladesh: Steps Toward a New National Water Plan*. Dhaka.

World Bank. 2000. *Bangladesh: Climate Change and Sustainable Development*. Report No. 21104-BD. Rural Development Unit, South Asia Region, World Bank, Dhaka.

World Bank. 2003. *Water Resources Sector Strategy*. Dhaka.

World Bank–WSP (Water and Sanitation Program). 2005. *Toward a More Effective Operational Response: Arsenic Contamination of Groundwater in South and East Asian Countries*. World Bank, Washington, D.C.

World Bank. 2005. *Bangladesh PRSP Forum Update: Recent Developments and Future Perspectives*. Dhaka.

World Bank. 2005. *Bangladesh Country Water Resources Assistance Strategy. Bangladesh Development Strategies*. Dhaka.

World Bank. 2006. *Country Environmental Analysis*. Dhaka.

World Bank. 2007. *Bangladesh: Strategy for Sustained Growth. Bangladesh Development Series*. Dhaka.

World Bank. 2007. *Bangladesh 2020. A Long-Run Perspective Study*. Dhaka and Washington, D.C.

Chapter 8

Agarwal, N.K. and Associates. 1991. *Energy Sectoral Review—Nepal Country Report*. Institute for Integrated Development Studies, Kathmandu.

Bajpai, U.S. 1986. Seminar report. In U.S. Bajpai (Ed.), *India and Its Neighborhood*. Lancer International, New Delhi.

Chaturvedi, M.C. 2011a. *Societal Environmental Systems Management* (in process of publication).

Chaturvedi, M.C. 2011b. *India's Waters: Environment, Economy and Development*. Taylor & Francis, Boca Raton, FL.

Chaturvedi, M.C. 2011c. *India's Waters: Advances in Development and Management*. Taylor & Francis, Boca Raton, FL.

Chitale, M.A. 1992. Cooperative development of the Indo–Nepal Water resources—prospects opportunities and challenges. Keynote address in *Proceedings of the Workshop on Cooperative Development of the Indo–Nepal Water Resources: Prospects, Opportunities and Challenges*, 20–30 May 1992, Centre for Water Resources Studies, Bihar College of Engineering, Patna.

Das, B.S. 1986. *Bhutan*. In U.S. Bajpai (Ed.), *India and Its Neighborhood*. Lancer International, New Delhi.

Dixit, A.K. 1991. *Water Resources and Hydrological Uncertainties—Nepal Country Report*. Institute for Integrated Development Studies, Kathmandu.

Government of India (GOI). 1972. *Report of the Irrigation Commission*, Vol. 2 and 3 (1), Ministry of Irrigation and Power, New Delhi.

Indo–Bangladesh Joint Rivers Commission Updated Proposals and Comments of Bangladesh and India on the Augmentation of the Dry Season Flows of the Ganges, Dhaka, May 1985.

Kaushik, P.D. 1992. Nepal's water resources and their impact on Indo–Nepalese relations. In *Proceedings of the Workshop on Cooperative Development of the Indo–Nepal Water Resources: Prospects, Opportunities and Challenges*, 20–30 May 1992, Centre for Water Resources Studies, Bihar College of Engineering, Patna.

Kumar S. and Sinha, A.K. 1992. Indo–Nepal water resources—a holistic approach. In *Proceedings of the Workshop on Cooperative Development of the Indo–Nepal Water*

Resources: Prospects, Opportunities and Challenges, 20–30 May 1992, Centre for Water Resources Studies, Bihar College of Engineering, Patna.

Kurve, M. 1981. Plea to declare Ganga an "international river." *The Times of India*, April 13, 1981.

Malla, G.L. 1985–1986. Regional co-operation in South Asia: A Nepalese perspective. *Strategic Studies Series*, No. 6 and 7.

Malla, S.K. 1991. *Hydropower Development, Nepal Country Report*. Institute for Integrated Development Studies, Kathmandu.

Muni, S.D. 1986. Nepal. In U.S. Bajpai (Ed.), *India and Its Neighborhood*. Lancer International, New Delhi.

Pradhan, B.K. and H.M. Shrestha. 1992. A Nepalese perspective of Himalayan water resources development. In D.J. Eaton (Ed.), *The Ganga Brahmaputra Basin*. LBJ School of Public Affairs, University of Texas, Austin.

Rao, A.R. and T. Prasad. 1992. Water resources development of the Indo–Nepal region. In Regmi, R.J. Indo–Nepal cooperation in mega-projects—learning from Itaipu. *Water Nepal*, 3, (1), 3–7.

Regmi, R.J. 1992. Indo–Nepal cooperation in mega-projects—learning from Itaipu. *Water Nepal*, 3, (1), 3–7.

Rogers, P., P. Lydon, and D. Seckler. 1989. *Eastern Waters Study: Strategies to Manage Flood and Drought in the Ganges–Brahmaputra Basin*. Agriculture and Rural Development Division, USAID, Washington, D.C.

Sain, K. 1978. *Reminiscences of an Engineer*. Young Indian, New Delhi.

Sharma, C.K. 1983. *Water and Energy Resources of the Himalayan Block*. Sangeeta Sharma, Kathmandu.

Thapa, B.D. and B.B. Pradhan. 1995. *Water Resources Development—Nepalese Perspective*. Konark Publishers Private Ltd., New Delhi.

Tiwari, A. 1992. Common man's woes. *India Today*, 15 August, pp. 74–85.

Upadhyay, S.N. 1991. Nepal's transboundary water resources—new perspectives on cooperation. Transboundary Resources Report. International Transboundary Resources Centre.

Verghese, B.C. and R.R. Iyer. 1993. *Harnessing the Eastern Himalayan Rivers*. Konark Publishers Pvt. Ltd., New Delhi.

Chapter 9

Abbas, B.M. 1982. *The Ganges Waters Dispute*. University Press, Dhaka.

Ahmad, Q.K., A.U. Ahmad, H.R. Khan, and K.B.S. Rasheed. 2001. GBM regional water vision: Bangladesh perspectives. In Q.K. Ahmad, A.K. Biswas, R. Rangachri, and M.M. Sainju (Eds.), *Ganga–Brahmaputra–Meghna Region—A Framework for Sustainable Development*. The University Press Ltd., Dhaka.

Bagchi, K. 1944. *The Ganges Delta*. Calcutta University Press.

Bagchi, K. (Ed.). 1972. The Bhagirathi–Hoogly basin. *Proceedings of an Interdisciplinary Symposium*, Sibeniranath Kanjilal, Calcutta.

Begum, K. 1988. *Tension over the Farakka Barrage: A Techno-Political Tangle in South Asia*. Steiner Verlag Wiesbaden Gmbh, Stuttgart.

Bhattacharya, S.K. 1975. Deltaic activity of Bhagirathi–Hoogly river system. *Proceedings of the American Society of Civil Engineers*, WWI 99, 69–87.

Chakravarty, N. 1986. Bangladesh. In U.S. Bajpai (Ed.), *India and Its Neighborhood*. Lancer International, New Delhi.

Chaturvedi, M.C. 1968. *Review of a Game Theory Approach to International Aspects of the Ganga Basin Study by Peter Rogers*. Government of India, Ministry of Irrigation, New Delhi. (mimeo)

Chaturvedi, M.C. 1985. Water in India's development. In M.C. Chaturvedi and P. Rogers (Eds.), *Water Resources Systems Planning—Some Case Studies for India*. Indian Academy of Sciences, Bangalore.

Chaturvedi, M.C. 1992. Water resources engineering considerations. In D.J. Eaton (Ed.), *The Ganga–Brahmaputra Basin—Water Resources Cooperation between Nepal, India and Bangladesh*. LBJ School of Public Affairs, University of Texas, Austin.

Chaturvedi, M.C. 1992. Transboundary river basin management and sustainable development in developing countries: Case study of Ganges–Brahmaputra–Meghna basin. International Symposium on Transboundary River Basin Management and Sustainable Development, Delft, The Netherlands, 18–22 May 1992.

Chaturvedi, M.C. 2011a. *Societal Environmental Systems Management* (in process of publication).

Chaturvedi, M.C. 2011b. *India's Waters: Environment, Economy and Development*. Taylor & Francis, Boca Raton, FL.

Chaturvedi, M.C. 2011c. *India's Waters: Advances in Development and Management*. Taylor & Francis, Boca Raton, FL.

Cotton, A. 1854. *Public Works in India*. W.H. Allen and Co., London.

Cotton, A. 1858. *A Memorandum on the Water Communication between Calcutta and Rajmahal*. Alipore Jail Press, Calcutta.

Crow, B. 1985. The making and breaking of agreement on the Ganga. In J. Lundqvist et al. (Eds.), *Strategies for River Basin Management*. D. Reidel, Dordrecht, Holland.

Crow, B. and A. Lindquist. 1990. Development of Rivers Ganges and Brahmaputra: The difficulty of negotiating a new line, development policy and practice. The Open University, Milton Keynes, UK, Working Paper No. 19.

Crow, B. 1995. *Sharing the Ganges—The Politics and Technology of River Development*. Sage Publication, New Delhi.

Eaton, D.J. (Ed.). 1992. *The Ganges Brahmaputra Basin: Water Resources Cooperation Between Nepal, India and Bangladesh*. LBJ School of Public Affairs, University of Texas, Austin.

Framji, K.K. 1975. The Farakka Barrage Project—the fulfillment of a dream. In Government of India, Souvenir, Farakka Barrage Project.

Government of India (GOI). 1961. *Preservation of the Port of Calcutta* (report). New Delhi.

Government of India (GOI) Ministry of External Affairs. 1976. *The Farakka Barrage*. New Delhi.

Harvard University, Centre for Population Studies. 1972 (June). Bangladesh land, water and power study. Final Report (Draft).

IMCB (International Movement to Save Bangladesh). 1994. Farakka—a death trap for Bangladesh. Seminar on Farakka Barrage, MIT, Cambridge, 18 June 1994.

Ippen, A.T. and C.F. Wicker. 1962. *The Hoogly River Problem*. Government of Pakistan. (mimeo)

Islam, M.R. 1987. *Ganges Waters Dispute: Its International Legal Aspects*. University Press, Dhaka.

Joglekar, J.V. 1955. Hydraulic investigations of the Hooghly to improve its naviga-bility. Congress of the International Association for Hydraulic Research, The Hague.

Leonard, H. 1865. *The River Hoogly: A Letter to the Undersecretary of State for India.* Eyre and Spootiswood for HMSO, London.

Mehta, J. 1992. Opportunity costs of delay in water resources management between Nepal, India and Bangladesh. In D.J. Eaton (Ed.), *The Ganges–Brahmaputra Basin—Water Resources Cooperation between Nepal, India and Bangladesh.* LBJ School of Public Affairs, University of Texas, Austin.

Mehta, J.S. 1993. Politics of riparian relations. In B.G. Verghese and R.R. Iyer (Eds.), *Harnessing the Eastern Rivers.* Konarak Publishers, New Delhi.

Ministry of Agriculture and Irrigation. 1984. *IUP (India Updated Proposal) for the Augmentation of Dry Season Flow of Ganga.* Delhi.

Ministry of Flood Control and Water Resources. 1985. *BUP (Bangladesh, Updated Proposal) for the Augmentation of the Dry Season Flow of the Ganges.* Ministry of Flood Control and Water Resources, Dhaka.

Mookerje, D. 1975. Farakka Barrage Project: A challenge to engineers. *Proceedings, Institution of Civil Engineers,* London, Part I, pp. 67–84.

Myrdal, G. 1971. *Asian Drama* (Abridgement by Seth S. King). Allen Lane, The Penguin Press, London.

Oaz, T.M. 1939. *Report on the River Hoogly and Its Headworks.* Calcutta. Bengal Secretariat Book Depot, Calcutta.

Rangachari, R. and R.R. Iyer. 1992. Indo–Bangladesh talks on the Ganga waters issue. In B.G. Verghese and R.R. Iyer (Eds.), *Harnessing the Eastern Rivers.* Konark Publishers, New Delhi.

Rangachari, R. and B.G. Verghese. 2001. Making water work to translate poverty into prosperity: the Ganga-Brahmaputra-Barak region. In Q.K. Ahamad, A.K. Biswas, R. Rangachari, and M.M. Sainju (Eds.), *Ganges–Brahmaputra Meghna Region—A Framework for Sustainable Development,* The University Press Ltd., Dhaka.

Sau, S.N. 1990. The economics of Calcutta–Heldia Port Complex. *Economic and Political Weekly,* 5–12 May 1990, pp. 1015–1026.

UNDP, India. 1972. *The National Water Grid.* New York.

Vernon-Harcourt, L.F. 1905. The River Hoogly. *Proceedings, Institution of Civil Engineers,* London, CLX.

Webster, A. 1946. Report on the Future Development of the Port of Calcutta. Commissioners for the Port of Calcutta, Calcutta. (Cotton 1854, Leonard 1864, Vernon, Harcourt 1896, 1905, Stevenson-Moore Committee 1916–1919, Wilcoks 1930, Oaz 1939 and Webster 1946).

World Bank, Bangladesh. 1972. *Land and Water Resources Sector Study.* Washington, D.C.

Chapter 10

Altekar, A.S. 1958. *State and Government in Ancient India.* Motilal Banarsidas, Delhi.

Avineri, S. (Ed.). 1968. *Karl Marx on Colonialism and Modernization.* Doubleday and Co., New York.

Bhagwati, G. 1994. India's economic state. *The American Enterprises,* March/April, 55–61.

Chaturvedi, M.C. 2002. Higher technical education: Challenges, opportunities and constraints. *Proceedings of the National Conference on the Management of Higher Education: 21st Century Challenges.* Anamaya Publishers, New Delhi.

Godgil, M. 1985. *Towards and Ecological History of India. Economic and Political Weekly*, Vol. XX, Nos. 45, 46, and 47, 1909–1918.

Government of India (GOI). 1976. *National Commission of Agriculture.* Ministry of Agriculture, New Delhi.

Government of India (GOI). 1992. *Eighth Five Year Plan.* Planning Commission, New Delhi.

Gulhati, N.D. 1965. Administration and Financing of Irrigation Works in India, Publication No. 77, Central Board of Irrigation and Power, Government of India.

Huque, C.E. and M. Rahman. 1994. *Water Resources Management.* In H. Zafaruqab, M.A. Tashim, and A. Chowdhury (Eds.), *Policy Issues in Bangladesh.* South Asian Publishers, New Delhi.

Mehta, P. 1989. *Bureaucracy, Organization Behavior, and Development.* Sage Publication, New Delhi.

Mohile, A.D. 2007. Government policies and programmes. In J. Briscoe and R.P.S. Malik (Eds.), *Handbook of Water Resources in India.* Oxford University Press, New Delhi.

Myrdal, G. 1971. *Asian Drama* (Abridgement by Seth S. King). Allen Lane, The Penguin Press, London.

Pai Panandiker, V.A. 1966. *Personnel System for Development Administration.* Popular Prakashan, Bombay.

Panikkar, K.M. 1959. *Afro-Asian States and Their Problems.* Allen and Unwin, London.

Sen, A. 1989. Indian development: Lessons and non-lessons. *Daedalus*, 118, 369–392.

Wittfogel, K.A. 1956. The hydraulic civilizations. In W.L. Thomas Jr. (Ed.), *Man's Role in Changing the Face of the Earth.* University of Chicago Press, Chicago.

Wittfogel, K.A. 1957. *Oriental Despotism—A Comparative Study of Total Power.* Yale University Press, New Haven.

Chapter 11

Alexandratos, N. (Ed.). 1995. *World Agriculture Towards 2010, An FAO Study.* John Wiley & Sons, Chichester, UK.

Bossel, H. 1998. *Earth at Cross Roads.* Cambridge University Press, Cambridge, UK.

Chaturvedi, M.C. 2001. Sustainable development of India's waters-some policy issues. *Water Policy*, 3, 297–320.

Chaturvedi, M.C. 2011a. *Societal Environmental Systems Management* (in process of publication).

Chaturvedi, M.C. 2011b. *India's Waters: Environment, Economy and Development.* Taylor & Francis, Boca Raton, FL.

Chaturvedi, M.C. 2011c. *India's Waters: Advances in Development and Management.* Taylor & Francis, Boca Raton, FL.

Conway, G. 1998. *The Doubly Green Revolution.* Cornell University Press, Ithaca, NY.

Cotton, A. 1867. On a communication between India and China by the line of the Burhampooter and Yang-tsze. *The Journal of the Royal Geographical Society*, 37, 231–239.

Food and Agriculture Organization of the United Nations (FAO). 1995. Assessment of the current world food security and medium term review. Item II of the Provisional Agenda, 20th Session, Committee on World Food Security, FAO, Rome, April 1995.

Foland, F.M. 1974. Agrarian unrest in Asia and Latin America. *World Development*, 2, (4 and 5), 56.

Gandhi, P.J. 2005. *Dr. Kalam's PURA Model and Societal Transformation*. Deep Publications, New Delhi.

United Nations Development Programme. 2006. *Beyond Scarcity: Power, Poverty and the Global Waters*. Human Development Report. UNDP, New York.

NCIWRDP (National Commission of Integrated Water Resources Development Planning). 1999. Government of India, New Delhi.

Weitz, R. 1971. *From Peasant to Farmer*. Columbia University Press, New York.

Chapter 12

Agenda 21. 1992. United Nations Conference on Environment and Development. Rio de Janeiro.

Ausbel, J.H. and H.D. Langford (Eds.). 1997. *Technological Trajectories and the Human Environment*. National Academy Press, Washington, D.C.

Bhatia, R. 2004. *Economic Impacts and Synergy Benefits of the Bhakra Multipurpose Dam, India: A Case Study*. World Bank, Washington, D.C.

Board of Sustainable Development (BSD), National Research Council (NRC). 1999. *Our Common Journey. A Transition toward Sustainability*. National Academy Press, Washington, D.C.

Bossel, H. 1998. *Earth at Cross Roads*. Cambridge University Press, Cambridge, UK.

Bower, B.T. and M.N. Hufschmidt. 1984. A conceptual framework for analysis of water resources management in Asia. *Natural Resources Forum*, 8, (4), 343–356.

Centre for Science and Environment (CSE). 1999. *The Citizen's Fifth Report*. CSE, New Delhi.

Cernea, M.M. 1994. The sociologist's approach to sustainable development. In I. Serageldin and A. Steers (Eds.), Making Development Sustainable—from Concept to Action, Environmentally Sustainable Development Occasional Papers, Series No. 2. World Bank, Washington, D.C.

Chaturvedi, M.C. 1976. *Water—Second India Studies* (published for the Ford Foundation). Macmillan, New Delhi.

Chaturvedi, M.C. 1987. *Water Resources System Planning and Management*. Tata McGraw-Hill, New Delhi.

Chaturvedi, M.C. 1991c. Water resources systems development and management in India—a policy analysis. National Seminar on Use of Computers in Hydrology and Water Resources, New Delhi. Central Water Commission, New Delhi, Dec 16–18.

Chaturvedi, M.C. 2001. Sustainable development of India's waters—some policy issues. *Water Policy*, 3, 297–320.

Chaturvedi, M.C. 2011a. *Societal Environmental Systems Management* (in process of publication).

Chaturvedi, M.C. 2011b. *India's Waters: Environment, Economy and Development*. Taylor & Francis, Boca Raton, FL.

Chaturvedi, M.C. 2011c. *India's Waters: Advances in Development and Management.* Taylor & Francis, Boca Raton, FL.

Chaturvedi, V. and M.C. Chaturvedi. 1977. Higher technical education—patterns, trends and implications for developing countires. *Journal of Higher Education*, 2, (3), 377–386.

Chaturvedi, M.C. and P. Rogers (Eds.). 1985. *Water Resources Systems Planning—Some Case Studies for India.* Indian Academy of Sciences, Bangalore.

Dreze, J. and A. Sen. 2002. *India—Development and Partcipation.* Oxford University Press, New Delhi.

Gardner, R.A.M. and A.J. Gerrard. 2003. *Runoff and Soil Erosion on Cultivated Rainfed Terraces in the Middle Hills of Nepal.* Elsevier Science Ltd., Amsterdam.

Government of India (GOI). 2002. *Tenth Five Year Plan—2002–2007.* Vol. 1. Planning Commission, New Delhi.

Haimes, V.Y. 1977. *Hierarchical Analysis of Water Resources Systems.* McGraw-Hill, New York.

Hall, W. A. and J.A. Dracup. 1970. *Water Resources Systems Engineering.* McGraw Hill, New York.

Harrison, L. and S. Huntington (Eds.). 2000. *Culture Matters—How Values Shape Human Progress.* Basic Books, New York.

Hassan, Q. 2004. A study of some aspects of sustainable development of India's waters. Ph.D. Thesis. Indian Institute of Technology, Delhi, New Delhi.

Hoekstra, A.Y. 1998. *Perspectives on Water—An Integrated Model Based Exploration of the Future.* International Books, Utrecht, The Netherlands.

IIPCC (Intergovernmental Panel on Climate Change). 2001. Climate Change 2001. Synthesis Report. A Synthesis of Working Groups I, II, and III to the Third Assessment Report of the Intergovernmental Panel on Climate Change. R.T. Watson and the Core Working Team, Cambridge University Press. Cambridge, UK and New York.

IUCN (The World Conservation Union), UNEP (United Nations Environmental programme), and WWF (World Wide Fund for Nature). 1991. *Caring for Earth: A Strategy for Sustainable Living.* UCN, UNEP, and WWF, Gland, Switzerland.

Kothari, M. 2000. Sustainable water resources management for a river basin. Ph.D. Thesis. Indian Institute of Technology, Delhi, New Delhi.

Landes, D.S. 2000. Culture matters. In L. Harrison and S. Huntington (Eds.). 2000. *Culture Matters—How Values Shape Human Progress.* Basic Books, New York.

Lenton, R. 2011. Integrated water resources management. In P. Rogers (Vol. Ed.), Management of Water Resources. In P. Wilderer (Ed.). *Treatise on Water Science*, Vol. 1, pp. 9–21. Elsevier, Amsterdam.

Loucks, D.P., J.R. Stedinger, and H.A. Haith. 1981. *Water Resources Systems Planning and Analysis.* Prentice Hall, Upper Saddle Rivers, New Jersey.

Loucks, D.P. and J.S. Gladwell (Eds.). 1999. Sustainability criteria for water resources systems. *International Hydrology Series.* Cambridge University Press, Cambridge, UK, Upper Saddle River, New Jersey.

L'vovick, M. and G.F. White. 1990. Use and transformation of terrestrial water system. In B.L. Turner II, W.C. Clark, R.W., Kates, J.F. Richards, J.T. Mathews, and W.B. Meyer (Eds.), *The Earth as Transformed by Human Action.* Cambridge University Press, Cambridge, UK.

Maass, A., M.M. Hufschmidt, R. Dorfman, H.A. Thomas, S. Marglin, and G.M. Fair (Eds.). 1962. *Design of Water Resources Systems.* Harvard University Press, Cambridge, MA.

Marglin, S.A. 1962. Objectives of water resources development. In A. Maass, M.A. Hurfschmidt, R. Dorfman, H.A. Thomas Jr., S.A. Marglin, and G.M. Fair (Eds.), *Design of Water Resource Systems.* Harvard University Press, Cambridge, MA.

North, D.C. 1996. Economic development in historical perspective: The Western world. In R.H. Myers (Ed.), *The Wealth of Nation in the Twentieth Century. The Policies and Institutional Development of Economic Development*. Stanford University Press, Stanford.

Putnam, R.D. 1993. *Making Democracy Work*. Princeton University Press, Princeton.

Rangachari, R. 2006. *Bhakra–Nangal Project—Socioeconomic and Environmental Impacts*. Oxford University Press, New Delhi.

Raskin, P., E. Hansen, and R. Margolis. 1995. Water and sustainability. *Polestar Series Report*, No. 4. Stockholm Environment Institute, Boston.

Rayner, S. 1994. A Wiring diagram for the study of land-cover change: A report of the working group A. In W.B. Meyer and B.L. Turner II (Eds.), *Changes in Land Use and Land Cover: A Global Pespective*. Cambridge University Press, Cambridge.

Richards, D.J. and G. Pearson. 1998. *The Ecology of Industry*. National Academy Press, Washington, D.C.

Rogers, P. and M.B. Fiering. 1986. Use of systems analysis in water resources. *Water Resources Research*, 22, (9), 146S–158S.

Rotmans, J. and H.J.M. de Vries. 1996. *Perspectives on Global Change: The TARGETS Approach*. Cambridge University Press, Cambridge.

Rouse, H. and S. Ince. 1957. *History of Hydraulics*. Dover, New York.

Savanije, H.H.G. and A.Y. Hoekstra. 2002. *Water Resources Management in Knowledge for Sustainable Development*. UNESCO and EOLSS Publishers, Paris.

Sen, A. 1999. *Development as Freedom*. Alfred A. Knopp, New York.

United Nations Development Programme. 2006. *Beyond Scarcity: Power, Poverty and the Global Waters*. Human Development Report. UNDP, New York.

Yamamura, K. 1996. Bridled capitalism and the economic development of Japan. In R.H. Myers (Ed.), *The Wealth of Nation in the Twentieth Century. The Policies and Institutional Development of Economic Development*. Stanford: Stanford University Press.

Young, G.J., J.C.I. Dooge, and J.C. Rodda. 1994. *Global Water Resource Issues*. Cambridge University Press, Cambridge, UK.

Wilson, H.M. 1892. Irrigation in India. U.S. Geological Survey, Water Supply and Irrigation Paper No. 87. US Department of Interior, Washington, D.C.

World Water Council. 2000. *World Water Vision—A Water Secure World*. World Water Council, Marseille.

World Bank. 1993. *Water Resources Management—A Policy Paper*. Washington, D.C.

World Bank. 2000. *World Development Report 1999/2000. Entering the 21st century*. Oxford University Press, New York.

World Commission on Environment and Development (WCED). 1987. *Our Common Future*. Oxford University Press, London.

Chapter 13

Agenda 21. 1992. United Nations Conference on Environment and Development. Rio de Janeiro.

Arya, Y.C. 1980. Dynamic system response of crops for improving management of irrigation. Ph.D. Thesis. Indian Institute of Technology, Delhi, New Delhi.

Asthana, B.N. 1984. System studies for regional environmental resources planning in a developing economy. Ph.D. Thesis. Indian Institute of Technology, Delhi, New Delhi.

Bhatia, P.K. 1984. Modeling integrated operation of multipurpose multireservoir water resources systems. Ph.D. Thesis. Indian Institute of Technology, Delhi, New Delhi.

Briscoe, J. and Malik, R.P.S. 2006. *India's Water Economy—Bracing for a Turbulent Future.* Oxford University Press, World Bank, New Delhi.

Central Water Power Commission. 1997. Storages in River Basins of India, Government of India, New Delhi.

Chaturvedi, M.C. 1973. Indian National Water Plan and Grid, First World Congress, International Water Resources Association, Chicago, Sept. 24–28.

Chaturvedi, M.C. 1981. Ganges–Water Machine. *Proceedings Indo-U.S. Scientists Workshop on Water Resources Development and Flood Mitigation*, New Delhi, March 30–April 1. Unpublished.

Chaturvedi, M.C. 1985. Water in India's development. In M.C. Chaturvedi and P. Rogers (Eds.), *Water Resources Systems Planning—Some Case Studies for India.* Indian Academy of Sciences, Bangalore.

Chaturvedi, M.C. 1987. *Water Resources Systems Planning and Management.* Tata McGraw Hill, New Delhi.

Chaturvedi, M.C. 1991a. *Water Management—Eastern Rivers Study.* Centre for Policy Research, New Delhi.

Chaturvedi, M.C. 1991b. *Environmental Systems Management—Eastern Rivers Study.* Center for Policy Research, New Delhi.

Chaturvedi, M.C. 1998. India Water Power Machine, Energy Policy of India, Workshop Proceedings, Indian National Academy of India, New Delhi.

Chaturvedi, M.C. 2001. Sustainable development of India's waters—some policy issues. *Water Policy*, 3, 297–320.

Chaturvedi, M.C. 2002. Integrating Water and Energy Sectors—Systems Planning and Case Study, IWRA Regional Symposium, New Delhi, 27–30 Nov. 2002.

Chaturvedi, M.C. 2011a. *Societal Environmental Systems Management* (in process of publication).

Chaturvedi, M.C. 2011b. *India's Waters: Environment, Economy and Development.* Taylor & Francis, Boca Raton, FL.

Chaturvedi, M.C. 2011c. *India's Waters: Advances in Development and Management.* Taylor & Francis, Boca Raton, FL.

Chaturvedi, M.C. and V. Chaturvedi. 2011. *Tryst with Destiny—Revolutionizing India.* Unpublished.

Chaturvedi, M.C., R. Revelle, and V.K. Srivastava. 1975. Induced Groundwater Recharge, Proceedings, Second World Congress, International Water Resources Association, New Delhi.

Chaturvedi, M.C. and V.K. Srivastava. 1979. Storage of surface flows through groundwater recharge. *Water Resources Research*, 15, 1156–1166.

Chaube, U.C. 1982. Two level multiobjective reconnaissance system study of a large water resources system. Ph.D. Thesis. Indian Institute of Technology, Delhi, New Delhi.

Central Groundwater Board (CGWB). 1996. *National Perspective Plan for Recharge by Utilizing Surplus Monsoon Runoff.* Unpublished.

Cotton, A. 1867. On a communication between India and China by the line of the Burhampooter and Yang-tsze. *The Journal of the Royal Geographical Society*, 37, 231–239.

Government of India (GOI). 2002. *Tenth Five Year Plan—2002–2007*, Vol. 1. Planning Commission, New Delhi.

Gopal, S. 1995. *Speeches of Jawaharlal Nehru*. Ministry of Information and Broadcasting, Government of India.

Gupta, D.K. 1984. Conjunctive Surface and Groundwater Development—Multilevel Multiobjective Analysis. Ph.D. Thesis. Indian Institute of Technology, Delhi, New Delhi.

Hagerstrand, T. and U. Lohm. 1990. Sweden. In B.L. Turner II, W.C. Clark, R.W. Kates, J.F. Richards, J.T. Mathews, and W.B. Meyer (Eds.), *The Earth as Transformed by Human Action*. Cambridge University Press, Cambridge, UK.

Hall, W.A. 1980. Cost differential of peaking and secondary power. Personal communication.

Hall, W. A. and J.A. Dracup. 1970. *Water Resources Systems Engineering*. McGraw Hill, New York.

Hassan, Q. 2004. A study of some aspects of sustainable development of India's waters. Ph.D. Thesis. Indian Institute of Technology, Delhi, New Delhi.

Jones, P.H. 1985. *Deep Aquifer Exploration Project—Upper Gangetic Plain, India*. Prepared for the World Bank. P.H. Jones Inc.

Khepar, S.D. 1980. System studies for microlevel irrigation water management. Ph.D. Thesis. Indian Institute of Technology, Delhi, New Delhi.

Kothari, M. 2000. Sustainable water resources development for a river basin. Ph.D. Thesis. Indian Institute of Technology, Delhi, New Delhi.

Landes, D.S. 2000. Culture matters. In L. Harrison and S. Huntington (Eds.), *Culture Matters—How Values Shape Human Progress*. Basic Books, New York.

National Commission for Integrated Water Resources Development Plan (NCIWRDP). 1999. Draft Report, Ministry of Water Resources, Government of India, New Delhi.

Lakshsminarayana, V. and R. Revelle. 1975. Ganga Water Machine. *Science*, 188, 541–549.

Nehru, J. 1947. *Discovery of India*. The Signet Press, Calcutta.

North, D.C. 2005. *Understanding the Process of Economic Change*. Princeton University Press, Princeton.

Pfister, C. and P. Messerli. 1990. Switzerland. In B.L. Turner II, W.C. Clark, R.W. Kates, J.F. Richards, J.T. Mathews, and W.B. Meyer (Eds.), *The Earth as Transformed by Human Action*. Cambridge University Press, Cambridge, UK.

Prasad, R.K. 1981. Finite element simulation of groundwater system. Ph.D. Thesis. Indian Institute of Technology, Delhi, New Delhi.

Rangachari, R. 2006. *Bhakra–Nangal Project—Socioeconomic and Environmental Impacts*. Oxford University Press, New Delhi.

Revelle, R. and T. Herman. 1972. Some possibilities for international development of the Ganges–Brahmaputra basin. Res. Report. Harvard University Centre for Population Studies, Cambridge, MA.

Revelle, R. and V. Lakshminarayana. 1985. An unconventional approach to integrated ground and surface water development. In M.C. Chaturvedi and P. Rogers (Eds.), *Water Resources Systems Planning—Some Case Studies for India*. Indian Academy of Sciences, Bangalore.

Ruttan, V.W. 1997. Sustainable growth in agricultural production: Poetry, policy and science. In S.A. Vosti and T. Reardon (Eds.), *Sustainability, Growth and Poverty Alleviation*. The John Hopkins Press, Baltimore and London.

Sarma, E.A.S. 1985. A study of energy policy for India. Ph.D. Thesis. Indian Institute of Technology, Delhi, New Delhi.

Singh, R. 1980. Study of some aspects of water resources and agricultural policy of India. Ph.D. Thesis. Indian Institute of Technology, Delhi, New Delhi.

Sohoni, V. 1976. Design of a small axial reversible flow pump. M.Tech. Thesis. Indian Institute of Technology, Delhi, New Delhi.

Spear, P. 1978. *A History of India*, Vol. 2. Penguin, London.

Srivastava, V.K. 1976. Study of induced groundwater recharge. Ph.D. Thesis. Indian Institute of Technology, Delhi, New Delhi.

Thangraj, C. 1988. Regional power systems planning. Ph.D. Thesis. Indian Institute of Technology, Delhi, New Delhi.

World Bank. 1998a–f. *India Water Resources Management Sector Review, Report on Initiating and Sustaining Water Sector Reforms: A Synthesis*. South Asia Rural Development Unit, Washington, D.C.

World Bank. 1986 (May). *Proceedings Groundwater Seminar and Technical Session*. World Bank, Washington, D.C.

Chapter 14

Ahmad, Q.K., N. Ahamad, and S. Rasheed (Eds.). 1994. *Resources, Environment and Development in Bangladesh with Particular Reference to the Ganges–Brahmaputra–Meghna Basins*. BUP/Academic Publishers, Dhaka.

Ahmad, Q.K., S.K. Malla, B.B. Pradhan, and B.G. Varghese (Eds.). 1994. *Converting Water into Wealth: Harnessing the Eastern Himalayan Rivers*. Academic Publishers, Dhaka.

Ahamad, Q.K., A.K. Biswas, R. Rangachari, and M.M. Sainju (Eds.). 2001. *Ganges–Brahmaputra–Meghna Region—A Framework for Sustainable Development*. The University Press, Dhaka.

Chaturvedi, M.C. 1987. *Water Resources Systems Planning and Management*. Tata McGraw Hill, New Delhi.

Chaturvedi, M.C. 2011a. *Societal Environmental Systems Management* (in process of publication).

Chaturvedi, M.C. 2011b. *India's Waters: Environment, Economy and Development*. Taylor & Francis, Boca Raton, FL.

Chaturvedi, M.C. 2011c. *India's Waters: Advances in Development and Management*. Taylor & Francis, Boca Raton, FL.

Chaturvedi, M.C. 2009. Revolutionizing the development and management of transboundary waters—shared waters and shared opportunities. World Water Day, 22 March 2009. Ministry of Water Resources, Government of India.

Chaturvedi, M.C. and P. Rogers (Eds.). 1985. *Water Resources Systems Planning—Some Case Studies for India*. Indian Academy of Sciences, Bangalore.

Crow, B. and A. Lindquist. 1990. Development of Rivers Ganges and Brahmaputra: The difficulty of negotiating a new line, development policy and practice. The Open University, Milton Keynes, UK, Working Paper No. 19.

Eaton, D.J. (Ed.). 1992. *The Ganges–Brahmaputra Basin—Water Resources Cooperation between Nepal, India and Bangladesh*. University of Texas, Austin.

Rao, K.L. 1975. *India's Water Wealth*. Orient Longans, New Delhi.

Verghese, B.G. and R.R. Iyer (Eds.) 1993. *Harnessing Eastern Himalayan Rivers*. Centre for Policy Research, Konarak Publishers, New Delhi.

Thapa, B.B. and B.B. Pradhan. 1996. *Water Resources Development—Nepalese Perspective*. Konark Publishers, New Delhi.

World Bank. 1989. *Bangladesh Action Plan for Flood Control*. Unpublished.

World Bank. 1998. *Bangladesh 2020*. The University Press Ltd., Dhaka.

World Bank. 1998. *Water Resources Management in Bangladesh: Steps Towards a New National Plan*. World Bank, Dhaka.

Chapter 15

Chaturvedi, M.C. 2011a. *Societal Environmental Systems Management* (in process of publication).

Landes, D.S. 1998. *The Wealth and Poverty of Nations—Why Some Are So Rich and Some So Poor*. W.W. Norton, New York.

NCIWRDP. 1999. *National Commission on Integrated Development Plan*. Ministry of Water Resources, Government of India.

World Resources 1996–1997. Oxford University Press, New York.

Index

Printed and bound by CPI Group (UK) Ltd, Croydon, CR0 4YY

21/10/2024

01777085-0019